Secondary Natural Products in Foods and Food Systems

Bryan Hanley

Secondary Natural Products in Foods and Food Systems

Springer

Bryan Hanley
Department of Preventative Dentistry
Academic Center for Dentistry Amsterdam
Amsterdam, The Netherlands

ISBN 978-3-032-00291-4 ISBN 978-3-032-00292-1 (eBook)
https://doi.org/10.1007/978-3-032-00292-1

This Springer imprint is published by the registered company Springer Nature Switzerland AG
The registered company address is: Gewerbestrasse 11, 6330 Cham, Switzerland

Preface

This book focuses upon the secondary natural products that occur in foods and food systems. The primary function of food is as a source of macro and micronutrients and this role of food has been extensively studied both in humans and in animals. In addition, however, foods (and food systems) also contain minor constituents with a range of properties. These constituents have been of interest to chemists on several levels. Many of the bioactive principles in nature have derived, directly or indirectly, from secondary natural products. These have been isolated, investigated and modified and their study has formed the basis of medicinal and organic chemistry. In some ways, these investigations have travelled full circle. They began with the application of reductionist principles to identify and characterise natural products, continued with their synthesis in the chemistry laboratory, and now, with the application of synthetic biology and the application of engineering principles to complex biological machinery, chemistry has returned to the optimisation of biology through manipulation driven by the development of exquisitely targeted techniques and a more complete understanding of how biological systems work and can be manipulated. While secondary natural products have been found to have a range of applications (medicinal, taste and flavour properties, for example), the value of these products to the producing species is, perhaps, less clear. In some cases, the molecules produced are toxic to competing species and therefore confer a growth/ecological advantage. In other cases, it has been suggested that some secondary natural products may act as feeding deterrents to inhibit predation. This raises a further question. If compounds are produced which deter consumption of a particular species, then the predating species must have a mechanism for being deterred. It is perhaps unclear what the evolutionary advantage this confers for the predatory organism. This is particularly pertinent in the case of foods and food systems.

In most cases, foods do not contain acutely toxic entities and this has a number of clear consequences. First of all, some of the compounds have multiple physiological targets. This multiplicity of biological activities can increase the biological potency of a single compound. Secondly, individual secondary natural products often occur as part of a mixture of bioactive entities and developing methods to determine synergistic (or antagonistic) effects can be challenging. Finally, many of the biological properties of secondary natural products in foods are of a chronic (long-term) nature. Most investigations of the response of biological systems to bioactive principles are driven by acute (and easily measurable) effects. The measurement of the consequences of lifetime exposure is much more challenging and the appreciation of a temporal axis requires that predictive models are developed.

Processing (including cooking) is an important factor in the use of foods and food systems and such operations may have a direct impact on secondary natural products. In addition to being dealt with in specific chapters, most of the chapters (2–10) contain a simple recipe which illustrates one or more principles of the chemistry of secondary natural products in a very practical setting. It is important to appreciate that this is not, by any means a recipe or cookbook. It is a chemistry book that will appeal to chemists who have a particular interest in the consequences of the occurrence of secondary natural products in foods and food systems.

Amsterdam, The Netherlands Bryan Hanley

Competing Interests The author has no competing interests to declare that are relevant to the content of this manuscript.

Contents

1 Secondary Natural Products in Foods and Food Systems 1
2 Secondary Natural Products and Disease 33
3 Analytical and Bioanalytical Methods 69
4 Bioassays 105
5 Biosynthesis—Approaches and Strategies 135
6 Epidemiology, Causation and Correlation 165
7 Taste and Other Receptors 197
8 Secondary Natural Products and Food Processing 227
9 Secondary Natural Products and Microbes 255
10 Processing, Varieties, Natural Products and Food Systems 289
11 Conclusions and the Future of Natural Product Research 317

About the Author

Dr. Bryan Hanley has worked both in academia and in industry in the UK, USA, and Europe. He is a chemist by training and has degrees from the University of St. Andrews and the University of Edinburgh. He has authored over 80 scientific papers and authored/edited four books. He was Visiting Professor at the University of Ulster (2013–2017).

He is also President of the International Life Sciences Institute, Europe, and was formerly President of the Industry and Technology Division Council of the Royal Society of Chemistry (UK) and Chairman of the Food Group of the Royal Society of Chemistry.

List of Figures

Fig. 1.1 Folate and homocysteine metabolism 3
Fig. 1.2 Affinity ordering system 5
Fig. 1.3 Classification of organic molecules by origin, functional group or biological activity 6
Fig. 1.4 Structure of resolvins 10
Fig. 1.5 Structures of quinine and allyl toluidine 12
Fig. 1.6 Enzymic hydrolysis of glucosinolates 22
Fig. 2.1 Structures of the chemical constituents of juniper berry oil 34
Fig. 2.2 The main bioactive constituents of coffee 37
Fig. 2.3 The glucosinolate/myrosinase system 40
Fig. 2.4 Structure of penicillin and cephalosporin 41
Fig. 2.5 Structures of yohimbine and atropine 46
Fig. 2.6 Annotated structure of glucosinolates 48
Fig. 2.7 Formation of S-containing compounds in allium species 49
Fig. 2.8 The biosynthesis of glucosinolates 50
Fig. 2.9 Formation of goitrin (vinyl-oxazolidinethione) 51
Fig. 2.10 Examples of bitter compounds with an NH–C=S moiety 51
Fig. 2.11 Formation of indole-3-carbinol 52
Fig. 3.1 Structure of capsaicin 70
Fig. 3.2 Structure of crocin 73
Fig. 3.3 Structure of crocetin 74
Fig. 3.4 Structure of Safranal 74
Fig. 3.5 Formation of safranal from picrocrocin 74
Fig. 3.6 Representative glucosinolate structures 76
Fig. 3.7 Enzymic hydrolysis of glucosinolates 76
Fig. 3.8 The anomeric approach to glucosinolate synthesis 78
Fig. 3.9 The nitronate route to glucosinolate synthesis 79
Fig. 3.10 The structure of estradiol 79
Fig. 3.11 Structures of isoflavones, coumestans and lignans 80
Fig. 3.12 Structure of zearalenone 80

Fig. 3.13 Structure of a representative heterocyclic aromatic amine (MeIQ) ... 97
Fig. 4.1 Structure of gingerol, shogaol and zingerone ... 107
Fig. 4.2 A postulated synthesis of quinine from aniline ... 116
Fig. 4.3 The synthesis of mauveine from o-toluidine ... 117
Fig. 4.4 Structures of representative alkaloids ... 118
Fig. 4.5 The Hoffman degradation reaction ... 119
Fig. 4.6 Subsequent hypothesised breakdown of indolyl-3-methyl isothiocyanate ... 125
Fig. 5.1 The cleaved structure of pectin ... 139
Fig. 5.2 The intermediacy of secologanin in biosynthesis ... 144
Fig. 5.3 The Diels Alderase reaction in nature ... 145
Fig. 5.4 Synthesis of 6-amino penicillanic acid ... 150
Fig. 5.5 Structure of 7-aminodeacetoxycephalosporanic acid ... 150
Fig. 5.6 Derivation of taxol from 10-deacetylbaccatin III ... 151
Fig. 5.7 Phases in the biosynthesis of glucosinolates ... 152
Fig. 5.8 Structure of alpha-linolenic acid ... 157
Fig. 6.1 Structure of hydroxy alpha sanshool ... 168
Fig. 6.2 Structure of trans-anethol ... 169
Fig. 6.3 Structure of Griseorhodin A ... 176
Fig. 6.4 Structure of Bottromycin A2—a representative RiPP ... 177
Fig. 6.5 Rearrangement of glucosinolate aglucone ... 179
Fig. 6.6 Products from Indole-3-carbinol ... 180
Fig. 6.7 Structure of brassinin ... 181
Fig. 6.8 Structure of goitrin and its precursor ... 181
Fig. 6.9 Structure of S-Alk(en)yl-L-cysteine Sulphoxides ... 181
Fig. 6.10 Structure of allicin ... 182
Fig. 6.11 Formation of onion lachrymatory factor ... 183
Fig. 6.12 Structure of ajoene ... 184
Fig. 6.13 Some phytochemicals with microbial quorum sensing inhibitor activity ... 187
Fig. 6.14 Structure of oxalates ... 189
Fig. 6.15 Structures of catechins, proanthocyanidins and epicatechins ... 190
Fig. 7.1 The Maillard reaction ... 200
Fig. 7.2 Formation of thiol compounds from Thiosulphinates ... 201
Fig. 7.3 Structures of some natural and synthetic bitter compounds ... 207
Fig. 7.4 Structure of denatonium ... 207
Fig. 7.5 Structure of 11-cis-retinal ... 212
Fig. 7.6 Structure of tetracene ... 213
Fig. 7.7 Structure of pentacene ... 213
Fig. 7.8 Ceric ammonium nitrate ... 214
Fig. 7.9 Chemical constituents of tyrian purple ... 216
Fig. 7.10 Perkin's putative synthesis of quinine ... 217
Fig. 8.1 The structure of vinyl oxazolidinethione (goitrin) ... 231
Fig. 8.2 Structures of 3-monochloropropane diol and acrylamide ... 233

Fig. 8.3 Structure of amylose 237
Fig. 9.1 The structure of crocin 258
Fig. 9.2 The structure of safranal 259
Fig. 9.3 The structure of fenchol 262
Fig. 9.4 Structure of representative sulphonamides 265
Fig. 9.5 The active agent of prontisil 265
Fig. 9.6 Structures of thiocillin 267
Fig. 9.7 Structure of acyl homoserine lactones 269
Fig. 9.8 Structure of fengycin 270
Fig. 9.9 Structure of lanthionine 274
Fig. 9.10 Formation of thiosulphinates 277
Fig. 9.11 Structure of ajoene 278
Fig. 10.1 Catabolism of caffeine 295
Fig. 10.2 The lossen rearrangement 299
Fig. 10.3 Structure of flavones 301

Chapter 1
Secondary Natural Products in Foods and Food Systems

Contents

1.1 Introduction 1
1.2 Macronutrient Analysis and Function 7
1.3 Micronutrient Analysis and Function 10
1.4 Food, Secondary Natural Products and Reductionism 11
1.5 Nutrition and Pharmacology 14
1.6 Metabolic Programming and Stratified Nutrition 16
1.7 Health Claims 20
References 25

Abstract The nature of food is that its primary purpose is to provide macro and micronutrients. These impact upon nutrition and primary metabolic pathways. Nutrition can be considered at various levels of granularity and can be reflective of population and individual health. This raises the question of how to measure health—apart from nutrient deficiency—and what is the role of non-nutritive food components? Nutrient deficiency has a number of clinical outcomes (depending on severity) and these can be measured. It is, however, more difficult to assess optimal inputs of nutrients and, in particular, non-nutrients. Clinical endpoints are difficult to define and changes in them challenging to ascribe to particular food components.

1.1 Introduction

Food and food systems are composed of chemicals and many of the properties of foods (and their constituents) is a function of the numbers and types of compounds and relationships and interactions between them. There are a number of classes of natural products present in foods and food systems including, in particular macronutrients (proteins, carbohydrates, lipids) and micronutrients (vitamins, minerals and other essential ingredients). These have been extensively studied—in part because the consequences of deficiencies in the intake of them are clinically measurable in a relatively short timescale therefore they can be classified by biological function. Not all sub-optimal intakes have the same consequences. In particular, there is a distinct difference between nutrient inadequacy (i.e. sub-optimal intake) and nutrient

B. Hanley, *Secondary Natural Products in Foods and Food Systems*,
https://doi.org/10.1007/978-3-032-00292-1_1

deficiency (severely reduced intake leading to a clinical disease outcome). Both deficiency and inadequacy can apply to macro and micronutrients—although the clinical manifestation is different [1]. In addition, some nutritional dysfunctions such as Crohn's disease can affect the absorption of nutrients and this may mimic the symptoms of inadequacy or deficiency. Recommended daily intakes (RDIs) have been calculated and are defined as the amount of a given nutrient each person should consume based upon factors such as age and gender. These are based on animal and human studies and have been codified into regulations. In most cases, adequate levels of nutrients are found in foods if an individual consumes a balanced diet however in some cases (for example, folate—see below) there is a requirement that is met either by supplementing foods or by external supplements. Certain cells in the body have particular requirements in terms of nutrients and not all cells or organs have equivalent needs. In the case of macronutrients, neuronal cells, for example require a good source of energy and the link between energy intake and cognitive function has been noted [2]. In the case of macronutrients both protein and lipid deficiencies can cause significant disease outcomes. This applies both to nutrient levels in general and, in some cases, specific members of the class. For example in the case of lipids, an adequate intake of long chain polyunsaturated fatty acids (LC-PUFAs) has been associated with a number of health endpoints. In particular, intake of docosahexaenoic acid, 22:6n-3 (DHA) intake (and that of related LC-PUFAs) has been correlated with improved cardioprotection, reduced inflammation and prevention of cognitive decline [3]. Clinical trials, while showing some promise, have yet to prove a convincing mechanistic link between types of LC-PUFA intakes and definitive clinical outcomes.

For essential micronutrients, the most common deficiencies occur due to compromised dietary levels of iron, folate, zinc, iodine and vitamin A. In addition to significant clinical diseases such as beri beri or pellagra, chronic deficiencies in some vitamins or micronutrients can have a range of subacute consequences. Vitamin and micronutrient deficiencies have been extensively studied however in some cases the association between optimal intakes of a particular dietary component and a biologically distant disease outcome can be difficult to prove. For example, while the link between vitamin D intake (and vitamin D receptor polymorphisms) and tuberculosis has been noted [4], it is difficult to prove causality and it has been suggested that vitamin D may improve symptoms rather than affecting infection rate per se [5]. Indeed, vitamin D adequacy was reviewed and even found not to be associated with the development or progression of tuberculosis from the latent to the active state [6]. Another example of a chronic disease state associated with an inadequate supply of a nutrient is the link between folate and vitamins B6 and B12 and cardiovascular disease. In this case, however, there is a sounder mechanistic basis for suggesting a causal link. Elevated plasma levels of homocysteine are associated with an increased risk of cardiovascular disease [7]. The link between vitamins B6 and B12 (and folate) and homocysteine metabolism provides a mechanistic basis for the cardioprotective effects of B vitamins (Fig. 1.1). Folate, vitamin B6 and B12 are implicated in the conversion of homocysteine to methionine or to cysteine thereby inhibiting the build up of homocysteine.

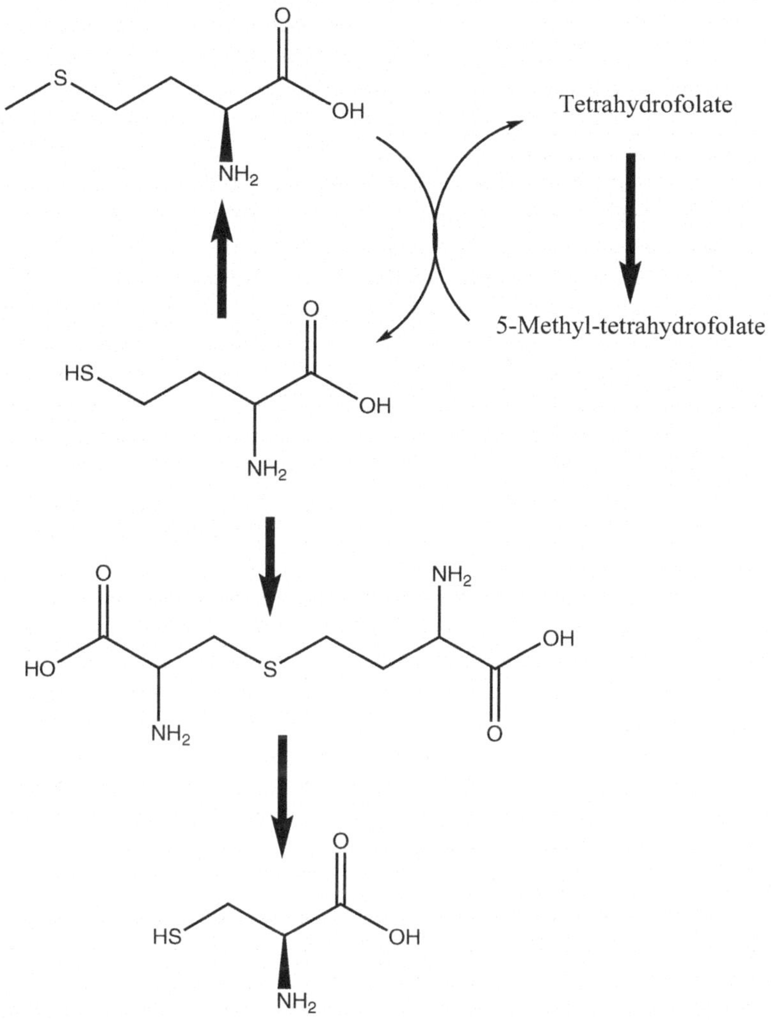

Fig. 1.1 Folate and homocysteine metabolism

In addition to inadequate intakes, vitamin and micronutrient deficiency can also be a result of malabsorption. In order to develop coherent interventions that may either prevent disease onset or improve health, it is important to have both associative and mechanistic information—evidence for an effect and a provable hypothesis as to how the effect takes place—and to understand the factors that may affect uptake as well as intake. Such information can be derived as a result of the application of reductionism to biological systems (see Sect. 1.2) and, while this may limit the study of some of the interrelationships that occur in whole living systems, it is often the most effective way of establishing a testable (empirical) hypothesis.

In addition to measuring the biological consequences of nutrients and micronutrients, it is essential to be able to firmly correlate these effects with specific levels of the compounds of interest. This clinical association is much easier to probe in the case of nutrients and micronutrients than it is for secondary natural products (as will be demonstrated later). However some of the principles that have been developed for measuring the levels and function of nutrients and micronutrients can also be applied to secondary natural products in foods and food systems. The analysis of macro and micronutrients in foods and also in clinical samples (serum, urine) has been a significant area of development. In order to understand the principles that underpin analysis, it is necessary to consider how the reductionist chemistry paradigm arose and, indeed, gave rise to the concept of elements and molecules with particular characteristics.

Traditionally, chemistry has adopted a reductionist approach to analysis. Indeed isolation and purification (removal of undesired components or impurities) lies at the core both of alchemy and the tradition of analytical chemistry [8]. Alchemists used a series of practical steps to transform substances and to remove some components and this remains a part of the core experimental practice of chemistry. With minerals and other inorganic substances (including gases), fragmentation by heating, distillation and displacement reactions shed light on the composition of materials and, later, on how individual substances combined as a function of their 'affinity' for one another. The concept of affinity was an important precursor to the development of chemical bond theory. While some aspects of the sharing of electrons and an understanding of bonds in complex systems has been adopted by quantum physics and biology respectively, chemistry remains at the intellectual centre of how atoms and molecules interact. By understanding how substances can be split into their component parts and combined to give either the original substance or a different substance, chemistry has adopted the dissociation (fragmentation) followed by association (synthesis) cycle to prove structures and relationships.

While this has, as noted above, been of particular value in the study of inorganic materials where, as will be demonstrated, components can be removed or substituted with other components, organic compounds are a more significant test of the cycle model. An inorganic substance can be split into its constituent parts until it cannot be subdivided any further (what remains are elements) and the proportions of the elements that combine together can be determined. The other aspect of inorganic chemistry (in addition to the dissociative (fragmentation) part) is the association (synthesis). As an example, the combination of a simple acid with a base to give a neutral salt is an exercise in synthesis that can be used to determine stoichiometry. By understanding how much of a given acid will combine with a known amount of a base to attain neutrality, the concept of valence can be introduced. Replacement of one component of a salt by a component with a greater affinity develops the concept of reactivity and the idea of the smallest unit that cannot be further divided describes the concept of elements. Using such an affinity ordering system (Fig. 1.2, shows a small part of a section of an affinity table), elements can be ordered both according to their affinity (reactivity) and the stoichiometric amount of them that will combine (equivalence).

Esprit Acide	Acide du sel marin	Acide nitreaux	Acide vitriolique
Sel alcali fixe	Etaine	Fer	Soufre minerale
Sel alcali volatile	Regule d'Antimoine	Cuivre	Sel alcali fixe
Eau	Cuivre	Plomb	Sel alcali volatile

Fig. 1.2 Affinity ordering system

Once similarities are observed between elements that are ordered into columns of affinity and equivalence and these are seen to occur regularly then the beginnings of the Periodic Table of elements can be seen. What is more, gaps in the Table can be observed and missing elements sought on the basis of the similarity in their properties to adjacent elements in the Table.

In the case of organic molecules, the situation is complicated by the level of reductionism that is of value in order to distinguish between different substances. This will be dealt with in more detail in Sect. 1.2 but, simply put, if an organic molecule is fragmented into its constituent parts then the chemist will quickly discover that all organic molecules consist primarily of just three elements (carbon, hydrogen and oxygen) combined in different ways—along with small amounts of some other elements. In addition, compounds with the same proportions of carbon hydrogen and oxygen may have very different properties. In other words, substances that have the same proportions of elements may, when isolated, be markedly different in appearance and properties. This raises the question of what level of decomposition is most useful in the study of organic compounds. In addition to being classified by their chemical or trivial name, organic compounds can be characterised by their origin (for example, mustard oil glycosides), by the functional groups they contain (for example, alkaloids, alcohols, aldehydes) or by their biological activity (for example, narcotics). (Fig. 1.3).

The idea of analysis (the resolution of anything complex into simple elements) being the etymological opposite of synthesis (the act or process of combining or composing different elements into a whole), as noted previously, underpinned much of inorganic chemistry and was also an important aspect of the development of knowledge about organic compounds. The proof of the composition of a material was thought to require both its breakdown into its constituent parts (analysis) followed by the reforming of the original material by combination (synthesis). The combination was more than simply mixing or blending—it also involved reaction and transformation. Assembly of the constituent parts was seen as a crucial component of the proof of a material. This is a theme that we will return to later after considering the principles that underpin macro and micronutrient analysis and how the methods developed can also be used in secondary natural product characterisation.

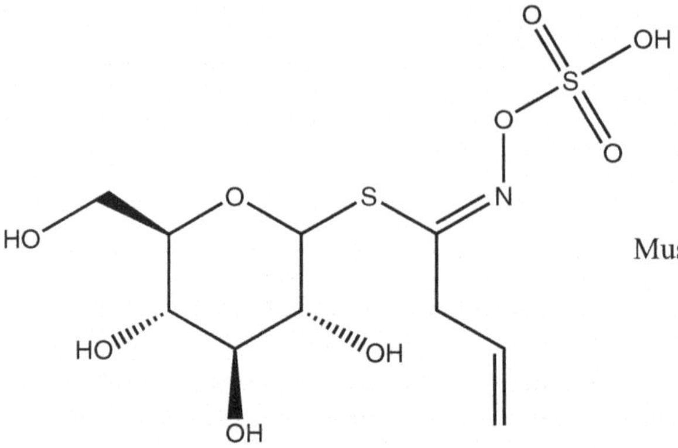

Mustard oil glycoside (sinigrin)

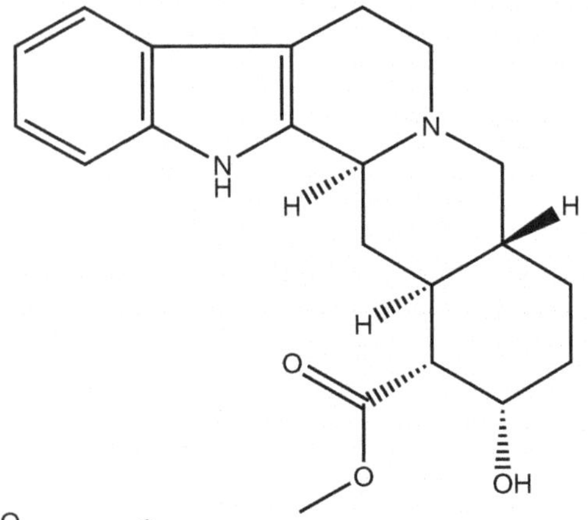

Alkaloid (yohimbine

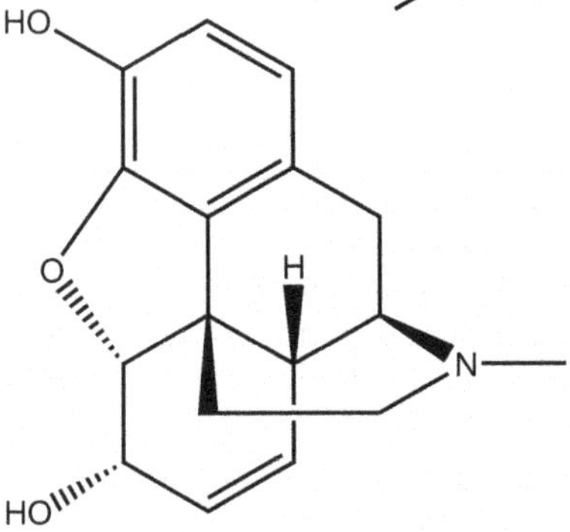

Narcotic (morphine)

Fig. 1.3 Classification of organic molecules by origin, functional group or biological activity

1.2 Macronutrient Analysis and Function

Analytical procedures for organic natural products (including nutrients) have been developed over many years. Ideally methods should be non-destructive (not require the complete destruction of the sample) and be capable of measuring more than a single macronutrient at a time. The potential of near infra red spectroscopy has been noted in this regard [9], although some variability in carbohydrate analysis has also been observed. Other spectroscopic methods have been used in particular instances including Raman spectroscopy for macronutrients and also in a food safety setting [10, 11]. Analysis of macronutrients has also included the use of machine learning to optimise analytical methods [12].

The analysis of macronutrients is linked closely to optimal (rather than sufficient to prevent deficiency) intake and this characteristic has both been an analytical goal and has been the subject of considerable investigation [13]. Macronutrients comprise proteins, carbohydrates and lipids. As will be mentioned further, macronutrients are, in some aspects, partially interchangeable but also have specific roles. For example, while all three major macronutrients can provide energy, proteins are the least effective in this regard and their primary metabolic function is as a source of amino acids and as providers of sulphur, nitrogen and some building blocks for the synthesis of biological molecules.

Carbohydrates are a significant energy source. As such, upon digestion (hydrolysis), they provide glucose which is the key component of carbohydrate energy metabolism. Carbohydrates also provide non-digestible components (fibre and some polysaccharides) which have a number of functions including inducing satiation and regulating the gut microbiome. Individual macronutrients should be part of a balanced diet. The effects of an unbalanced diet can include both macronutrient deficiencies and excess intake and the consequences of this can include wasting, deleterious effects on energy metabolism and, at the other extreme, increased weight and adiposity. In common with other nutrients, recommended daily allowances (RDAs) for macronutrients are based upon the level that prevents a deficiency disease from being established rather than the *optimal* level of intake for health. Protein, lipid and carbohydrate deficiency diseases are relatively rare however are not unknown and an unbalanced diet may have consequences. The analysis of individual macronutrients can be simple or complex—depending on the sensitivity and degree of detail required. For carbohydrates, for example, analysis can either be by difference (although this may fail to exclude non-carbohydrate components such as lignin) or may be based upon the size of the constituent carbohydrates based on molecular size (degree of polymerization or DP) into sugars (DP 1–2), oligosaccharides (DP 3–9) and polysaccharides (DP > 10). The way of reporting macronutrient composition depends upon the nature of the requirement for the analysis and this may correspond to the function of the nutrient in vivo. For example, in the case of carbohydrate analysis, the division between digestible and non-digestible carbohydrates is an important biological/physiological distinction and a range of more specific chromatographic and enzymic methods that can differentiate between these types of carbohydrate have largely replaced older methods (e.g. colorimetric methods for total carbohydrate, various methods to determine total sugars of a particular type—for example reducing sugars). Various texts are available that describe analytical methods and enable the selection of the most appropriate of these [14]. The progress in the analysis of carbohydrates can be considered to reflect the development of chemistry in general. From a state of measuring total carbohydrates by difference (i.e. anything that cannot be classified in some other way is considered to be a carbohydrate) to colourimetric assays for total carbohydrates and then to the quantification of different sub-classes within the classification using more and more specific methods, carbohydrate analysis is a mixture of analysis by the number of repeat units and structure (DP) and characterisation by biological function (digestibility). Similarly chemistry has progressed from allencompassing alchemical definitions of substances and processes as largely bulk materials with visibly quantifiable properties to an awareness of atoms, elements and molecules to the further subdivision of the discipline into classification based on structures (in the broadest sense, organic and inorganic) and on states of being (physical chemistry).

In the case of another set of macromolecules, proteins, in a similar way to lipids (see below) and carbohydrates, the macromolecular analysis also can be considered to progress from the generic to the specific. A range of techniques have been developed depending on the level of granularity required [15]. The simplest methods are the Kjeldahl and Dumas methods. Both of these measure the amount of released nitrogen and so are not specific for protein. In addition, they do not provide information on the type of protein present. In the Kjeldahl method, the analytical sample is broken down by incubation with sulphuric acid. The resulting ammonia is distilled and collected and the amount determined by titration. The Kjeldahl method has been used as the reference protocol for the analysis in a range of foods [16]. The Dumas method also relies on the release of nitrogen but in this case it is by combustion. The two methods have been compared and both are used as reference methods for total protein determination [17]. As is the case with carbohydrates, the degree of granularity and detail required for protein analysis depends upon the use to which the data is being put. For example, insect protein has been suggested to have some useful processing characteristics and high levels of desirable nutrients [18], however the possibility of the protein content in insects being over-estimated due to the presence of non-protein nitrogen and a subsequent error in the N to protein conversion ratio has been noted [19]. Such errors may impact upon the calculations about the usefulness of different protein sources. In addition to the crude measurement of the amount of protein present as estimated by measuring nitrogen content, a range of criteria are used to assess protein quality [20]. These can include detailed analysis of the amino acid content—particularly the essential amino acids which must be obtained from the diet (histidine, isoleucine, leucine, lysine, methionine, phenylalanine, threonine, tryptophan, and valine), protein digestibility and nitrogen utilisation. All of these factors can impact upon the assessment of protein quality and use. In addition, certain population groups (e.g. infants, the elderly) have particular protein nutritional requirements. A consideration of this lies at the centre of the development of stratified nutrition which will be discussed in more detail later in this chapter.

The third category of macronutrients are those classified as fats or lipids. Lipids are the most energy dense of the macronutrients however, in addition to being a potential energy source, they are important functional components of several other aspects of living systems. They are important in the synthesis of a range of essential components of living systems (including some hormones) and in the uptake and absorption of fat-soluble vitamins. Dietary lipids can be divided into a number of subtypes which will not all be considered in detail. These groups include triglycerides, phospholipids, sterols and fatty acids. This last group are further sub-divided depending on the degree of unsaturation. Unsaturation changes the functional properties of fatty acids and they can therefore be considered to be secondary natural products with defined properties rather than simply macronutrients. From a nutritional/metabolic/functional standpoint, both the degree and position of unsaturation is important. The most biologically important fatty acids are long chain polyunsaturated fatty acids (LC-PUFAs) and some details of their functional properties will be mentioned later. These are classified as omega-3, omega-6 or omega-9 with the number referring to the position of the first double bond in the LC-PUFA counting from the methyl end of the molecule. The potential health effects of different LC-PUFAs has already been mentioned (see above) and these functional properties have been extensively studied [21]. The analysis of total fat or lipid content includes glycerides, sterols, phospholipids and other minor components. Fats can be extracted *in toto* by Soxhlet extraction using a suitable non-polar solvent. Acid hydrolysis is included as a step to release bound fats. Once the fat is solubilised, the sample is dried to constant weight. Care must be taken to prevent oxidation of the fat since this can increase the weight of the fat as a result of addition of oxygen. A range of methods including, in particular chromatographic and spectroscopic methods can be used to quantify individual lipid species [22]. Although the total amount of lipids may be relatively easy to quantify, there is considerable heterogeneity in the types of lipids present. The analysis of the complete set of lipids from a given source is referred to as the lipidome (by analogy with the genome) [23]. The use of mass spectrometry has facilitated a significant advance in the characterisation of lipids from biological systems [24] however less precise methods can distinguish between different levels of saturation. From a health standpoint, there is a presumed link between saturated fat intake and the risk of developing cardiovascular disease although this simple relationship has been questioned [25] and further studies and the re-analysis of available clinical data is warranted. In the case of LC-PUFAs, the two most significant classes are the omega-6 and the omega-3. Fatty acids can be biosynthesised both anaerobically (through a polyketide route) or aerobically through successive elongation and desaturation. One of the major biological effects of lipids appears to be in the modulation of inflammation.

A number of chronic disease states involve inflammation—in some cases through the mediation of inflammasomes [26] and, while omega-3 fatty acids and some of their metabolic products (for example resolvins—Fig. 1.4) have been implicated in anti-inflammatory effects and while some mechanistic rationales have been proposed (including alteration of cell membrane phospholipid fatty acid composition, disruption of lipid rafts, inhibition of the activation of the pro-inflammatory nuclear transcription factor kappa B), this remains an area of some controversy [3].

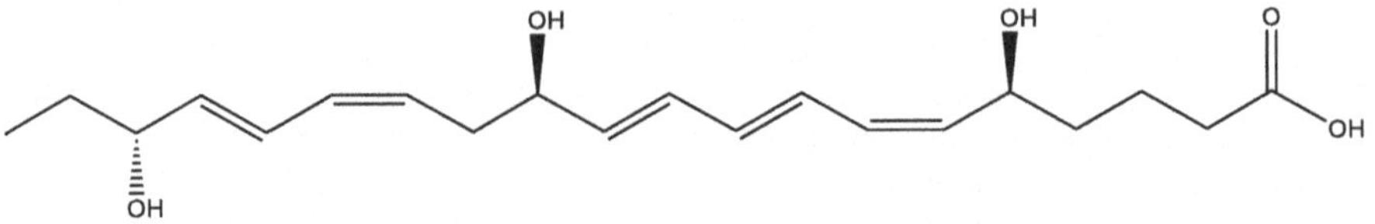

Fig. 1.4 Structure of resolvins

The impact of macronutrients is not limited to effects on energy utilisation and primary metabolism. They are also, to varying degrees, implicated in structural effects, as precursors of some secondary natural products and, as mentioned, may have an impact on clinical outcomes such as inflammation. In addition to the effects of classes and individual compounds, the biosynthesis of both macro and micro nutrients has been extensively studied and the interrelationships between them (and, indeed secondary metabolites) in terms of biosynthesis and biological activity has been investigated. In addition, and perhaps in some ways in contrast to secondary metabolites, the structure, production and use of primary metabolites by living systems are strictly controlled since the consequences of any errors or deficiencies are significant and can be rapid.

1.3 Micronutrient Analysis and Function

The analysis of macronutrients was underway in the nineteenth century. For micronutrients, their occurrence at much lower levels means that analysis was much more challenging. Indeed, in some cases, a functional rather than a chemical analysis was favoured. In addition, methods to assess micronutrient status has been of considerable interest. In this case there has been considerable interest in the development of methods to measure multiple micronutrients including functional markers to assess micronutrient status. The European Society for Clinical Nutrition and Metabolism has coordinated efforts to tabulate and assess the analysis of clinical samples for key micronutrients. Their prioritisation is generally based upon those micronutrients for which a deficiency is more common and more serious in terms of clinical outcome [27]. A number of techniques have been developed to measure micronutrient levels in foods specifically including, in particular, chromatographic methods, often coupled to mass spectrometry or other detection and structure confirmation methods. This has facilitated the production of models that can calculate and estimate micronutrient status based on dietary micronutrient intake data of different populations [28]. As with macronutrients, methods which can measure multiple analytes simultaneously—using either chemical or functional techniques—are favoured. The concept of micronutrient density has been introduced and used to build a global food composition database that can be used by nutritionists and NGOs such as the FAO to build intake assessments and estimate the nature and impact of deficiencies [29]. Methods continue to be developed and both ease of operation and sensitivity remain significant analytical goals. For example, the use of capillary electromigration (CE) [30], multiplex immunoassay [31] and surface plasmon resonance [32] have all been used to analyse for vitamins and other micronutrients.

The molecular heterogeneity of micronutrients means that, unlike macronutrients, methods to analyse all members of a group of micronutrients in complex systems (foods, clinical samples etc.) are neither simple to develop nor to automate to facilitate rapid throughput screens.

1.4 Food, Secondary Natural Products and Reductionism

While, as noted above, there have been many studies of the impact of nutrients and micronutrients on primary metabolites and their involvement in key metabolic pathways such as the tricarboxylic acid cycle, there has also been a growing interest in metabolically non-essential secondary natural products. These are generally produced in much smaller amounts and have a more diverse range of structures which in turn leads to a wider range of functional uses (see below). While their principal use is often to protect the producing organism from predation (consumption and ingestion) or to provide it with an advantage in the competition for resources, they have, in many cases, been found to have additional biological functionality for the consuming organism. This includes anti-microbial activity, anti-inflammatory activity, binding to specific receptors and a range of deleterious and hypothesised beneficial effects. The study of these products reflects the growth in sophistication of techniques for their isolation and characterisation (biological and chemical). The principles of characterisation by degradation and the identification of specific structural features noted previously and the subsequent synthesis of compounds to prove structures is an important aspect of the growth of natural product chemistry.

At the beginning of the nineteenth century, chemists had only a limited number of options for either characterisation or synthesis. For example, when Perkin tried to synthesise quinine, he attempted to prepare the molecule on the basis of the relative equivalent amounts of C, H, N and O in the desired product and in his starting materials. He had no way of probing the relative relationships between functional groups in a molecule. Perkin was tasked with synthesising quinine using coal tar as a starting material. Perkin's approach was based upon the knowledge of the molecular formula of quinine ($C_{20}H_{24}N_2O_2$). He hypothesised that, if he took two molecules of allyl toluidine ($C_{10}H_{13}N$) and added oxygen using one of the few reagents available to him (potassium dichromate) then he would produce quinine and water.

It is clear from the relative structures of quinine and allyl toluidine (Fig. 1.5) that this reaction would not give the anticipated product. Indeed, it was not until the twentieth century that a successful synthesis of quinine was achieved [33]. Even then, there was some debate as to whether the synthesis had in fact been achieved earlier (in 1945) by Woodward [34]. What is clear is that the reductionist approach (measurement of the proportions of the different constituent atoms in a molecule)—which was all that was available to Perkin—gave a misleading impression of how organic molecules were related to each other.

While knowledge of the proportions of the constituent atoms could be a valuable tool, this level of reductionism was unlikely to enable chemists to carry out a complete synthesis of complex molecules based upon limited molecular knowledge. While Perkin was mistaken in his appreciation of how allyl toluidine and quinine might be related, nonetheless his approach was to start with a molecule that was already quite complex (even if he did not know the level) and to try to modify it by arithmetic addition to reach an even more complex system. The adoption of this strategy suggests that organic chemists were perhaps greatly underestimating the numbers of different potential structures that might exist in nature.

In addition, the very idea of the existence of a similarity between molecules created by chemists and those which occur in nature was only established in the early part of

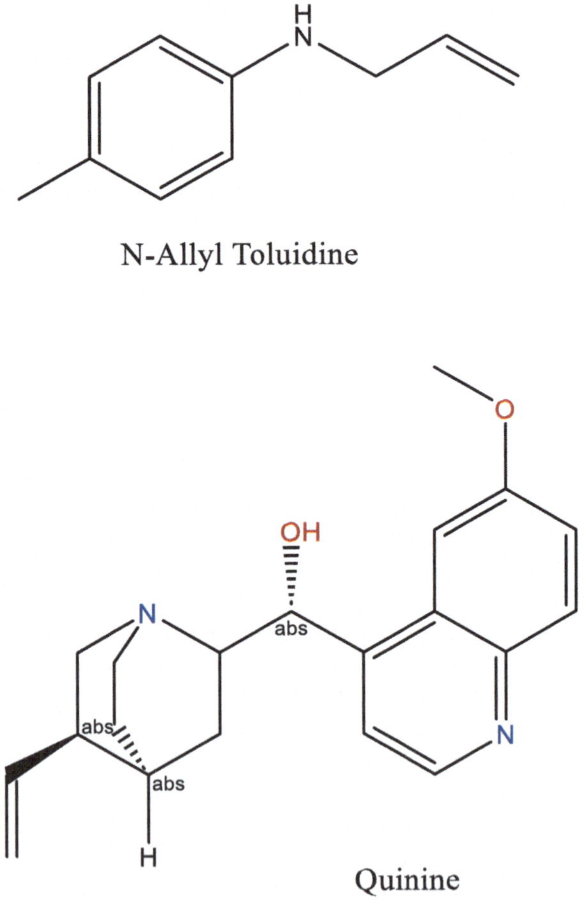

Fig. 1.5 Structures of quinine and allyl toluidine

the nineteenth century. It was only in 1828 that Wohler synthesised urea from potassium cyanate thereby revealing that natural products did not possess some unique property related to their origin (vitalism). This was contrary to the widely accepted view of Berzelius and others however the overturning of the vitalist theory was an important paradigm shift that enabled chemists like Perkin to think in terms of the constituent parts of natural products being composed solely of what can be measured and not possessing some 'extra', vital component. The idea of vitalism persisted in the nineteenth century, however and remained an important consideration as organic and natural product chemistry became more clearly defined as distinct (but interlinked) areas of investigation. While the destructive nature of structure confirmation (combustion analysis)—only gave as a primary piece of data, the formula but, importantly, not the structure. The knowledge base of structures (and the interrelationships between natural products) remained incomplete. Since many molecules that occur in nature have the same formula but are clearly different from each other, then this awareness and how different molecules might be interconverted (using schemes such as that developed by Perkin in his attempt to synthesise quinine) became of primary importance however there were still a considerable number of conceptual and practical obstacles that had to be overcome. In the absence of spectroscopic methods or ways of probing structures directly, the properties of substances became increasingly important. Melting point, boiling point and solubility became key attributes that could be measured and act as a comparison point between compounds produced by researchers in different laboratories. This reversion to a known attribute greatly increased the amount of synthetic chemistry that could be carried out but by the middle of the nineteenth century, molecular structures were still, for the most part, obscure. One other crucially important property of some natural products was optical activity. The capability of some substances to rotate the plane of polarisation of plane polarised light was first described by Jean-Baptist Biot in 1815. The association of this attribute with crystal shape (and molecular asymmetry) was not realised or proposed until much later by Pasteur [35]. The use of the term molécules organiques to describe components of living systems has its origins in the eighteenth century however only a small number of organic molecules had been characterised (alcohol, ether, tartaric acid, formic acid and acetic acid were chief among them). The primary methods of determining the constitution of substances were, as noted above, by combustion analysis as pioneered by Lavoisier (in 1794) and others and this enabled crude composition analysis to be carried out. Chemistry remained largely a descriptive practice and many natural products were named by a combination of their derivation and their properties. It is worth noting that Wohler's synthesis of urea was not only a refutation of vitalism (and of the previously unsurmountable division of organic and inorganic compounds), but it was also a demonstration that compounds could have identical constituents and, by changing the relative positions of these constituents, could form different substances. One of the most significant isolations of natural products also occurred at the beginning of the nineteenth century when Serturner isolated morphine from opium [36]. The isolation of morphine (and later other alkaloids) was significant for a number of reasons. First of all, it led to the development of methods to isolate classes of compounds (alkaloids) which had sufficient molecular similarity to respond to certain procedures in the same way (despite their overall structural complexity). Secondly, it demonstrated that some secondary natural products occurred in significant amounts (and without other competing substances being present) that they could be isolated and purified. This helped to define the level of reductionism that is of greatest value in natural product chemistry. Finally, isolation of purified compounds enabled pharmacologists to define the effective (and sub-toxic) dose of naturally occurring bioactive compounds. Another bioactive compound, quinine was isolated from the bark of the Cinchona tree by Pelletier and Caventou in 1817. As with the isolation of other alkaloids, isolation of the pure compound enabled the therapeutically effective dosage to be defined [37]. This is particularly important in the case of quinine since it is toxic and the therapeutic window is small. Opium and quinine alkaloids were two of the most scientifically and societally important products that arose from the application of natural product chemistry in the nineteenth century and helped to define the approach of isolating useful biologically active compounds from nature that continued throughout the remainder of the nineteenth and twentieth centuries.

Secondary natural products in foods have traditionally been employed to enhance the experience of consumption rather than for a direct physiological effect. In addition both macro and micronutrients have been found to be both biologically and sensorially significant. Most notably the sensation of sweet taste of and the mouthfeel of fat are both considered to be positive factors in foods [38]. In addition, in some species, nutrient sensing pathways operate. These can become dysregulated in cases of metabolic disease [39]. This can occur at the whole organ and systemic level (for example glucose detection by the pancreas) [40] but also at a cellular level [41]. The development of knowledge about extra-oral taste receptors with multiple biological functions [42] suggests that secondary natural products, as agonists (or antagonists) of taste receptors may potentiate a number of biological effects in addition to being important in the taste perception of foods.

1.5 Nutrition and Pharmacology

Traditionally, although they may both deal with bioactive constituents, there has been a clear scientific (and philosophical) division between nutrition and pharmacology. Nutrition has been defined as the assimilation and use of food constituents by living systems to allow them to grow and thrive. Pharmacology is specifically the study of drugs and their effect on living systems and, particularly in the treatment of dysfunction. The philosophical distinction between something that takes a state of unbalanced disrepair (disease) and rebalances it and something that creates the conditions for healthy growth is clear. This simple division—food provides nutrition while drugs cure disease, has become less clear cut. In the case of macro and micronutrients (as we have seen), deficiencies or even sub (or supra) optimal intakes can result in disease which can be reversed by changing the nutrient status. An example of a crossover effect between nutrients and drugs occurs with GLP-1 agonists. Obesity is a disease (with a number of subsequent co-morbidities) that is caused by overnutrition, and the treatments have included suppression of energy intake, enhanced energy expenditure and increased excretion of caloric foods [43]. The development of GLP-1 agonists as a treatment for obesity has changed the paradigm and created a pharmacological solution to a nutrient-based disorder [44]. In the case of micronutrients, a deficiency in vitamin B-1 (thiamine) can cause neurological dysfunction while excess vitamin A can have a number of deleterious clinical outcomes including headache and dizziness. In both cases, optimisation of intake will alleviate symptoms. LC-PUFAs are, as already mentioned, an example of a member of the class of macronutrients (lipids) which may have a number of specific physiological functions that are unrelated to metabolic health. The division between nutrients and pharmacological/clinical effects is again blurred. The study of pharmacology and the development of drugs has been mostly oriented towards specific disease states. For example, antibiotics will perform the classic 'Silver Bullet' exercise described by Ehrlich by killing harmful microorganisms while being benign (largely) to the host infected species. Further discussion of the concept of the 'magic bullet' is carried out below. Drugs that target specific receptors (e.g. beta blockers) are designed to target specific physiological functions often by acting as receptor or enzyme agonists (activators) or antagonists (blockers). Such specificity is an important aspect of pharmaceuticals—indeed it is a pre-requisite for approval. This encapsulates a challenge for secondary natural products which often have multiple structural features and/or multiple biological targets. The side-effects of drugs are generally the focus of minimisation strategies since these effects are often unwelcome and can limit the therapeutic window. The situation becomes more complex when the condition being treated is either chronic or multifactorial. There are a number of chronic conditions that can develop over a long period of time and for which no targeted pharmacological intervention is possible. Some of these noncommunicable diseases (NCDs) are lifestyle conditions and the development of chronically acting bioactive constituents have been suggested to be of benefit. While in some cases, the solution is to optimise micronutrient status, in others administration of non-nutritive bioactive food constituents may be a more efficacious route. In the case of micronutrients, the condition that is being treated is often not a consequence of deficiency but physiologically unrelated to the primary biological target of the micronutrient. For example, one of the primary targets of selenium is the thyroid and an adequate intake of selenium is important for thyroid health [45]. Other effects of selenium have been suggested to be a consequence of its antioxidative capability—including improvements in cognitive heath, asthma, arthritis, prostate cancer and infertility—however many of these benefits remain unproven [46]. Such claims are often difficult to establish unequivocally since the effects are chronic and there are many confounders so studies to provide proof are often lengthy and require a large intervention cohort over long periods of time to demonstrate statistical significance and this is expensive and the results are often inconclusive.

In the case of secondary natural products, proof of functionality can be even more difficult than for nutrients or micronutrients. At one level, epidemiological studies can provide correlative indicators (but not necessarily proof). For example, there is some indication from dietary intake studies that some cruciferous vegetable may have constituents that confer some degree of neuroprotection for consumers [47]. The challenge is to convert what is, in essence, an observational study into something that has a mechanistic underpinning and can be tested empirically. General comments regarding 'anti-oxidative' or 'anti-inflammatory' effects are not easy to test and, in contrast with pharmacological agents, the effects are both attributable to a range of secondary natural products and any compound may affect a number of biological pathways. Despite the difficulty in carrying out meaningful and definitive studies, considerable efforts have been made to confirm the usefulness of some bioactive compounds in the treatment of chronic diseases. In many cases, interactions with specific pathways have been noted. For example, among other mechanisms, the compound postulated to have a significant biological effect derived from *Brassica* vegetables, sulforaphane, is found to have an effect on the activation of the transcription factor Nrf2 [48]. While this may represent a plausible mechanism, polyphenolic compounds which occur extensively in plants have also been found to impact Nrf2 [49]. Furthermore the flavonoid, resveratrol, which has been extensively studied, has been shown to have a positive effect on the Nrf2 pathway and is postulated to prevent age-related decline in pathways associated with Nrf2 [50]. Not only do individual secondary products apparently compete (and may even synergise) in their effects on clinically relevant pathways, some individual compounds may have an effect on multiple pathways. For example, sulforaphane can form the corresponding dithiocarbamate through reaction with hydrogen sulphide [51]. The reaction is reversible in vivo and may lead to the local release of hydrogen sulphide [52]. The dithiocarbamate may, therefore, function as a prodrug for hydrogen sulphide in addition to its postulated effect on the Nrf2 pathway. The importance of hydrogen sulphide as a gasotransmitter in vivo is increasingly being recognised [53] and so the biological activities of sulforaphane may be linked to this property of the molecule rather than any innate biological relevance per se. Dithiocarbamates have a range of medicinal properties as well [54]. In addition to being a different way of considering the biological activity of secondary natural products, the idea that molecules may be broken down in vivo to reveal a previously unrealised biological profile is, in some ways (as mentioned previously), a mimic of the methods used by organic chemists to determine the structure of individual natural products in the nineteenth century by 'useful' degradation. It is also a more advanced and thoughtful way of considering the biological activity of secondary natural products than the structure:activity relationship approach or the bioactivity 'fishing' approach that are both more commonly used.

1.6 Metabolic Programming and Stratified Nutrition

Metabolic programming is an important concept that lies at the interface between nutrition and pharmacology [55]. The concept of metabolic programming suggests that there are specific windows of opportunity during which nutrient intake (or its absence) can impact upon longer term health effects. This implies that the extent of phenotypic plasticity is different at different stages in development and is therefore susceptible to alteration as a result of environmental factors at a given time point. If phenotypic change is a consequence of genotypic (and epigenotypic) change then phenotypic plasticity is the means by which these effects are realised. As such, it has some similarity to the investigation of chronic disease outcomes mentioned previously. An impact at a given point will, depending on the level of phenotypic plasticity, have a longer term, subsequent effect. The impact of nutrients has been most extensively studied but secondary natural products may also produce effects depending on susceptibility at a given time point. There are a number of concepts that underpin metabolic programming.

- In order for metabolic programming to be measured and its significance determined, predictive biomarkers should preferably be developed. Without these the study becomes largely one of correlation over a complete life course with the impact of all the confounders implicit in this approach.
- Mechanistic or hypothesis-driven studies are more useful since they can help to determine which variables should be altered to confirm (or disprove) a given hypothesis.
- The use of longitudinal studies is a powerful (but costly) tool for determining diet/disease relationships however these can be difficult to set up and to interpret [56].

One of the most significant (and simplest) associations that have been related to metabolic programming is that of low birth weight with an increased risk of adult cardiovascular disease and diabetes [57]. This led to the hypothesis of the developmental origins of health and disease and to the idea that pre- and post-natal interventions may influence the progression of later metabolic diseases. The link between metabolism (and metabolic programming) and the development of immune cell-mediated responses has also been noted [58]. The idea of metabolic programming is, of course, only an option if theories regarding how germ cells operate, which were developed in the nineteenth century, are correct and other theories that were proposed are deficient. For example, using microscopy and significant numbers of developmental studies based on animals, the homunculus theory of germ cells and, later, the homunculus theory of neurology [59] are both incorrect [60]. The impact of, in particular macronutrients on development has remained an important and testable aspect of pre and peri-natal nutrition. There is some degree of confusion as to the difference between metabolic programming and developmental windows. LC-PUFA intake at crucial developmental windows may affect fetal development [61] and a link has been suggested to brain development [62] and intake of docosahexaenoic acid (DHA) has been associated additionally with visual acuity [63]. These developmental effects lie outside the strict definition of *metabolic* programming however they are frequently cited since they can be tested and coherent hypotheses proposed. It should be noted, however, that some studies on brain and visual development have been challenged [64] and the precise timing of the developmental window remains a matter of debate.

In the case of micronutrients, the effect of early exposure and metabolic health programming has been implicated in later disease through epigenetic mechanisms [65].

A further example of metabolic programming which may be a result of early life exposure lies with the gut (and other) microbiomes. While the sterile state of the pre-natal fetal milieu has been assumed, prenatal transfer of microbes to the prenatal fetus from the mother has been suggested. In essence, this hypothesis suggests that the maternal microbiome may influence the microbiome of the pre-natal fetus and post-natal infant [66] and this, in turn affects the epigenesis of the infant microbiome with concomitant effects on health [67].

A further possible effect on programming and its impact on later health arises as a consequence of the metabolism of xenobiotics by endogenous enzyme systems (phase I and phase II enzymes). Both systems comprise both inducible and constitutive enzymes. Phase I metabolism, which is carried out by Cytochrome P450 enzymes, generally leads to activation of the xenobiotic [68]. Typically phase I enzymes will make a xenobiotic more water soluble and may also lead to the biological activation of the compound by inserting (or making available) groups such as OH, NH_2 or SH. For example, some members of the class of pyrrolizidine alkaloids are metabolised by phase I enzymes to give bioactive species that can bind to biological macromolecules including DNA [69]. Phase II enzymes generally catalyse conjugation type reactions (typically glucuronidation or sulphation, for example) and increase the water solubility of the xenobiotic prior to excretion. While substantial inter-individual differences in the levels and activities of phase I and phase II enzymes has been noted, not all of these differences can be explained by differences in genotype or exposure to known inducers and therefore one possibility is that early exposure to secondary natural products may have a metabolic programming effect on the enzyme systems. Using human cells in culture, it was noted that the effect of phytochemicals on phase II enzyme expression in infant human primary skin fibroblast cells and adult cells was different and that metabolic plasticity and response to xenobiotics may be different in infants and adults [70].

Stratified nutrition is a concept that rose from the development of ideas of personalised nutrition (see below) and follows the same concept as metabolic programming—interventions will have different levels of significance at different stages of growth and development. Different individual responses to nutritional interventions may provide a basis for developing nutrition that is tailored towards the individual and these are discussed in more detail below. One of the obstacles to this highly granulated, ‘personalised’ approach is that a diet designed for a specific individual is likely to be expensive and adherence may also be a challenge. The extent of inter-individual variation (particularly in terms of primary metabolism which must fulfil specific functions) is questionable. In addition, the numbers of genetic (and, indeed epigenetic) polymorphisms that would need to be quantified and their various interlinked consequences determined is significant and they are difficult to analyse and therefore it is challenging to come to meaningful conclusions [71]. In terms of secondary natural products, differences in receptor polymorphisms, constitutive or induced phase I and phase II metabolism discussed above and different microbiomes may be responsible for a significant proportion of inter-individual differences observed. Personalised nutrition, even if it is too granular to be practically investigated, is one way of trying to recognise that dietary factors have an important role to play in a number of chronic, non-infectious diseases including some types of cancer, metabolic diseases such as type II diabetes and circulatory diseases including cardiovascular disease and stroke.

Targeted nutritional advice or intervention can be considered with varying degrees of granularity and this defines the size of the population that makes up the cohort. Apart from generalised, populationlevel nutritional advice, two other more specific approaches will be described. While personalised nutrition, which is discussed in more detail below, is probably too specific to be of value or significantly impactful at a population level, the use of stratified nutrition has been developed which is also more tailored than general nutritional guidelines by segmenting at a sub-population level. Stratified nutrition seeks to connect individuals with broad shared characteristics (age, gender, medical history) with beneficial dietary advice. It deals with large groups of individuals organised into cohorts with comparable susceptibilities to disease and it assumes that responses to dietary interventions will be broadly similar. A pharmacological equivalent would be the prophylactic application of statins to people who have familial hypercholesteremia. Stratified nutrition typically deals with broad-based characteristics of groups whose unifying characteristic contains a number of generic characteristics that may respond to dietary changes. For example, dietary phytoestrogens have been suggested to be of benefit to some perimenopausal females [72] however such effects are not universally experienced and there is debate about efficacious dose. Personalised or precision nutrition is even more individualised. As a consequence, the number of factors that have to be measured to obtain a unique health profile of the individual will increase and more biomarkers must be developed and validated. In addition, personalised nutrition will also depend on genotype and phenotype (see above) and so will target the individual but only that specific individual. This creates another challenge. If a particular food or dietary constituent is effective in a personalised nutrition setting then a health claim for a component that applies to the general population cannot be supported by the individualised evidence (see Sect. 1.7 for a more general discussion of health claims). Nutrigenetics is a field of study within personalised nutrition that attempts to correlate the (generally phenotypic) response to food with specific genetic variance. Depending on the extent and occurrence of the genetic variation, the information generated may also impact upon stratified groups and not just the individual.

While variations on the theme of personalised and stratified nutrition can be difficult to quantify and interpret, the more personalised the approach, the more likely it is that advice will be adopted by the individual. The shift from a highly reductionist approach that focuses on the individual to a population-based intervention that will have significantly greater societal impact is important [73].

The simple (and simplistic) notion of applying pharmacological standards of evidence gathering to nutritional, dietary interventions is at best mistaken. The difference between a pharmacological intervention designed to alleviate a specific disease state and a nutritional intervention designed to improve health status is scientifically and philosophically profound. The concept of food as medicine, which has been popularised by some in the nutrition community [74], is essentially a bogus concept since it takes, as a starting point, an essentially disease-free group. The need for significant reforms to nutritional epidemiology has been recognised [75]. Indeed, the idea that randomised control trials (RCTs) might provide a way of bridging the gap between data on the nutritional status of individuals and that of population groups has been criticised [76]. Attempts to build evidence-based models and the constraints inherent in personalised nutrition are an example of two approaches with widely conflicting objectives and, concomitantly, irreconcilable methods. Nutritional science seems to be incapable of deciding whether to fully embrace reductionism and precision or to focus on societal challenges. Resolving these conflicting paradigms will define the future of diet and health and the role, if any, nutritional science will play.

1.7 Health Claims

As noted above, the importance of diet and dietary factors in health is widely appreciated but more difficult to quantify. This is, in part, a consequence of the history of the link between diet and disease and in the difference between nutrients and non-nutrients. The spectrum of effects can be considered in terms of the following.

- Overt toxicity.
- Disease caused by a deficiency in an essential dietary component (macro or micro nutrient).
- The difference between adequacy and an optimal diet.
- Non-nutritive components which enhance health.

Toxic substances are not a normal part of the diet. In some cases, at least, the toxicity of a food component is reflected in its adverse taste characteristics (in particular bitterness). This will discourage consumption therefore development of sensitivity to bitterness can help both the producer (as a consumption deterrent) and the consumer (avoidance of toxin ingestion). Many such principles are non-nutritive. Some substances confer an environmental advantage to the producing species. One example of this is the production of anti-microbial species by some fungi. These compounds are specifically directed against bacterial pathways and are, therefore, selective toxins. Indeed, selective toxicity is a key concept for successful species. The concept of selective toxicity was, in particular, a key aspect of how Paul Ehrlich developed pharmacological compounds. He coined the term 'magic bullet' to describe the administration of a substance that would selectively kill pathogens but would not harm the host. The concept of a magic bullet (Zauberkugeln) most likely derives from the opera Der Freischultz by Carl Maria von Weber which was the first Romantic opera produced in Germany. The term magic bullet was used to describe the organarsenical compound, Salvarsan which was the first effective treatment for syphilis. While overtly toxic species are not part of the diet, some compounds are able to be tolerated and may actually have a useful function. For example, studies on free living chimpanzees suggest that a proportion of the plants they consume have distinct pharmacological profiles including anti-microbial, anti-inflammatory and other activities (for example against gut parasites) [77]. This example demonstrates that toxicity is dependent upon both species and dose. It is also a function of benefit versus harm. A further example of the use of toxins and a positive aspect to their ingestion is the accumulation of exogenous toxins by some species in order to protect themselves against predation by other species. The sequestration of plant toxins by insects has been observed [78]. In order for the strategy to be effective, the host insect must have a tolerance mechanism to protect itself from the toxin and it must also retain (or even enhance) the bioactivity of the toxin. An alternative approach is for the insect (or other species) to store the toxin precursor and to only activate it when required. The cabbage aphid (*Brevicoryne brassicae*) is able to accumulate glucosinolates. When these compounds are hydrolysed by a thioglucosidase enzyme (myrosinase) they produce toxic, volatile isothiocyanates (Fig. 1.6) and other products.

The cabbage aphid can, upon being attacked, produce an endogenous myrosinase and thereby produce isothiocyanates [79]. Isothiocyanates are toxic to a number of aphid predators [80] and also have bacteriostatic and fungistatic activity [81].

Prevention of disease caused by inadequacy in the levels of a given macro or micronutrient can form the basis of a type of health claim. Under many regulatory principles, indeed, health claims are considered in conjunction with nutrition claims. These also include foods for specific groups with particular needs (for example, the elderly, infants etc.). In the EU, nutrition and health claims legislation was introduced in 2006 (Regulation (EC) No 1924/2006) and was effective from 2007. The EU Register of Nutrition and Health Claims lists all permitted nutrition claims and all authorised and non-authorised health claims. The distinction between nutrition and health claims is noteworthy. A nutrition claim is any claim that suggests or implies that a food has particular beneficial nutritional properties (and has a relevant nutritional composition). A health claim refers to any relationship between food and health. In an important change from previous regulatory frameworks, diseaserelated claims are allowed if they can be scientifically substantiated. One of the problems, which was referred to previously, is that the burden of proof generally requires human clinical studies of the type more usually carried out for bioactive pharmaceutical preparations. These are not only expensive, the nature of the disease prevention or protective effect is often relatively small so, in order to achieve statistical significance, very large cohort sizes must be used. While some health and some nutrition claims have been authorised, many are rejected and the cost of compiling a dossier for submission to the European Food Safety Agency (EFSA) can be substantial. Nonetheless mechanistically confirmed claims have been authorised and the development of New Alternative Methods (NAMs) may reduce the volume of confirmatory documents required to make a health claim. A list of deficiency diseases correlated with specific micronutrient deficiencies is given in Table 1.1. In some cases (for example vitamin D and calcium) adequacy in more than one nutrient is required to prevent disease. Genetic polymorphisms may also impact on susceptibility to some diseases.

It is important to recognise the difference between adequacy and optimal intakes. Traditionally insufficiency in the intake of a particular macro or micronutrient has

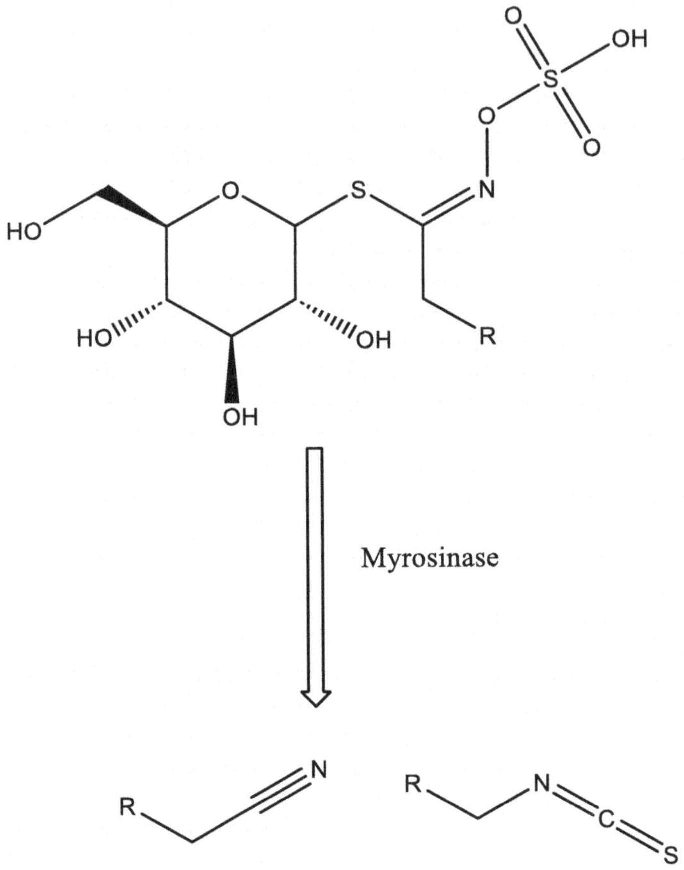

Fig. 1.6 Enzymic hydrolysis of glucosinolates

been recognised by the development of clinical symptoms associated with the disease states given in Table 1.1. Deficiency may also exacerbate other conditions. Optimal intakes are more difficult to define. This is particularly the case for nutrients which have a therapeutic window and can be toxic in high doses.

Hypervitaminosis is rare however, it has been observed, for example, in the case of vitamin D [82]. The optimal daily intake is defined as the level of a given nutrient for optimal health rather than the level required to prevent the onset of a deficiency disease. A number of other definitions are used however many of these apply to the majority of the population (for example, 97.5% is a common benchmark) and are not intended for use as individual intake recommendations. This challenge has been approached by scoring various dietary nutrient intakes for an individual and then providing an index by which individuals can adjust their nutrient intake to achieve an optimal score [83]. Other tools are available including those which, analogously

Table 1.1 Some common micronutrient deficiency diseases and their cause

Micronutrient	Deficiency disease or condition
Vitamin A (retinol)	Night blindness
Vitamin B1 (thiamine)	Beri-beri
Vitamin B2 (riboflavin)	Growth retardation
Vitamin B12 (cyanocobalamin)	Anaemia
Vitamin C (ascorbic acid)	Scurvy
Vitamin D (calciferol)	Rickets
Vitamin K (phylloquinone)	Bleeding after injury
Calcium	Brittle bones
Phosphorus	Dental health, bone health
Iron	Anaemia
Iodine	Thyroid dysfunction
Copper	Growth retardation

to energy density, can measure nutrient density with the aim of providing information about nutrient dense foods.

While macro and micronutrients that enhance health are able to be investigated since they normally correspond to a specific biological mechanism and have a biological target than can be monitored—either directly or through the use of clinical biomarkers and their absence gives rise to quantifiable effects, the situation with non-nutritive (secondary) natural products is more challenging. In most cases there are no measurable indicators of the health endpoint that is affected, and the establishment of a dose/effect relationship can be difficult. One example of this are the plethora of claims and counter claims that relate specific dietary components to the development or prevention of cancer. In some cases, the conclusions are mis-represented in order to create a headline that is easily understood by the consumer. One example of this is the headlines generated by reports in the science literature and in the scientific (and popular) press of the link between consumption of red and/or cooked meat and the risk of colorectal cancer. Although the link was 'weak' it nonetheless exists and therefore the International Agency for Research on Cancer (IARC) classified processed meat in Group 1 of their classification system (carcinogenic to humans). This system fails to take into account the magnitude of the increased risk. The lifetime chance of a non-smoker developing lung cancer is 1%. The lifetime chance of a smoker developing lung cancer is over 20%. There is a large and significant difference between the two groups. In the case of bowel cancer, the chance of developing the disease is about 6%. Consumption of an extra 50 g per day of processed meat every day will increase that chance to 7% (i.e. one extra death). So while both cigarette smoking and consumption of processed meat both meet the IARC requirement for classification in Group 1, the difference in the magnitude of the increase in risk is huge. Despite the nature of the data, some nutritionists will still want to use it as a basis for changing dietary habits. One of the issues with such trials that there is a lack of good data. For pharmaceutical type trials, the scientific standard is random controlled trials (RCTs). In the case of food components, the time required for an effect to be statistically significant is long and that increases greatly the chances of non-compliance and the introduction of confounders. Looking at the inverse of increased risk—lowered incidence of cancer—the situation is also somewhat confusing. Although decreased risk of cancer has been hypothesised due to a number of factors (including lowered inflammation), prospective studies do not appear to provide a convincing link between lowered incidence of cancer and fruit and vegetable consumption [84], however higher fruit and vegetable consumption appears to be correlated with all cause mortality including, in particular, cardiovascular disease [85]. Overall, the link between consumption of fruit and vegetables and the risk of developing cancer remains obscure and generally unproven [86]. A number of researchers have favoured particular types of study (case-controlled or cohort studies) or particular mechanisms (reduction of inflammation has been particularly popular) however the promulgation of intentional or unintentional bias has created a situation where definitive answers as to which secondary natural product may have an effect on the incidence of cancer is impossible to assess. Other clinical endpoints may be more productive but have been less well studied due to the popularity of 'anti-cancer' effects among scientists, clinicians and funding bodies. One example of this is the increased incidence of cognitive decline and, in particular, Alzheimer's disease in ageing populations. There have been a number of associative studies that suggest fruit and vegetable consumption appears to reduce the risk of developing Alzheimer's disease however, as with other disease endpoints, correlation does not prove causality and, rather than looking for correlations, studies should be designed to attempt to disprove simplistic correlations and demonstrate rigorous and definitive links with an absence of confounders. Unfortunately, many scientists who are wedded to the concept of seeking correlations are insufficiently rigorous in their deliberations and either examine pre-existing data (always an easy option since it does not require the generation of anything new) or carry out experiments that are designed specifically to find correlations. In one area, in particular, however, mechanistically-derived hypotheses may be able to be demonstrated. One of the key clinical markers of Alzheimer's disease is the occurrence of beta amyloid plaques in the brain. These are accompanied by (or may lead to formation of) tangles of a hyperphosphorylated protein called tau. Some of the most recent, antibody-based therapies that have been developed, seek to eliminate beta amyloid plaques [87] and such therapies show some promise without being able to be described as a definitive 'cure'. Methods to inhibit or remove tau tangles has also shown some promise (albeit only thus far in animal models) and it is hoped that removal of these aberrant proteins will have clinical significance [88]. While immune-mediated responses may show some promise. Some secondary natural products may also target beta-amyloid and tau tangles. Broccoli juice appeared to inhibit the action of beta amyloid on neuronal cells [89]. In addition *Brassica* vegetables appear to contain neuroprotective agents [47]. So far as mechanisms are concerned, *Brassica* vegetables contain isothiocyanates, some of which can inhibit tau hyperphosphorylation and formation of beta amyloid plaques [90]. In addition isothiocyanates can react with hydrogen sulphide to form dithiocarbamates [91]. This reaction is reversible in vivo and therefore the dithiocarbamates may release both the isothiocyanate and hydrogen sulphide. Both parts of the molecule are neuroprotective [92] and so may inhibit the progress of Alzheimer's disease. If such a 'small molecule' solution can be shown to be effective then both dietary and pharmacological solutions to Alzheimer's disease may be potentiated by a secondary natural product.

The physiological importance of macro and micronutrients is clearly demonstrable and the consequences of deficiencies in these is both quantifiable and, in many cases, relatively easily reversible by supplementation. Measuring the effects (beneficial and deleterious) of secondary natural products is much more challenging. In many cases the pharmacological models (including RCTs) are not fit for purpose and alternative methods (including IT-driven models) may be of value. In many cases the effects of secondary natural products are chronic rather than acute (for example effects on nutrient uptake or on specific organs such as the thyroid or on biological systems such as hormone pathways) and these will be examined in more detail in Chapter Two. Cause and effect, if not underpinned by a mechanistic understanding, is challenging to demonstrate—particularly when relatively small effects that are subject to confounders are being considered and this is part of the challenge for natural products. How to combine reductionist chemistry with the effects on complex biological systems? Part of this challenge is encompassed by the early Kantian concept of the contrast between intrinsic and relational properties [93]. To what extent do the properties of secondary natural products depend on their relationship to complex living systems? We have already touched on this topic in the context of toxicity and the 'magic bullet' and this will be explored in more detail in succeeding chapters.

References

1. Kiani AK, Dhuli K, Donato K, Aquilanti B, Velluti V, Matera G, Iaconelli A, Connelly ST, Bellinato F, Gisondi P, Bertelli M. Main nutritional deficiencies. J Prev Med Hyg 2022;63(suppl.3):E93E101.
2. Altayyar M, Nasser JA, Thomopoulos D, Bruneau M Jr. The implication of physiological ketosis on the cognitive brain: a narrative review. Nutrients. 2022;14(3):513. https://doi.org/10.3390/nu14030513.PMID:35276871;PMCID:PMC8840718.
3. Zárate R, El Jaber-Vazdekis N, Tejera N, Pérez JA, Rodríguez C. Significance of long chain polyunsaturated fatty acids in human health. Clin Transl Med. 2017;6(1):25. https://doi.org/10.1186/s40169-017-0153-6. Epub 2017 Jul 27. PMID: 28752333; PMCID: PMC5532176.
4. Zhang Y, Zhu H, Yang X, Guo S, Liang Q, Lu Y, Chen X. Serum vitamin D level and vitamin D receptor genotypes may be associated with tuberculosis clinical characteristics: A casecontrol study. Medicine (Baltimore). 2018;97(30): e11732. https://doi.org/10.1097/MD.0000000000011732. PMID:30045341;PMCID:PMC6078749.
5. Salahuddin N, Ali F, Hasan Z, et al. Vitamin D accelerates clinical recovery from tuberculosis: results of the SUCCINCT Study [Supplementary Cholecalciferol in recovery from tuberculosis]. A randomized, placebo-controlled, clinical trial of vitamin D supplementation in patients with pulmonary tuberculosis'. BMC Infect Dis. 2013;13:22. https://doi.org/10.1186/1471-2334-13-22

6. Cao Y, Wang X, Liu P, et al. Vitamin D and the risk of latent tuberculosis infection: a systematic review and meta-analysis. BMC Pulm Med. 2022;22:39. https://doi.org/10.1186/s12890-022-01830-5.
7. Peng HY, Man CF, Xu J, Fan Y. Elevated homocysteine levels and risk of cardiovascular and allcause mortality: A meta-analysis of prospective studies. J Zhejiang Univ Sci B. 2015;16(1):78–86. https://doi.org/10.1631/jzus.B1400183. PMID:25559959;PMCID:PMC4288948.
8. Chalupa R, Nesměrák K. Analytical chemistry reveals secrets of alchemy. Monatsh Chem. 2021;152:1019–32. https://doi.org/10.1007/s00706-021-02813-8.
9. Perrin MT, Festival J, Starks S, Mondeaux L, Brownell EA, Vickers A. Accuracy and reliability of infrared analyzers for measuring human milk macronutrients in a milk bank setting. Curr Dev Nutr. 2019;3(11):nzz116. https://doi.org/10.1093/cdn/nzz116. PMID: 31723725; PMCID: PMC6838652.
10. Petersen M, Yu Z, Lu X. Application of Raman spectroscopic methods in food safety: A review. Biosensors (Basel). 2021;11(6):187. https://doi.org/10.3390/bios11060187. PMID: 34201167; PMCID: PMC8229164) and mid infra red spectroscopy.
11. Parat S, Groh-Wargo S, Merlino S, Wijers C, Super DM. Validation of mid-infrared spectroscopy for macronutrient analysis of human milk. J Perinatol. 2017;37(7):822–6. https://doi.org/10.1038/jp.2017.52. Epub 2017 Apr 20 PMID: 28425978.
12. Towett EK, Drake LB, Acquah GE, Haefele SM, McGrath SP, Shepherd KD. Comprehensive nutrient analysis in agricultural organic amendments through nondestructive assays using machine learning. PLoS ONE. 2020;15(12): e0242821. https://doi.org/10.1371/journal.pone.0242821.
13. Lonnie M, Hooker E, Brunstrom JM, Corfe BM, Green MA, Watson AW, Williams EA, Stevenson EJ, Penson S, Johnstone AM. Protein for life: Review of optimal protein intake, sustainable dietary sources and the effect on appetite in ageing adults. Nutrients. 2018;10(3):360. https://doi.org/10.3390/nu10030360. PMID:29547523;PMCID:PMC5872778.
14. Huber KC, BeMiller JN. Carbohydrate analysis. In: Ismail BP, Nielsen SS, editors. Nielsen's food analysis. Food science text series. Springer, Cham; 2024. https://doi.org/10.1007/978-3-031-50643-7_19
15. Jain A, Jain R, Jain S. Protein analysis in food. In: Basic techniques in biochemistry, microbiology and molecular biology. Springer Protocols Handbooks. New York, NY: Humana; 2020. https://doi.org/10.1007/978-1-4939-9861-6_41
16. Lynch JM, Barbano DM. Kjeldahl nitrogen analysis as a reference method for protein determination in dairy products. J AOAC Int. 1999;82(6):1389–98. PMID: 10589493.
17. Wiles PG, Gray IK, Kissling RC. Collaborators: Routine analysis of proteins by Kjeldahl and Dumas methods: Review and interlaboratory study using dairy products. J AOAC Int. 1998;81(3):620–32. https://doi.org/10.1093/jaoac/81.3.620.
18. Kim TK, Yong HI, Kim YB, Kim HW, Choi YS. Edible insects as a protein source: A review of public perception, processing technology, and research trends. Food Sci Anim Resour. 2019;39(4):521–40. https://doi.org/10.5851/kosfa.2019.e53. Epub 2019 Aug 31. PMID: 31508584; PMCID: PMC6728817.
19. Janssen RH, Vincken JP, van den Broek LA, Fogliano V, Lakemond CM. Nitrogen-to-protein conversion factors for three edible insects: *Tenebrio molitor*, *Alphitobius diaperinus*, and *Hermetia illucens*. J Agric Food Chem. 2017;65(11):2275–78. https://doi.org/10.1021/acs.jafc.7b00471. Epub 2017 Mar 14. PMID: 28252948; PMCID: PMC5364430.
20. Tome D. Criteria and markers for protein quality assessment—A review. Br J Nutr. 2012;108(S2):S222–9. https://doi.org/10.1017/S0007114512002565.
21. Gladine C, Roy NC, Rigaudière JP, Laillet B, Da Silva G, Joly C, Pujos-Guillot E, Morio B, FeilletCoudray C, McNabb WC, Chardigny JM, Comte B. Increasing intake of long-chain n-3 PUFA enhances lipoperoxidation and modulates hepatic gene expression in a dose-dependent manner. Br J Nutr. 2012;107(9):1254–73. https://doi.org/10.1017/S0007114511004259. Epub 2011 Sep 14 PMID: 21914239.

22. Ellefson WC. Fat analysis. In: Nielsen SS, editors. Food analysis. Food science text series. Cham: Springer; 2017. https://doi.org/10.1007/978-3-319-45776-5_17
23. Swinnen JV, Dehairs J. A beginner's guide to lipidomics. Biochem (Lond) 2022;44(1):20–4. https://doi.org/10.1042/bio_2021_181.
24. Züllig T, Trötzmüller M, Köfeler HC. Lipidomics from sample preparation to data analysis: a primer. Anal Bioanal Chem. 2020;412(10):2191–2209. https://doi.org/10.1007/s00216-019-02241-y. Epub 2019 Dec 10. PMID: 31820027; PMCID: PMC7118050.
25. Johnston BC, Zeraatkar D, Steen J, et al. Saturated fat and human health: a protocol for a methodologically innovative systematic review and meta-analysis to inform public health nutrition guidelines. Syst Rev. 2023;12:39. https://doi.org/10.1186/s13643-023-02209-1.
26. Yao J, Sterling K, Wang Z, et al. The role of inflammasomes in human diseases and their potential as therapeutic targets. Sig Transduct Target Ther. 2024;9:10. https://doi.org/10.1038/s41392-023-01687-y.
27. Berger MM, Shenkin A, Dizdar OS, Amrein K, Augsburger M, Biesalski HK, Bischoff SC, Casaer MP, Gundogan K, Lepp HL, de Man AME, Muscogiuri G, Pietka M, Pironi L, Rezzi S, Schweinlin A, Cuerda C. ESPEN practical short micronutrient guideline. Clin Nutr. 2024;43(3):825857. https://doi.org/10.1016/j.clnu.2024.01.030. Epub 2024 Jan 30. PMID: 38350290.
28. Wang X, Dou Z, Feng S, et al. Global food nutrients analysis reveals alarming gaps and daunting challenges. Nat Food. 2023;4:1007–17. https://doi.org/10.1038/s43016-02300851-5.
29. Beal T, Ortenzi F. Priority micronutrient density in foods. Front Nutr. 2022;7(9): 806566. https://doi.org/10.3389/fnut.2022.806566. Erratum.In:FrontNutr.2022Apr25;9:908592.doi:10.3389/fnut.2022.908592.Erratumin:FrontNutr.2023Apr11;10:1195752.doi:10.3389/fnut.2023.1195752.PMID:35321287;PMCID:PMC8936507.
30. Cheung RH, Marriott PJ, Small DM. CE methods applied to the analysis of micronutrients in foods. Electrophoresis. 2007;28(19):3390–413. https://doi.org/10.1002/elps.200700100. PMID: 17847130.
31. Brindle E, Stevens D, Crudder C, Levin CE, Garrett D, Lyman C, Boyle DS. A multiplex immunoassay method for simultaneous quantification of iron, vitamin A and inflammation status markers. PLoS ONE. 2014;9(12): e115164. https://doi.org/10.1371/journal.pone.0115164. PMID:25525806;PMCID:PMC4272301.
32. Çimen D, Denizli A. Development of rapid, sensitive, and effective plasmonic nanosensor for the detection of vitamins in infact formula and milk samples. Photonic Sens. 2020;10:316–32. https://doi.org/10.1007/s13320-020-0578-1.
33. Stork G, Niu D, Fujimoto RA, Koft ER, Balkovec JM, Tata JR, Dake GR. The first stereoselective total synthesis of quinine. J Am Chem Soc. 2001;123(14):3239–42. https://doi.org/10.1021/ja004325r. Erratum.In:JAmChemSoc2001Sep5;123(35):8644.FujimotoA[correctedtoFujimotoRA] PMID: 11457058.
34. Seeman JI. The Woodward-Doering/Rabe-Kindler total synthesis of quinine: setting the record straight. Angew Chem Int Ed Engl. 2007;46(9):1378–413. https://doi.org/10.1002/anie.200601551. PMID: 17294412.
35. Gal J. Pasteur and the art of chirality. Nat Chem. 2017;9(7):604–5. https://doi.org/10.1038/nchem.2790. Epub 2017 May 29 PMID: 28644483.
36. Huxtable RJ, Schwarz SK. The isolation of morphine–first principles in science and ethics. Mol Interv. 2001;1(4):189–91 PMID: 14993340.
37. Hosztafi S. The discovery of alkaloids. Pharmazie. 1997;52(7):546–50 PMID: 9266593.
38. Cox DN, Hendrie GA, Lease HJ, Rebuli MA, Barnes M. How does fatty mouthfeel, saltiness or sweetness of diets contribute to dietary energy intake? Appetite. 2018;1(131):36–43. https://doi.org/10.1016/j.appet.2018.08.039. Epub 2018 Aug 31 PMID: 30176297.
39. Sung Y, Yu YC, Han JM. Nutrient sensors and their crosstalk. Exp Mol Med. 2023;55:1076–89. https://doi.org/10.1038/s12276-023-01006-z.
40. MacDonald PE, Joseph JW, Rorsman P. Glucose-sensing mechanisms in pancreatic beta-cells. Philos Trans R Soc Lond B Biol Sci. 2005;360(1464):2211–25. https://doi.org/10.1098/rstb.2005.1762. PMID:16321791;PMCID:PMC1569593.

41. Glucose-sensing mechanisms in eukaryotic cells. Rolland, Filip et al. Trends Biochem Sci. 26(5):310–17.
42. Behrens M, Lang T. Extra-oral taste receptors-function, disease, and perspectives. Front Nutr. 2022;4(9): 881177. https://doi.org/10.3389/fnut.2022.881177. PMID:35445064;PMCID:PMC9014832.
43. Obesity: causes, consequences, treatments, and challenges. J Mol Cell Biol. 2021;13(7):463–65. https://doi.org/10.1093/jmcb/mjab056. PMID: 34673982; PMCID: PMC8530518.
44. Popoviciu MS, Păduraru L, Yahya G, Metwally K, Cavalu S. Emerging role of GLP-1 agonists in obesity: a comprehensive review of randomised controlled trials. Int J Mol Sci. 2023;24(13):10449. https://doi.org/10.3390/ijms241310449. PMID:37445623;PMCID:PMC10341852.
45. Ventura M, Melo M, Carrilho F. Selenium and thyroid disease: From pathophysiology to treatment. Int J Endocrinol. 2017;2017:1297658. https://doi.org/10.1155/2017/1297658. Epub 2017 Jan 31. PMID: 28255299; PMCID: PMC5307254.
46. Bai S, Zhang M, Tang S, Li M, Wu R, Wan S, Chen L, Wei X, Feng S. Effects and impact of selenium on human health, a review. Molecules. 2024;30(1):50. https://doi.org/10.3390/molecules30010050. PMID:39795109;PMCID:PMC11721941.
47. Beretta HV, Torres-Palazzolo C, Ramírez DA, Camargo AB. Brassica vegetables: Rich sources of neuroprotective compounds. In: Gargiulo PÁ, Mesones Arroyo HL, editors. Psychiatry and neuroscience update. Cham: Springer; 2021. https://doi.org/10.1007/978-3-03061721-9_24
48. Russo M, Spagnuolo C, Russo GL, Skalicka-Woźniak K, Daglia M, Sobarzo-Sánchez E, Nabavi SF, Nabavi SM. Nrf2 targeting by sulforaphane: A potential therapy for cancer treatment. Crit Rev Food Sci Nutr. 2018;58(8):1391–405. https://doi.org/10.1080/10408398.2016.1259983. Epub 2017 Jul 21 PMID: 28001083.
49. Sharifi-Rad J, Seidel V, Izabela M, Monserrat-Mequida M, Sureda A, Ormazabal V, Zuniga FA, Mangalpady SS, Pezzani R, Ydyrys A, Tussupbekova G, Martorell M, Calina D, Cho WC. Phenolic compounds as Nrf2 inhibitors: potential applications in cancer therapy. Cell Commun Signal. 2023;21(1):89. https://doi.org/10.1186/s12964-023-01109-0. PMID:37127651;PMCID:PMC10152593.
50. Kim EN, Lim JH, Kim MY, Ban TH, Jang IA, Yoon HE, Park CW, Chang YS, Choi BS. Resveratrol, an Nrf2 activator, ameliorates aging-related progressive renal injury. Aging (Albany NY). 2018;10(1):83–99. https://doi.org/10.18632/aging.101361. PMID:29326403;PMCID:PMC5811244.
51. Tian M, Hanley AB, Dodds MW, Yaegaki K. Chewing gum containing allyl isothiocyanate from mustard seed extract is effective in reducing volatile sulfur compounds responsible for oral malodor. Am J Dent. 2013;26(4):180–4 PMID: 24693626.
52. Rose P, Moore PK, Zhu YZ. H2S biosynthesis and catabolism: new insights from molecular studies. Cell Mol Life Sci. 2017;74(8):1391–1412. https://doi.org/10.1007/s00018-016-2406-8. Epub 2016 Nov 14. PMID: 27844098; PMCID: PMC5357297.
53. Faris P, Negri S, Faris D, Scolari F, Montagna D, Moccia F. Hydrogen sulfide (H2S): As a potent modulator and therapeutic prodrug in cancer. Curr Med Chem. 2023;30(40):4506–32. https://doi.org/10.2174/0929867330666230126100638. PMID: 36703602.
54. Shinde SD, Sakla AP, Shankaraiah N. An insight into medicinal attributes of dithiocarbamates: Bird's eye view. Bioorg Chem. 2020;105: 104346. https://doi.org/10.1016/j.bioorg.2020.104346. Epub 2020 Oct 7 PMID: 33074122.
55. Hanley AB, Dijane J, Fewtrell M, Grynberg A, Hummel S, Junien C, Koletzko B, Lewis S, Renz H, Symonds M, Gros M, Harthoorn L, Mace K, Samuels F, van Der Beek EM. Metabolic programming—mechanisms, opportunities and challenges. Br J Nutr. 2010;104 Suppl 1:S1–25.
56. Sempos CT, Flegal KM, Johnson CL, Loria CM, Woteki CE, Briefel RR. Issues in the long-term evaluation of diet in longitudinal studies. J Nutr. 123:406–12.
57. Fall CHD, Kumaran K. Metabolic programming in early life in humans. Philos Trans R Soc Lond B Biol Sci. 2019;374(1770):20180123. https://doi.org/10.1098/rstb.2018.0123. PMID:30966889;PMCID:PMC6460078.

58. Møller SH, Wang L, Ho PC. Metabolic programming in dendritic cells tailors immune responses and homeostasis. Cell Mol Immunol. 19:370–83 (2022). https://doi.org/10.1038/s41423-021-00753-1
59. Penfield W, Boldrey E. Somatic motor and sensory representation in the cerebral cortex of man as studied by electrical stimulation. Brain. 1937;60:389–440.
60. Catani M. A little man of some importance. Brain. 2017;140(11):3055–61. https://doi.org/10.1093/brain/awx270. PMID:29088352;PMCID:PMC5841206.
61. Kabaran S, Besler HT. Do fatty acids affect fetal programming? J Health Popul Nutr. 2015;33:14. https://doi.org/10.1186/s41043-015-0018-9.
62. Kuratko CN, Barrett EC, Nelson EB, Salem N Jr. The relationship of docosahexaenoic acid (DHA) with learning and behavior in healthy children: A review. Nutrients. 2013;5(7):2777–810. https://doi.org/10.3390/nu5072777. PMID:23877090;PMCID:PMC3738999.
63. Shindou H, Koso H, Sasaki J, Nakanishi H, Sagara H, Nakagawa KM, Takahashi Y, Hishikawa D, Iizuka-Hishikawa Y, Tokumasu F, Noguchi H, Watanabe S, Sasaki T, Shimizu T. Docosahexaenoic acid preserves visual function by maintaining correct disc morphology in retinal photoreceptor cells. J Biol Chem. 2017;292(29):12054–064. https://doi.org/10.1074/jbc.M117.790568. Epub 2017 Jun 3. PMID: 28578316; PMCID: PMC5519357.
64. de Jong C, Kikkert HK, Fidler V, Hadders-Algra M. The groningen LCPUFA study: No effect of postnatal long-chain polyunsaturated fatty acids in healthy term infants on neurological condition at 9 years. Br J Nutr. 2010;104(4):566–72. https://doi.org/10.1017/S0007114510000863.
65. Maguolo A, Gabbianelli R, Maffeis C. Micronutrients in early life and offspring metabolic health programming: A promising target for preventing non-communicable diseases. Eur J Clin Nutr. 2023;77:1105–12. https://doi.org/10.1038/s41430-023-01333-4.
66. Smith-Brown P, Morrison M, Krause L, Davies PS. Mothers secretor status affects development of childrens microbiota composition and function: A pilot study. PLoS ONE. 2016;11(9): e0161211. https://doi.org/10.1371/journal.pone.0161211. PMID:27644050;PMCID:PMC5028039.
67. Lee HS. The interaction between gut microbiome and nutrients on development of human disease through epigenetic mechanisms. Genomics Inform. 2019;17(3):e24. https://doi.org/10.5808/GI.2019.17.3.e24. Epub 2019 Sep 26. PMID: 31610620; PMCID: PMC6808642.
68. McDonnell AM, Dang CH. Basic review of the cytochrome p450 system. J Adv Pract Oncol. 2013;4(4):263–8. https://doi.org/10.6004/jadpro.2013.4.4.7. PMID:25032007;PMCID:PMC4093435.
69. Couet CE, Hopley J, Hanley AB. Metabolic activation of pyrrolizidine alkaloids by human, rat and avocado microsomes. Toxicon. 1996;34:1058.
70. Warwick E, Cassidy A, Hanley B, Jouni ZE, Bao Y. Effect of phytochemicals on phase II enzyme expression in infant human primary skin fibroblast cells. Br J Nutr. 2012;108(12):215865. https://doi.org/10.1017/S0007114512000554. Epub 2012 Mar 19. PMID: 22424477.
71. Webster AK, Phillips PC. Epigenetics and individuality: from concepts to causality across timescales. Nat Rev Genet. 2025. https://doi.org/10.1038/s41576-024-00804-z.
72. Chen MN, Lin CC, Liu CF. Efficacy of phytoestrogens for menopausal symptoms: A metaanalysis and systematic review. Climacteric. 2015;18(2):260–9. https://doi.org/10.3109/13697137.2014.966241. Epub 2014 Dec 1. PMID: 25263312; PMCID: PMC4389700.
73. Barrow M, Bell L, Bell C. Transforming personalized nutrition practice. Nutr Rev. 2020;78(12):1046–51. https://doi.org/10.1093/nutrit/nuaa012.
74. Food as medicine. Translating the evidence. Nat Med. 2023;29:753–4. https://doi.org/10.1038/s41591-023-02330-7.
75. Brown AW, Aslibekyan S, Bier D, Ferreira DaSilva R, Hoover A, Klurfeld DM, Loken E, Mayo-Wilson E, Menachemi N, Pavela G, Quinn PD, Schoeller D, Tekwe C, Valdez D, Vorland CJ, Whigham LD, Allison DB. Toward more rigorous and informative nutritional epidemiology: The rational space between dismissal and defense of the status quo. Crit Rev Food Sci Nutr. 2021. https://doi.org/10.1080/10408398.2021.1985427.

76. Hébert JR, Frongillo EA, Adams SA, Turner-McGrievy GM, Hurley TG, Miller DR, Ockene IS. Perspective: Randomized controlled trials are not a panacea for diet-related research. Adv Nutr. 2016;7(3):423–32. https://doi.org/10.3945/an.115.011023. PMID:27184269;PMCID:PMC4863268.
77. Freymann E, Carvalho S, Garbe LA, Dwi. Ghazhelia D, Hobaiter C, Huffman MA, et al. Pharmacological and behavioral investigation of putative self-medicative plants in Budongo chimpanzee diets. PLoS ONE. 2024;19(6):e0305219.https://doi.org/10.1371/journal.pone.0305219
78. Erb M, Robert CA. Sequestration of plant secondary metabolites by insect herbivores: Molecular mechanisms and ecological consequences. Curr Opin Insect Sci. 2016;14:8–11. https://doi.org/10.1016/j.cois.2015.11.005. Epub 2015 Dec 12 PMID: 27436640.
79. Kazana E, Pope TW, Tibbles L, Bridges M, Pickett JA, Bones AM, et al. The cabbage aphid: A walking mustard oil bomb. Proc Royal Soc B. 2007;274:2271–7. https://doi.org/10.1098/rspb.2007.0237.
80. Sun R, Jiang X, Reichelt M, Gershenzon J, Vassão DG. The selective sequestration of glucosinolates by the cabbage aphid severely impacts a predatory lacewing. J Pest Sci. 2021;94:1147–60. https://doi.org/10.1007/s10340-020-01319-2.
81. Sotelo T, Lema M, Soengas P, Cartea ME, Velasco P. In vitro activity of glucosinolates and their degradation products against brassica-pathogenic bacteria and fungi. Appl Environ Microbiol. 2015;81:432–40. https://doi.org/10.1128/AEM.03142-14.
82. Marcinowska-Suchowierska E, Kupisz-Urbańska M, Łukaszkiewicz J, Płudowski P, Jones G. Vitamin D Toxicity-A clinical perspective. Front Endocrinol (Lausanne). 2018;20(9):550. https://doi.org/10.3389/fendo.2018.00550. PMID:30294301;PMCID:PMC6158375.
83. Busgang SA, Malin AJ, Gennings C. My nutrition index: a method for measuring optimal daily nutrient intake. BMC Nutr. 2022;8(1):16. https://doi.org/10.1186/s40795-022-00497-9. PMID:35189956;PMCID:PMC8862522.
84. George SM, Park Y, Leitzmann MF, Freedman ND, Dowling EC, Reedy J, Schatzkin A, Hollenbeck A, Subar AF. Fruit and vegetable intake and risk of cancer: a prospective cohort study. Am J Clin Nutr. 2009;89(1):347–53. https://doi.org/10.3945/ajcn.2008.26722. Epub 2008 Dec 3. Erratum in: Am J Clin Nutr. 2010 Oct;92(4):1001. PMID: 19056579; PMCID: PMC2647712.
85. Wang X, Ouyang Y, Liu J, Zhu M, Zhao G, Bao W et al. Fruit and vegetable consumption and mortality from all causes, cardiovascular disease, and cancer: Systematic review and doseresponse meta-analysis of prospective cohort studies. BMJ 2014;349:g4490. https://doi.org/10.1136/bmj.g4490
86. DeClercq V, Nearing JT, Sweeney E. Plant-based diets and cancer risk: What is the evidence? Curr Nutr Rep. 2022;11:354–69. https://doi.org/10.1007/s13668-022-00409-0.
87. Avgerinos KI, Manolopoulos A, Ferrucci L, et al. Critical assessment of anti-amyloid-β monoclonal antibodies effects in Alzheimer's disease: A systematic review and meta-analysis highlighting target engagement and clinical meaningfulness. Sci Rep. 2024;14:25741. https://doi.org/10.1038/s41598-024-75204-8.
88. Mukadam AS, et al. Cytosolic antibody receptor TRIM21 is required for effective tau immunotherapy in mouse models. Science. 2023;379:1336–41. https://doi.org/10.1126/science.abn1366.
89. Masci A, Mattioli R, Costantino P, Baima S, Morelli G, Punzi P, Giordano C, Pinto A, Donini LM, d'Erme M, Mosca L. Neuroprotective effect of *Brassica oleracea* sprouts crude juice in a cellular model of Alzheimer's disease. Oxid Med Cell Longev. 2015;2015:781938. https://doi.org/10.1155/2015/781938. Epub 2015 Jun 9. PMID: 26180595; PMCID: PMC4477226.
90. Li, C. et al. Sulforaphane suppresses Aβ accumulation and tau hyperphosphorylation in vascular cognitive impairment (VCI). J Nutritional Biochem. 2024:109803.
91. Tian M. et al. Chewing gum containing allyl isothiocyanate from mustard seed extract is effective in reducing volatile sulfur compounds responsible for oral malodor. Am J Dent. 26 4;2013:180–4.
92. Giovinazzo D, Bursac B, Sbodio JI, Nalluru S, Vignane T, Snowman AM, Albacarys LM, Sedlak TW, Torregrossa R, Whiteman M, Filipovic MR, Snyder SH, Paul BD. Hydrogen

sulfide is neuroprotective in Alzheimer's disease by sulfhydrating GSK3β and inhibiting Tau hyperphosphorylation. Proc Natl Acad Sci U S A. 2021;118(4): e2017225118. https://doi.org/10.1073/pnas.2017225118.PMID:33431651;PMCID:PMC7848711.
93. Langton R. 'Kant's Rejection of Reducibility', Kantian humility: Our ignorance of things in themselves. Oxford;2001; online edn, Oxford Academic, 1 Nov. 2003). https://doi.org/10.1093/0199243174.003.0006

Chapter 2
Secondary Natural Products and Disease

Contents

2.1 Recipe Gin & Tonic and Coffee 33
2.2 Chronic and Acute Disease 38
2.3 Quinine, Caffeine and Bitter Taste Receptors 42
2.4 Classes of Secondary Natural Products 43
2.5 Secondary Natural Products and Their Effects upon Health and Disease 53
2.6 The 'Magic Bullet' and Pharmacology 57
2.7 Mutagenicity and Carcinogenicity of Secondary Natural Products 59
References 61

Abstract The difference between acute and chronic effects and the impact they have can be challenging to determine. Secondary natural products do not have a direct nutritive impact and this can make the consequences of their ingestion difficult to determine. One of the measurable outcomes from secondary natural product ingestion is bitterness which can be measured both in terms of binding to a receptor and through a physiological sensation. The naming of natural products can be based on derivation, chemical or biological characteristics and these can be demonstrated. The biological activity may be a consequence of the compounds as produced or their metabolites or breakdown products. Examples of these can be found and of synergies between different compounds and metabolites. The 'Magic Bullet' concept can apply to some bioactive secondary natural products but is less useful in foods and food systems. The complexity of the potential effects of secondary natural products is illustrated by studies of mutagenicity and carcinogenicity.

2.1 Recipe Gin & Tonic and Coffee

Gin & tonic and coffee are interesting from a natural product chemistry perspective since they contain bitter compounds (quinine and caffeine) which, in addition to being bitter, are also pharmacologically active. Gin also traditionally contains a mixture of herbs—most commonly including juniper berries. Juniper berry oil typically contains monoterpenes including α-pinene, myrcene, sabinene, limonene and β-pinene (Fig. 2.1). The combination of bioactive constituents in gin combined with

B. Hanley, *Secondary Natural Products in Foods and Food Systems*,
https://doi.org/10.1007/978-3-032-00292-1_2

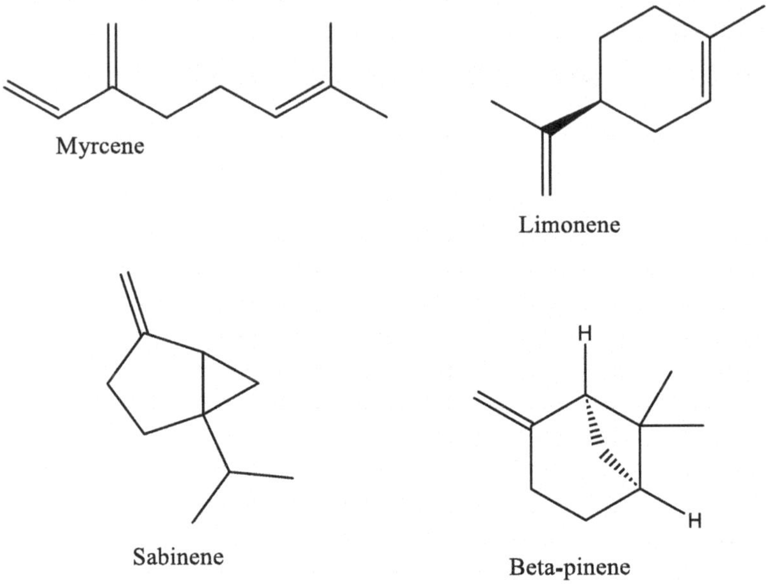

Fig. 2.1 Structures of the chemical constituents of juniper berry oil

the quinine that is used in tonic water suggests that gin & tonic is a rich source of bioactive phytochemicals.

Recipes for the perfect gin & tonic generally involve one part gin to three parts tonic but other ratios have been used depending on individual taste, the type of gin being used and the desired level of alcohol. It is worth being aware that the optimal alcohol level for a refreshing drink may depend on the time of year and the type of gin used. However, there is more to the perfect gin & tonic than simply mixing the two constituents in the correct ratios. Preparing a gin & tonic has some aspects of the tea ceremony.

The first thing is to select your gin. Some of those available are rather heavily influenced by juniper and this can make them somewhat overpowering. There are two principal methods of infusing gin with flavour compounds. A range of flavour notes can be introduced depending on the essential herbs that are added. These can impart floral, fruity (particularly citrus), spicy (cinnamon is favoured by some), piney and herbal flavours.

Steeping extracts the flavours from the juniper and botanicals through a process of soaking the botanicals in the unflavoured natural spirit before distillation. The spirit is distilled and the condensate which is collected is gin.

The alternative method is vapour infusion in which the botanicals are placed in baskets above the base spirit. Heat is applied and as the liquid vaporises it extracts essential flavours from the botanicals. The infused vapour condenses and is collected. The complexity of the flavour of the finished product will depend upon the botanicals used and the method of extraction of the flavour compounds.

Of course if you have access to a gas chromatograph, you can analyse the volatile components of gin and combine this information with taste testing to find your preference. If you employ advanced detection techniques [1] then the greater the power of your analysis and the more components you will be able to identify and characterise. It is, of course, also possible to carry out a series of taste tests to determine your personal favourite however it is important to standardise the taste conditions and there will remain a degree of subjectivity to the choices made. There are a number of online guides that will provide advice on the most appropriate protocols however with all of them, it is best to include a 'retest' phase. Once you have ranked your three favourite gins, go back the following day and check your ranking with a fresh palate. This method is time (and gin) consuming however diligence and optimisation of your choice will be beneficial in the long run. Storage of gin is a much-neglected aspect of gin & tonic preparation. The major volatiles in gin are pinenes. These are unsaturated molecules and will, over time oxidise to give diols and aldehydes. These have a distinctly different flavour profile compared to pinenes. While the oxidation is a relatively slow process, the alcohol (ethanol) in gin can also oxidise giving acetaldehyde and, ultimately, acetic acid. The reaction of acetic acid with residual ethanol can give ethyl acetate which has a strong flavour profile. Other off flavours may also be formed [2]. It is, therefore, preferable to keep gin at a low temperature in the dark since chemical reactions proceed more slowly at low temperatures and light-induced photo oxidation can also occur [3]. The alcohol content of gin (about 40%) will prevent it from freezing in a domestic freezer so this type of storage is also possible.

Once the most appropriate gin has been selected, an appropriate glass should be selected and filled with ice. It should be left for several minutes to cool the glass and the ice discarded and replaced. A small amount (about 10% of the total) of tonic water should be added and the measure of gin. The glass should then be filled with the rest of the tonic. It is usual for a slice of lemon or lime (or even cucumber) to be added but, unlike in a gin martini, these add little by way of flavour compounds and are, therefore unnecessary.

The choice of tonic water is largely down to personal preference. It is worth noting, however, that the amount of quinine generally present in tonic water is considerably (tenfold or so) below the therapeutic dose. If tonic water was (as urban legend suggests) used as a prophylactic measure to counteract malaria, then the levels of quinine present (and the concomitant bitterness) must have been substantially greater than that in modern tonic water. While adverse reactions to the quinine in tonic water are rare, they are not unknown [4], and instances of exanthema have been observed (exanthema is a rash commonly caused by viral infections but can also be a reaction to bioactive compounds—usually drugs). The possibility of an adverse effect from long term quinine exposure has been studied [5] and some association was noted although the quinine exposure (100 mg/day) was high compared to dietary levels. Toxic effects due to quinine consumption have been observed [6]. Quinine is also prescribed for leg cramps and an association with kidney failure in an older population has been noted [7]. The possible link between the biological activity of quinine (and caffeine) and their function as bitter agents is explored further in Sect. 2.3.

There is a long and significant history associated with the consumption of coffee. Coffee was first consumed in Ethiopia before spreading globally. Unlike gin & tonic—which may have been developed initially as a drug (quinine) delivery system, the bioactive ingredients in coffee are an integral part of the popularity of the drink. Coffee was (and still is) consumed for its stimulatory activity. Coffee contains a number of bioactive compounds including caffeine and a range of phenolics which between them may contribute to the health promoting properties of coffee [8]. Some of the components of coffee have been implicated in protective effects against chronic, non-communicable diseases. While a substantial number of species of coffee plant are known, two of them, *Coffea arabica* (Arabica) and *Coffea canephora* (Robusta) make up the vast majority of the commercial crop. A range of techniques are used to extract or solubilise secondary natural products from coffee and these form the basis of making a coffee beverage [9]. In general, the solvent of choice is water although a range of techniques using different solvents and different detection methods have been used [10]. The main bioactive constituents in coffee are chlorogenic acids and derivatives, trigonelline, nicotinic acid, 5-hydroxymethylfurfural, theobromine, theophylline, and caffeine and these are shown in Fig. 2.2.

In addition to extraction to produce a balanced cup of coffee, specific extraction methods are used to remove caffeine (and similar stimulatory compounds containing a xanthine core) and to produce completely soluble, instant coffee.

The production of instant coffee relies upon two techniques to remove the solvent (generally water) after extraction of the soluble components—either spray drying or freeze drying. Prior to this final step, the stages involved in processing coffee (starting with green coffee beans) are roasting, grinding, extraction into boiling water then evaporation and packaging. Spray drying leads to more loss of aroma compounds than freeze drying [11] but it is cheaper. Drying also decreases the rate of microbial spoilage and improves the shelf life of coffee. For some premium brands of instant coffee, some volatiles that are released during the process are added back at the end. Roasting will also inactivate any enzymes in the coffee beans and prevent any degradation of the plant material, however the roasting process can also lead to the formation of acrylamide and strategies have been developed to minimise formation of this unwanted Maillard product [12]. Caffeine is a potent bioactive constituent of coffee. Although it is a (quite) mild stimulant, it has both positive and negative aspects. While caffeine can improve neurocognition [13], it is also toxic (although the small amounts of caffeine in coffee do not pose a significant risk) [14]. Some of the sub-acute symptoms of caffeine ingestion include nausea, palpitations and headache however there is significant interindividual variability in sensitivity [15]. This accounts for the considerable interest in the process of decaffeination and the production of coffee with minimal amounts of caffeine. One of the problems with the removal of caffeine is that a number of other compounds that contribute to the aroma and taste of coffee may be removed as well and even caffeine itself has a flavour profile that contributes to the organoleptic properties of the drink (see Sect. 2.3). While decaffeinated coffee is only about 10–12% of the total coffee that is consumed, this is a significant market and is predicted to grow therefore there is interest in ensuring that the decaffeination process is able to be further developed and optimised. As

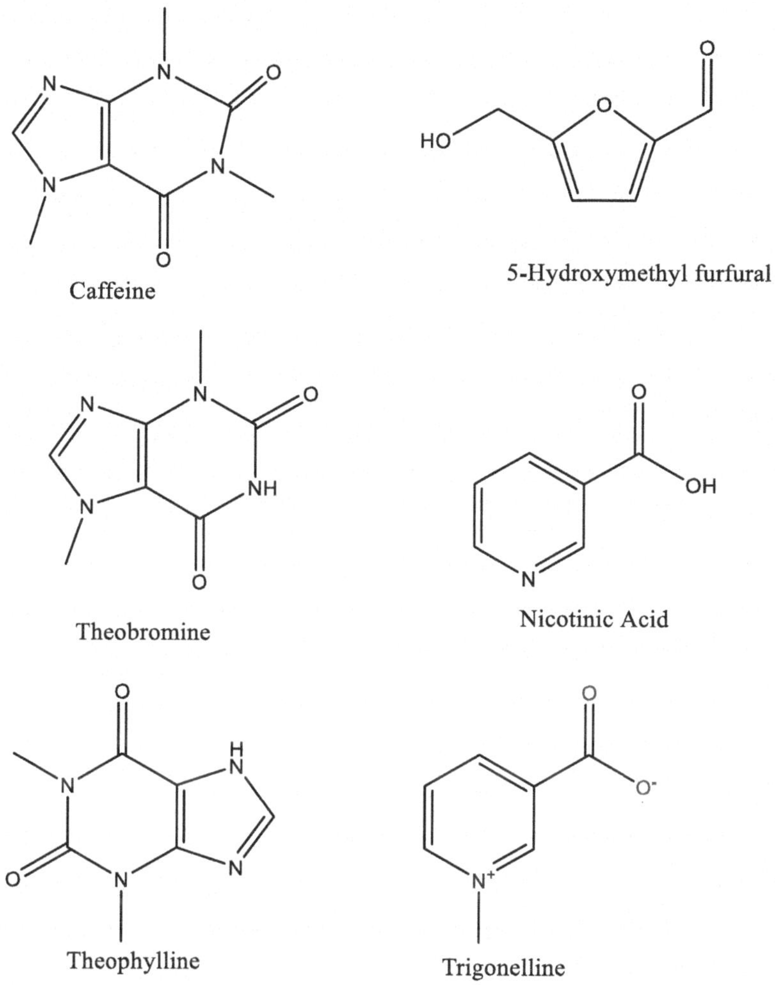

Fig. 2.2 The main bioactive constituents of coffee

noted previously, caffeine can be removed by extraction with an organic solvent (ethyl acetate and methylene chloride are the most commonly used of these), by water, either alone (in which case some of the other phenolic and other organic compounds have to be isolated and added back to the decaffeinated coffee), using activated charcoal to remove the caffeine or by using supercritical carbon dioxide [16]. The differences in the chemical composition of regular and decaffeinated coffee have been explored [17]. It was found that pyrazines, in particular, were reduced by the decaffeination process and this had a significant impact upon the taste profile of the final beverage. While the methods to make coffee from beans has, to an extent been standardised, the effects of various parameters (e.g. roasting, particle size after grinding, water quality) has been studied [18]. Making coffee from roasted beans is a solid/liquid extraction process and is therefore affected by several parameters that are understood to greater or lesser degrees. Overall, because of advances in analytical chemistry and the adoption of a reductionist chemistry approach, the effect of variation in the parameters associated with brewing coffee on the extraction or production of specific chemicals is better understood, the interplay between physicochemical variables is less extensively explored.

There is very little that can be manipulated in the production of instant coffee. For that reason, the process that will be described is for the production of coffee from beans. In addition, while there are a number of coffee machines on the market including those that make use of pre-filled coffee 'capsules', the simplest and most cost effective way of making fresh coffee is to use a French Press or cafetiere. The process is relatively simple and can be summarised as follows. Ground coffee is placed in a cylindrical jug. Boiling water is added and the mixture left to brew for 4 min. The grounds are then separated by pushing a plunger with a built-in filter and the filtered coffee poured from the top of the jug. While it is possible to use pre-ground coffee, this loses some flavour over time (even if kept in a sealed container such as a Kilner jar under refrigeration) and so it is better to grind the roast coffee beans freshly as required. The steps in the process are as follows.

- ***Weigh out the required amount of coffee beans. Weighing is better that using a scoop to estimate the amount of coffee since measurement by volume may vary by about 10%. The amount of coffee is a matter of personal preference but a ratio of about 20 g of coffee beans to 500 ml of water is a good starting point. The type (level of roast and origin) of the beans is also a matter of personal preference but a medium roast from one of the major coffee producing regions is a good place to begin.***
- ***Grind the coffee beans either in a purpose made grinder or in a spice grinder and place in the jug. It is also possible to grind beans by hand using a pestle and mortar but this a timeconsuming process and difficult to standardise. Add hot (almost boiling) water. Most experts suggest a water temperature of about 92–95C although since the water will cool as you pour it, there is no need to be obsessive about this.***
- ***Leave the mixture to stand for 3 min then stir once with a spoon. Leave a further minute then depress the plunger slowly and pour the coffee.***

2.2 Chronic and Acute Disease

Many secondary natural products which are found in foods and in food systems exhibit an effect on chronic rather than on acute disease states. This is a consequence of the biological activity and the low levels found to occur and was mentioned in Chap. 1 in the context of the difference between nutrition and pharmacology. This will also be explored further in Sect. 2.5. Both chronic and acute disease are defined in terms of measurable clinical endpoints but the definition of each also tends to reflect the rate of progression of the disease or condition. Acute states progress quickly while those that are chronic have a slower evolution. One of the problems with using the rate of disease progression as a defining metric is that this can mask both the severity and the site of the disease. For example, a cut to a finger can be considered to be acute (it occurs quickly) but in many cases it is neither severe nor physiologically of great significance and will often heal without any external treatment. Alzheimer's disease, on the other hand, usually takes a considerable time to develop but once beta-amyloid plaques and tau tangles have accumulated, the process is largely irreversible and the impact is severe. Acute does not necessarily equate to physiologically critical while chronic does not mean insignificant. According to the Center for Disease Control (CDC) in the US, a chronic condition is one that lasts for a year or more and affects both health and quality of life. Many acute conditions at both an individual and societal level arise as a consequence of infectious disease. Indeed the term 'non-communicable disease' (NCD) has been used as an analogy for chronic disease based upon the timeline of the disease and the nature of the process. Clearly since there is both a mortality and a morbidity aspect to many NCDs and susceptibility increases according to socioeconomic factors, the economic aspect of NCDs has been recognised in many studies [19] In brief, the prevalence and impact of chronic diseases or NCDs—including, in particular, diabetes, stroke, cardiovascular disease and cancer—is linked to economic inequality. In addition, chronic disease is often associated with a lack of fitness and resistance to co-morbidities [20].

In the context of secondary natural products, low level, chronic effects are observed with consumption of a number of them however precise and predictive clinical data and appropriate experimental models are not commonly available. For example, the fungal metabolite, ochratoxin A is associated with a number of disease endpoints including Balkan endemic nephropathy and it is tempting to see causality in what is, in fact, possibly a coincidence and the normal exposure levels and the lack of long term toxicity data suggest that ochratoxin A is unlikely to be a significant factor in any nephrotic disease state [21] in the short or medium term although the risk of consequences of weakened kidney function or the additive effects of ingestion of combinations of toxins (e.g. fungal toxins together with other toxins such as the plant-derived toxin, aristolochic acid) cannot be discounted [22]. The use of animal (particularly rodent)-based models is still the most widely accepted test method for measuring sub-acute and chronic toxicity although the cost and relevance of the models to humans are challenging factors. International standards to limit exposure to some natural toxins are developed by the Codex Alimentarius Commission.

Secondary natural products confer some advantage to the organism that produces them. This can include toxicity to competing species, protection against microbial or insect attack and/or taste aversion directed against predators. This creates a biological paradox—how does the species that produces the secondary natural product protect itself against continual exposure to the bioactive compound? In some cases, the constitutively produced secondary natural products which are present at sub-lethal levels, are supported by stress-induced formation either of higher levels of a particular secondary product at elevated levels or of a new spectrum of products [23]. In other cases, the toxic secondary natural product is stored in an inactive form and the

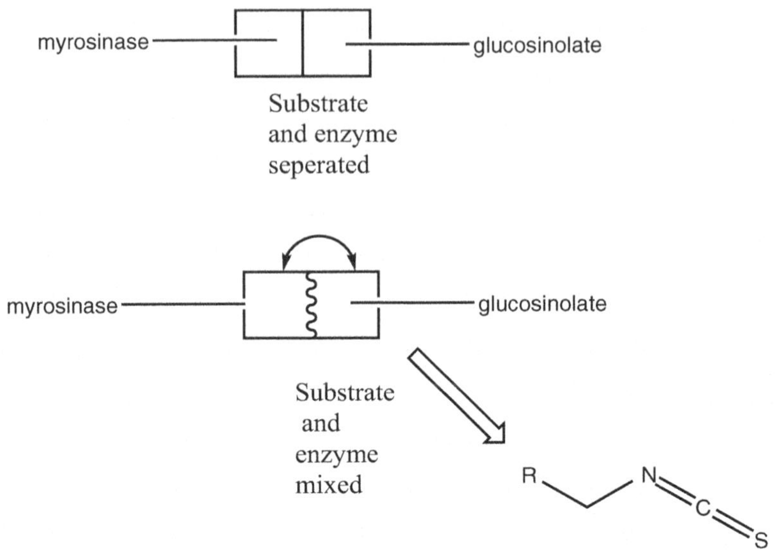

Fig. 2.3 The glucosinolate/myrosinase system

bioactive compound is released when the host organism is damaged. An example of this is the glucosinolate/myrosinase system (Fig. 2.3).

In this case the glucosinolate is stored in one part of the organism and a thioglucosidase in adjacent parts. When the tissue is disrupted (for example, as a result of herbivore or insect damage), the two components are mixed and the resultant isothiocyanate is the toxic entity that is released. Both the enzyme and the substrate are required and this has been demonstrated by the removal of the cells that contain the enzyme (myrosin cells) [24]. In many cases, the effects of secondary natural products are directed towards biological pathways that the host either does not use or are nonessential. An exception to this rule occurs in the case of 3-nitropropanoic acid. This toxin can occur as a free compound or as a glycoside which is hydrolysed in the gut of consuming animals to give the acid. The free acid is a suicide inhibitor of succinate dehydrogenase and therefore inhibits a ubiquitous energy pathway (the tricarboxylic acid cycle). As such, it will be toxic both to the producing species and to any consuming organism. Many species that produce 3-nitropropanoic acid also produce a nitronate mono-oxygenase that converts the toxin to malonic semialdehyde thereby providing a degree of protection for the host species [25]. In some cases, the toxin produced is directed against pathways that are specific to the target species and are therefore benign to the producing organism. One of the most significant examples of this occurs with penicillins and cephalosporins (Fig. 2.4).

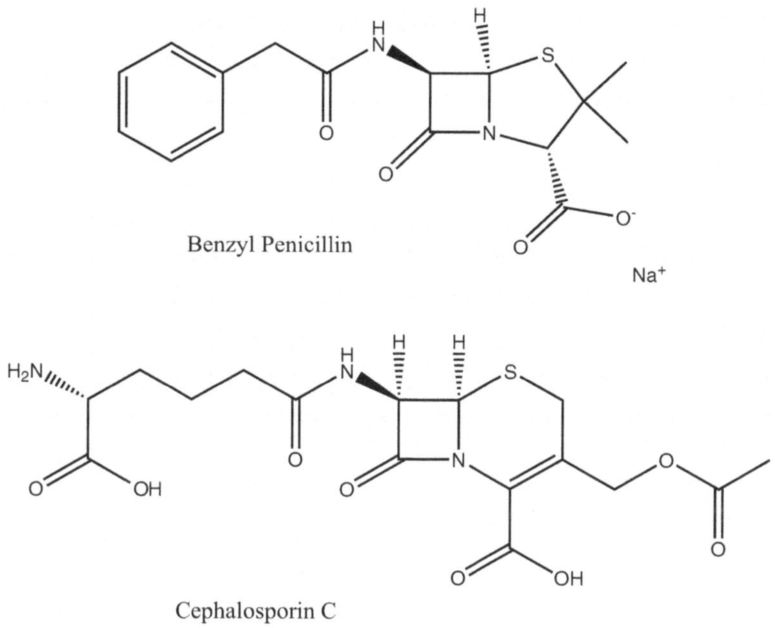

Fig. 2.4 Structure of penicillin and cephalosporin

Both of these compounds, which are produced by fungal species, work by inhibiting the transpeptidase that catalyses the cross-linking of bacterial peptidoglycan cell walls [26]. Since nonbacterial species do not make use of the peptidoglycan system of cell wall synthesis, penicillins and cephalosporins demonstrate selective toxicity.

In addition to intrinsic biological protection against the effects of toxins, protection against some toxic secondary natural products in food products can be achieved by processing [27]. For example, cassava contains cyanogenic glycosides which can produce hydrogen cyanide if consumed. The toxic effects of hydrogen cyanide are observed upon consumption at acute and sub-acute levels. Peeling the rood followed by grating enhances both the removal of cyanogenic glycosides and their hydrolysis by a co-occurring glycosidase enzyme. Heating, boiling in water, drying and fermentation are also effective methods for reducing cyanogenic glycoside content [28].

Other techniques such as microwave, ultrasound and high-pressure processing have been used to reduce the levels of some secondary natural products. The challenge, with all methods, is to reduce the levels of toxins selectively without degrading the nutritional value of the food. Methods such as supercritical extraction can, as was observed for decaffeination of coffee, be effective however the high initial capital investment will be a barrier except for higher value or high-volume foods.

2.3 Quinine, Caffeine and Bitter Taste Receptors

Although quinine and caffeine have very different biological profiles, they both bind to bitter receptors—although transmission of bitter taste is not the sole purpose of bitter taste receptors. They are found in a number of extra-oral sites including the pancreas, brain and alimentary tract. This suggests that taste-receptor mediated activity may not be purely associated with the sense of taste. Unlike some other taste receptors (salt and sour), there are multiple bitter taste receptors—in humans there are 25 different members of the class. They are G Protein Coupled Receptors (GPCRs) and are designated TAS2 type receptors followed by an integer. Studies have demonstrated that caffeine binds to a number of bitter receptors including TAS2R7, TAS2R10, TAS2R14, TAS2R43, and TAS2R46 [29]. In addition to binding to a range of bitter receptors, the affinity of some compounds is also affected by polymorphisms within individual receptors. For example, polymorphisms in TAS2R43 are associated with coffee preference and perceived bitterness. Some of the bitter receptors for which caffeine is a ligand (and other bitter receptors) have been found in a range of extra-oral sites and this may have biological significance. For example competence stimulating peptides from some bacteria act as ligands for TAS2R14 and this may have biological significance [30]. A role for taste receptors in the immune response to microbial quorum sensing systems has been suggested [31] and caffeine has been shown to have an immunomodulatory effect which may be potentiated by interaction with bitter receptors [32]. The role of coffee consumption on human health has been extensively reviewed [33] and some of these effects may be a consequence of binding to extra-oral bitter receptors.

In the case of quinine, one biological effect which has been documented is the reduction in calorie intake and this may occur as a result of interaction with bitter taste receptors in the gut [34] through enhancing cholecystokinin (CCK) release, a known satiety-inducing hormone [35]. Quinine is known to stimulate nitric oxide production in a way that is consistent with extra oral TAS2R stimulation [36] and the role of specific members of the class of bitter receptors has been explored [37]. Quinine, along with some other bitter compounds with different pharmacological profiles, binds to TAS2R7 [38] and this particular receptor has been implicated in the response to metal ions (analogous to the calcium-sensing receptor) in both oral and extra-oral tissue [39]. This might suggest that defining the biological activity of bitter compounds based solely on their bitter taste receptor binding profile in oral tissue may be inappropriate. An in vitro assay of the binding of a number of compounds to a range of TAS2R bitter receptors has been carried out [29]. The implication of this study is that the spectrum of binding of bitter compounds to different bitter receptors will influence their perceived taste characteristics and may also give some indication as to their biological activity. This also suggests that bitter compounds may have a range of different biological activities which are, at least in part, dependent on the extra-oral site and role of bitter receptors.

2.4 Classes of Secondary Natural Products

The discovery of secondary natural products evolved alongside the development of reductive chemistry methodologies and the realisation that natural products were the same as non-natural (synthetic) materials. The rejection of vitalism allowed natural product structure determination to develop alongside the lysis-synthesis paradigm since there was no unique property that was possessed by material that derived from a living source. In the eighteenth century, only a small number of compounds had been isolated from natural sources and these were divided into organic and inorganic (mineral) substances. While investigative methods were still relatively scarce, an appreciation grew that some individual organic substances could, in some cases, be transformed into other materials by living systems and this became the basis for fermentation and other techniques. The transformations that the alchemists sought became less viable and knowledge grew as to the atomic nature of matter and the difference between elements and molecules. By the middle of the nineteenth century, analysis had shown that organic materials contained principally carbon, hydrogen and oxygen. Additionally some also contained nitrogen and other chemical entities in smaller amounts. While an understanding of chemical functional groups was not fully developed, the idea that substances could be combined to give new compounds was appreciated—even if the rules that governed such combinatorial approaches were still obscure. In addition, Lavoisier stated that both analysis (breaking a substance into identifiable constituent parts) and synthesis (creation of a molecule that was identical to that which was isolated) were required to confirm a structure. Overlaid upon this were the traditional alchemical methods of distillation, sublimation and extraction as well as combustion and these techniques remained the principal tools that were available to chemists.

The extraction and purification of bioactive principles significantly predates knowledge of chemical structures. In many cases, the presence of active principles was (at least in part) ascribed to external agencies (e.g. phases of the moon or the position in the sky of particular constellations) [40]. While this may seem fanciful, it derives from an attempt to correlate the properties of substances with factors that can be observed and measured rather than to some invisible (and indivisible) property (atoms and molecules). This also led to the development of therapies that may seem strange including those based upon ‘the mosses of dead men’s skulls’. Lichens or other growths on the skulls of men who had suffered a sudden or violent death were considered to have considerable curative properties [41]. Some opportunistic fungal pathogens, including those of the *Mucormycetes* family have been found as infective agents in the brain and these can cause both sudden death and may be a source of fungal-derived antibiotics including polyketides. Some isolated instances of cures for normally lethal bacterial infections by extracts of ‘mosses of dead men’s skulls’ which contained polyketide antibiotics may have formed the basis for a belief in this particular folk remedy.

The classification of secondary natural products was driven initially by common structural features that gave rise to particular properties. These have also helped in the development of specific extraction and isolative techniques [42]. These include solubility in alcohol or in water and acidic or basic properties. Another key feature of compounds (which eventually led to the development of 'affinities' and the concept of valency) was neutralisation. Mixing an acid and a base led to the formation of a neutral 'salt' and by knowing how much of a certain acid was required to neutralise a given base, the strength of the base (in acid equivalents) could be determined. Another key principle was that of functional groups. If organic compounds were reduced to their components, then it was found that many of them had the same empirical formula (proportions of carbon, hydrogen and oxygen) but had very different properties. This led to the development of ideas concerning the structures of compounds rather than simply their atomic composition being their chief defining characteristic.

Natural products with specific characteristics began to be isolated in the eighteenth century. Since knowledge of chemical structures was not available, compounds were named after their derivation and broad chemical properties. This resulted in compounds being described as, for example 'sugar acid' (aldonic, aldaric, and uronic acids) and 'formic acid' (derived from the Latin for ant—formica—from the family Formicidae).

One of the most significant classes of functionally classified secondary natural products are alkaloids. These were first described by Meissner in 1819 who isolated plant-derived 'alkali-like' principles (Pflanzenalkali). The name alkaloid was given to this type of compound that Meissner isolated since they were very different in properties from mineral alkalis [43]. Meissner also proposed that in any treatise on plant chemistry, the alkaloids would precede plant acids (sauren). Interestingly the purification of the first alkaloid (morphine) by Serturner in 1804 predated the naming of the class of compounds. While alkaloids as components of plant extracts had been known for a considerable time, their isolation was an essential precursor to their safe and effective use in pharmacology since knowing how much was being administered was a vital characteristic of their use in medicines. The isolation of alkaloids is aided by their basic properties. By varying the pH, the solubility of alkaloids in aqueous and immiscible organic solvents can be manipulated. This is the basis of the Stas-Otto method for alkaloid isolation [44]. In many cases, there was a requirement for toxic alkaloids to be released from the tissue of people who had been purposely poisoned and this led to the development of many clinical chemistry modifications of the basic procedure. While the free bases of alkaloids are often difficult to crystallise, the corresponding salts are frequently easier to handle and fractional crystallisation can sometimes be a viable option [45]. Nonetheless, despite the sometimes considerable technical challenges, alkaloids such as atropine, strychnine and theophylline were isolated in crystalline form before the advent of chromatographic methods of separation.

One important factor in the isolation and purification of natural products is the stability of the compounds to the chosen methods. Extraction with hot solvent (methanol and ethanol—sometimes with an aqueous component added) is the method of choice and it is both convenient and technically unchallenging although heat labile compounds are often broken down with a concomitant deleterious effect on the yield of isolated material. Various distillation techniques were also common in the nineteenth century but were only able to be used for heat stable compounds. As a consequence of this there has been a growth in some less harsh technologies and this has led to the development of techniques including microwave-assisted and supercritical fluid extraction methods. Chromatographic techniques have also been developed to enable the fractionation of molecular types with solid–liquid adsorption chromatography being the most general method of choice. It has been found that some natural products are unstable under some fractionation and analytical conditions (see [46]) for an example and this has limited the use of alumina and silica in some cases. A number of other separation techniques such as partition chromatography, countercurrent chromatography, gel filtration chromatography and ion exchange chromatography have also been developed however some of these are used for specific types of molecules only or have other drawbacks that limit their effectiveness. One area that appears to be of particular value is preparative molecular imprinting technology [47]. This method can be specifically tuned for specific structural characteristics and is, therefore, less dependent upon the trial-and-error approach of traditional chromatography.

As isolative techniques improved, the types of secondary natural products that could be isolated increased and compounds present in smaller and smaller amounts could be purified. In the nineteenth and early twentieth centuries, structure elucidation followed the Lavoisier principles of degradative analysis and synthesis as coupled confirmatory techniques. In the absence of advanced analytical (particularly spectroscopic) methods, there was a reliance upon degradation and sub-structure analysis together with an increased awareness of the biosynthetic derivation of natural products which enabled compounds with similar derivations to be classified together. The lack of tools to aid direct structural identification led to the accumulation of experimental data and structural assignment by inference. Alkaloids generally have complex structures and in many cases, their isolation preceded their structural identification by a considerable period. This led to alkaloids (and other natural products) being named according to any details of their biological or chemical derivation that could be surmised (see below) rather than being grouped according to their core structure (as occurred later). For example, the indole alkaloid yohimbine (Fig. 2.5) was isolated from the bark of the tree *Pausinystalia johimbe* and atropine (Fig. 2.5) was derived from *Atropa belladonna.*

The derivational approach means that alkaloids have been classified according to their biosynthetic derivation although this creates some anomalies and some alkaloids cannot be classified in this way. It can give rise to classification of some of the major classes based upon the amino acid from which they derive.

Indole alkaloids—tryptophan.

Piperidine alkaloids—lysine.

Pyrrolidine alkaloids—ornithine.

Phenylethylamine alkaloids—tyrosine.

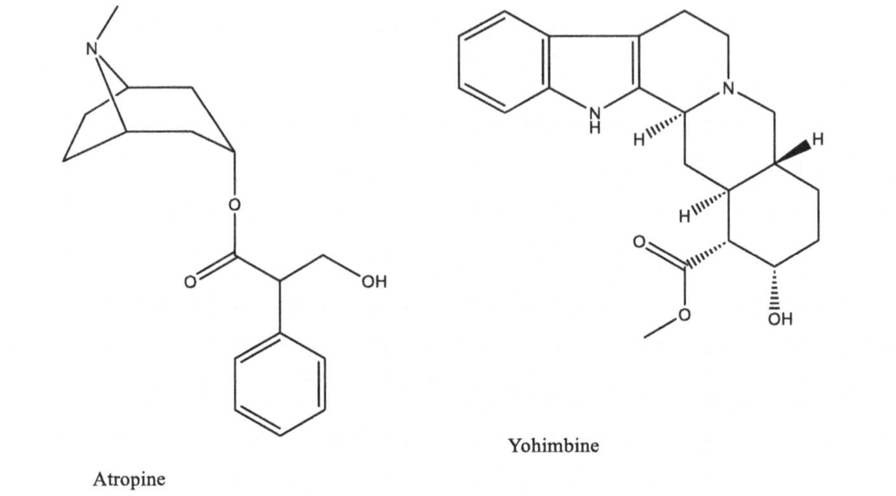

Fig. 2.5 Structures of yohimbine and atropine

Imidazole alkaloids—histidine.

As structural awareness grew and there were advances in chemical structure determination, the classification of alkaloids was based more upon similarities in their core chemical structure (e.g. steroidal and terpene alkaloids) than on their botanical derivation. Other classifications based on biological/functional properties (e.g. analgesics, hallucinogens etc.) was also possible.

Other secondary natural products are known including those derived from other (non-nitrogenous) core chemical structures or biosynthetic pathways. Perhaps the simplest classification is a mixture of structural and biosynthetic considerations. This divides natural products into a limited number of groups—alkaloids, phenylpropanoid-derived compounds, polyketides, amino acid-derived molecules, fatty acids and terpenoids. There are, of course, sub-divisions within these categories. The concept of the derivation of secondary natural products being derived from primary metabolites has been the subject of considerable biosynthetic and mechanistic interest. In the case of polyketides, for example, the condensation of acetyl CoA with malonyl CoA is an important initiating step and is the link between primary and secondary metabolism. Amino acid-derived natural products are common and also represent an intersection between primary and secondary metabolism. Post-translational modification is an important factor in the development of the final natural products and can have a high level of significance in defining their physiochemical and biological properties.

There is perhaps a wider philosophical point to be made about natural products. Clearly organisms are considerably invested in creating these products and it is generally assumed that the compounds confer some advantage, as inferred previously, for the host species. In some cases this may manifest in a bitter taste or some other adverse property for a consuming species to deter predation. Natural products may prevent microbial attack (as it the case for plant-derived anti-microbial compounds) or may be selectively toxic to species which compete for an ecological niche or for resources. It is unclear, however, why such a plethora of compounds of different structural types are made. A limited number of bioactive entities would be required to confer the purported growth or survival advantage and the diversion of such a large amount of primary metabolites into secondary natural product production would appear to be an inefficient use of resources. In addition, in many cases, secondary natural product formation is increased when an organism is placed under stress and it might be considered more likely that building up a store of precursors of primary metabolites for use in growth and survival situations would be of greater value.

One example of a class of secondary natural products that divert amino acids away from primary metabolism and may simultaneously confer a growth advantage for plants that produce them are glucosinolates. These compounds occur widely in food plants—in particular in plants of the *Brassicaceae* family and, in addition to being plant protection agents, also exemplify the principles of the use of an inactive 'pro-drug' approach where a compound is created as an inactive form and is then transformed into a biologically active species and used in nature. In general glucosinolates occur as mixtures derived from different (and differently modified) amino acids in vegetative tissue however some seeds contain just a single member of the class and this facilitates isolation of enough of the compound to use as a standard and for biological testing [48] however the complex nature of the different structures of individual compounds in the class, the varied occurrence of glucosinolates and the difficulty in isolating large amounts of pure material means that many studies have been carried out using whole plant material or extracts with the attendant problem of confounders affecting the interpretation of results. This is perhaps less of a problem for nutritional studies which focus more on the effects of diets on primary metabolism than pharmacological or mechanistic investigations for which pure material is required in order to generate meaningful results. Having introduced the concept of glucosinolates, a more detailed consideration of the compounds and their significance now follows.

Glucosinolates are amino acid-derived secondary natural products that occur in a number of plants of the order *Capparales* and, as noted above, particularly in the family *Brassicaceae.* They are one of the relatively limited number of S-containing natural products and consist of a thioglucose moiety and a sulphonated oxime (Fig. 2.6).

Glucosinolates are relatively biologically inert but are hydrolysed by a co-occurring thioglucosidase enzyme, myrosinase. The cleavage of the glucose moiety results in a Lossen type rearrangement and formation of a number of products including highly reactive isothiocyanates. The original structure of glucosinolates was proposed by Gadamer in 1897 [49], however the nature and formation of mustard oil which comprises largely allyl isothiocyanate was described earlier [50]. The structure proposed for the precursor of the isothiocyanate was logical in that it explained well the formation of the isothiocyanate since the important aspects of that moiety were present (Fig. 2.8). It was only considerably later that the correct structure was confirmed by isolation, analysis [51] and synthesis [52] and by an explanation of

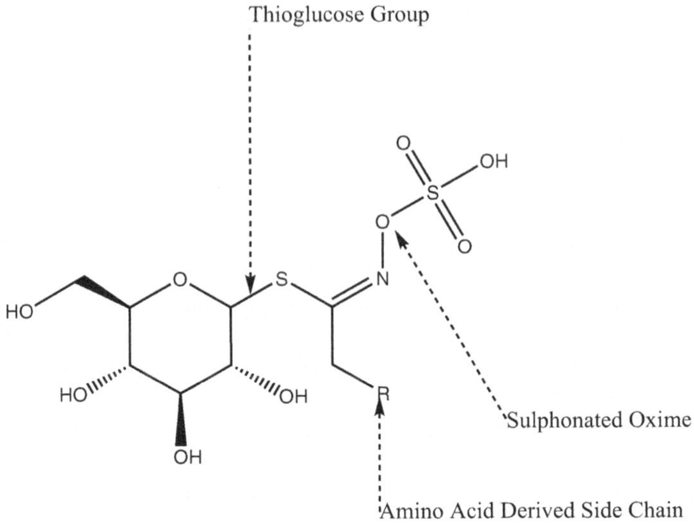

Fig. 2.6 Annotated structure of glucosinolates

the mechanism of formation of the isothiocyanate via a Lossen type rearrangement [53] which was originally described in 1875. Perhaps one of the most significant aspects of the biology of glucosinolates is a consequence of the different natures of the glucosinolates themselves and their enzymically-produced metabolites. Glucosinolates are water soluble salts and are physically separated from the thioglucosidase enzyme (located in myrosin cells) in plant cells. It is only upon cellular disruption (damage) that the enzyme and substrate are mixed and the isothiocyanate (and other organic compounds derived from the side chain) are formed. The range of compounds formed is largely a consequence of the large number of different glucosinolates that are synthesised by modification of the core amino acid structures. In contrast, while the Scontaining compounds of the genus *Allium* are stored using a similar approach of separating an enzyme from its biologically relatively inert secondary natural product, there are only a limited number of inert substrates and most of the chemical diversity arises from reactions between the hydrolysis products (Fig. 2.7).

In summary, most of the structural diversity of the glucosinolate metabolites derives from variability in the side chain of the parent glycoside while most of the structural diversity of S compounds of the genus *Allium* derives from the reactivity of the post-hydrolysis compounds. The diversity in starting materials is reflected in the biosynthesis of glucosinolates (Fig. 2.8).

Other methods that are used by host plants to protect themselves from the consequences of production of bioactive secondary natural products have been noted previously but the glucosinolate and *Allium* systems are both complex and require a considerable biosynthetic effort to not only produce the required products but also to ensure that they are employed to maximum effect. Producing a range of products may give

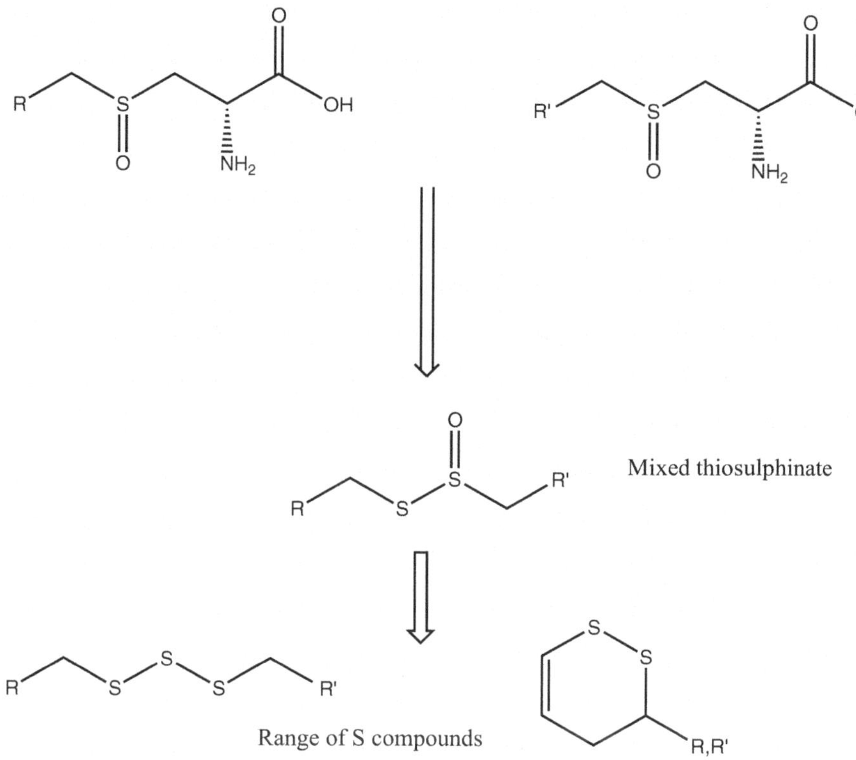

Fig. 2.7 Formation of S-containing compounds in allium species

a host species additional protection against predation through synergistic or complementary mechanisms however this may not invariably be the case. For example, while synergy between sweeteners has been noted [54], a similar effect has not been noted for bitterness. Some compounds are known to bind to multiple bitter receptors and synergistic effects have been noted in the activity of isothiocyanates when used in an assay of cancer cells [55] however this is a very specific instance and it is unclear if additional protection is afforded against predators by mixtures of isothiocyanates, for example. One notable example of a potential synergistic effect has however been noted in the anti-fungal activity of glucosinolate-derived isothiocyanates [56]. The volatile oils from some *Allium* plants demonstrated anti-fungal activity when administered alone and demonstrated synergy with ketoconazole [57]. It is unclear if volatiles from *Allium* species have any significant synergistic anti-fungal activity when acting with each other.

The ubiquity of glucosinolate- producing plants suggests that the strategy of producing multiple secondary natural products from primary metabolite starting materials is, in this case, successful. Different plants face different evolutionary challenges and the presence of multiple secondary natural product options could be of value in adapting to these. Two further variations may, in different ways, provide

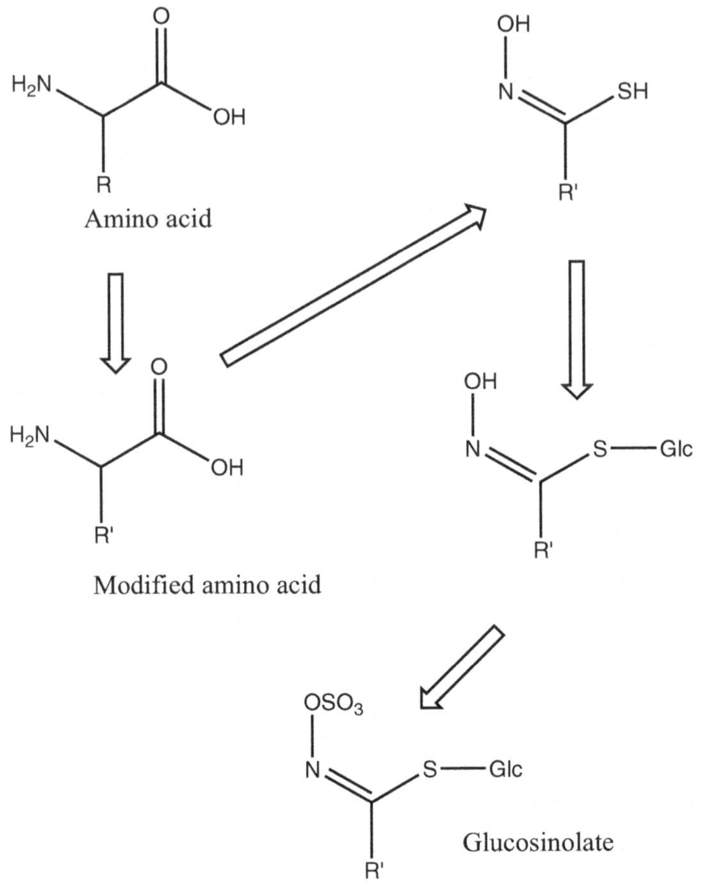

Fig. 2.8 The biosynthesis of glucosinolates

some advantage to host species that contain glucosinolates. These are the spontaneous formation of the bioactive agent goitrin (vinyl oxazolidinethione) and the production of tryptophan-derived glucosinolates, glucobrassicin and related, ring functionalised analogues. The hydroxy-butenyl glucosinolate trivially known as progoitrin because of the antithyroid activity of its principal breakdown product forms an unstable isothiocyanate after plant tissue damage caused by, for example, predation, which spontaneously cyclises to give the goitrogenic product, vinyl oxazolidinethione (Fig. 2.9).

This compound contains the NH–C=S moiety that is responsible for the pronounced bitterness of a range of tastants (Fig. 2.10).

This property is due principally to binding of these compounds to the TAS2R38 bitter receptor [58]. The sensation of bitter taste (particularly the polymorphism associated with 'super bitter' tasters, may be a function of factors other than simply

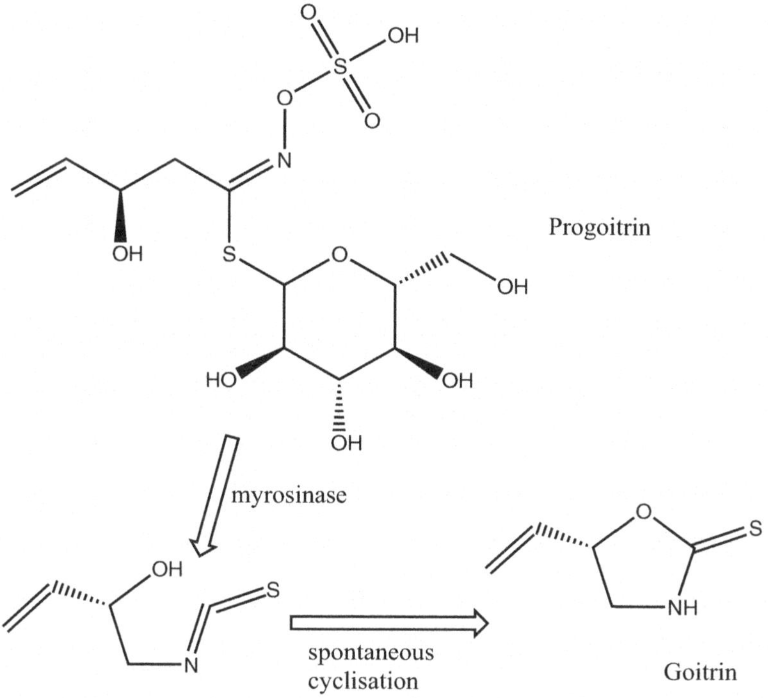

Fig. 2.9 Formation of goitrin (vinyl-oxazolidinethione)

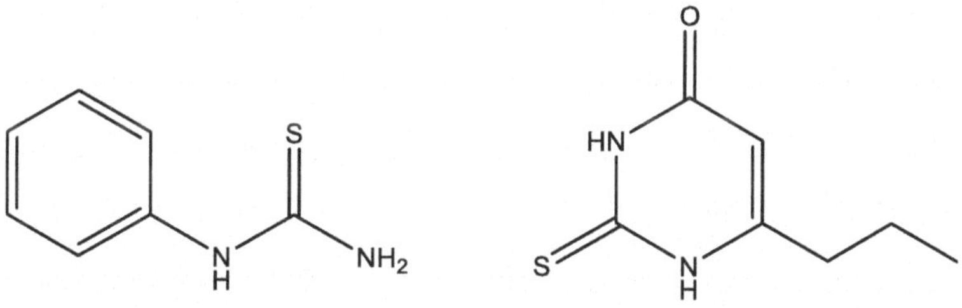

phenylthiocarbamide **6-n-propylthiouracil**

Fig. 2.10 Examples of bitter compounds with an NH–C=S moiety

binding to TAS2R38 [59]. It has been suggested that some of the parent glucosinolates also have a bitter taste [60] however this is at a lower level than vinyl oxazolidinethione and, indeed, while the intact glucosinolate, sinigrin appears to have some intrinsic bitter quality, progoitrin is only bitter after it has been hydrolysed and undergone cyclisation to give the oxazolidinethione [61]. Clearly the formation of the intensely bitter cyclisation product will confer an advantage to those plants that contain the precursor glucosinolate since they will be less likely to undergo predation.

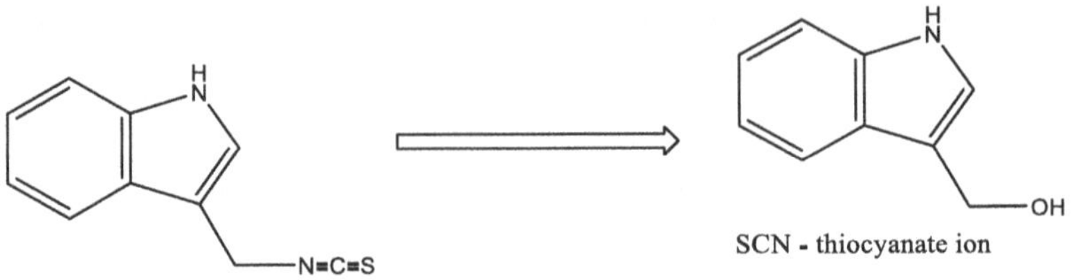

Fig. 2.11 Formation of indole-3-carbinol

The second group of glucosinolates whose post-hydrolysis breakdown is atypical is those that are derived from tryptophan and contain an indole nucleus. The isothiocyanates formed in this case is unstable to nucleophilic attack and, in the presence of water, for example, they will expel thiocyanate ion and form indole-3-carbinol (Fig. 2.11).

This intermediate is relatively stable however a number of further reactions can take place to give a range of indole products. While some of these have been implicated in a number of biological benefits of consumption of *Brassica* vegetables, thiocyanate ion is goitrogenic [62] and consumption of large amounts of *Brassica* vegetables may have an impact on thyroid function however effects are likely to be small and countered by adequate iodine intake [63]. The biosynthesis of indole glucosinolates such as glucobrassicin and phytoalexins like brassinin and wasalexins are linked [64]. This complex series of interactions leads to production of auxins from glucobrassicin via the metabolic intermediates, indole-3-acetaldoxime and indole-3-acetonitrile as well as compounds with antimicrobial, antifungal and insecticidal properties such as brassinin [65]. In simple terms, it appears that indole glucosinolates, give rise to plant protective factors and plant growth factors which, unlike other glucosinolates, are not based upon the production of volatile isothiocyanates but operate by different mechanisms. The structural links between sulphur-containing indole compounds in *Brassica* and related species (indole alkaloids and compounds with a functionalised indole ring) encourages chemists to try to identify the types of reactions that might be involved. The characterisation of enzymes that may catalyse such reactions provides a biological underpinning for chemically-based speculation [66]. In terms of biological activity, the anti-proliferative effects of sulphur-containing indole compounds from *Brassica* species has probably been subject to the greatest amount of study however the link between the consumption of *Brassica* vegetables and the prevention of inflammatory-based neurodegenerative diseases [67] and the possible formation of dithiocarbamates as a component of the inter-relationships between indole glucosinolates, transient intermediates such as isothiocyanates and indole phytoalexins, suggests that the protective effect of *Brassica* vegetable consumption may be a consequence of the formation of these and similar compounds [68].

2.5 Secondary Natural Products and Their Effects upon Health and Disease

Some limited mention of the biological properties of secondary natural products (particularly quinine and caffeine) has already been introduced. While the delineation of pharmacological properties was an important factor that fuelled the study and isolation of natural products and provided considerable impetus for natural product discovery programmes of work—particularly in the nineteenth and twentieth centuries—the beneficial, longer term effects of natural products and mixtures of natural products has been comparatively little studied. These longer-term effects are more typically caused by the low level exposures that occur as a consequence of the consumption of food containing secondary natural products. The two main reasons for this lack of concerted studies are the difficulty in carrying out long term studies of the type that most people are exposed to and the challenge of deconvoluting causative factors from confounders.

For these and other reasons, many investigations into the effects of secondary natural products have been initiated by observational studies on population groups. Studying isolated populations has a number of advantages. First of all, they are more likely to be less genetically heterogenous than a general mixed population. Secondly dietary habits are likely to be more similar between individuals (perhaps for social or religious reasons) than is the case for a more mixed population. Finally changes to diets caused by access to a wider selection of foods may highlight inter-generational changes and this may also alter factors such as epigenetic markers.

The study of the factors that affect disease prevalence in populations and among groups is called epidemiology. Although prevention and treatment of disease is not part of epidemiology as such, it is an important factor in providing evidence for the link between a specific factor and a disease or health issue. In the case of many dietary influences, the chronic nature of the effects that are measured means that longitudinal epidemiological studies are necessary. These measure the same factors over a long period of time in order to confirm the validity of a relationship. The numbers of epidemiological studies carried out in a particular area and with a particular dietary component can depend on a number of factors including the importance of the health or disease issue, the status of the link to diet and the availability of suitable predictive biomarkers that can be used. The use of biomarkers is an important aspect of measuring both beneficial and deleterious effects of food components. Biomarkers are often divided into three types—exposure, effect and susceptibility. Biomarkers of exposure can be measured by the cohort in a study being asked to keep a food diary and the components of interest can be measured after analysis of a representative sample of the food. This is not, strictly speaking, a measurement of exposure since it does not measure absorption distribution metabolism and excretion (ADME). While analysis of representative food samples is often possible, the accuracy of the measurement and the presence of analytical confounders (for example matrix effects) can be problematic. An additional approach is to measure the analyte of interest in serum, blood or urine and to correlate this with intake. While this provides a partial validation, it does not directly measure metabolism or (unless other studies are carried out) exposure of the target organ or system to the analyte. In addition, measurement of exposure biomarkers is generally carried out at a single timepoint and this may not be representative of the actual exposure. For example, the bioavailability of phytoestrogens is influenced by uptake and, to a significant degree, by metabolism. Both of these are subject to considerable inter-individual variation and so any data on 'typical' exposures to dietary phytoestrogens should be treated with caution [69]. The use of dietary biomarkers to measure exposure to specific dietary components has been reviewed [70] and a set of validation criteria developed. These include dose response, time response, robustness, reliability, stability, analytical performance, and interlaboratory reproducibility. The focus was on habitual food intake rather than short term intervention studies and this is, therefore, markedly different from pharmacological testing of bioactive agents. In addition, individual constituents may come from a number of foods rather than just a single source. In order for exposure results to be meaningful, the source of the dietary intake must be confirmed so that a dose/response correlation can be made. In addition, the biomarker (or the sum of relevant biomarkers) must be specific for the food constituent being monitored. This means that due consideration must be given to metabolic and excretion pathways and factors that influence them. While specificity of exposure is important—particularly for secondary natural products, it is also possible to measure intake of general classes of foods such as fats, meat-derived protein and carbohydrates. In some cases, gut microbial metabolism may be a significant confounder and this may be subject to multiple influencing factors. Processing contaminants such as heterocyclic amines (from cooked meat) have also been proposed as a topic for study. While these are not generally categorised as natural products, they are produced as a result of the application of a process—just as glucosinolate-derived isothiocyanates are produced as a result of an extraneous factor leading to cellular disruption. One of the most interesting members of this class is acrylamide which has been found to be present in a range of foods and is of concern because it is a genotoxic carcinogen. Until a study by the University of Stockholm, the source of acrylamide in foods was thought to be as a contaminant from the chemical industry. The discovery that acrylamide could be 'naturally' present in food and not purely as a contaminant came as a result of environmental monitoring of a building project. During digging of a tunnel near Hallandsåsen in Sweden, some of those working on the construction had high levels of acrylamide and it was also found in fish and cows downstream of the construction work. This raised some concerns and investigators at the University of Stockholm sought to establish exposure levels. In order to reach a true figure, they had to measure background levels in non-exposed individuals. These studies established that a small but significant level of acrylamide could be formed in some cooked foods [71]. The mechanism of formation was later established to be the result of a Maillard reaction between the amino acid asparagine and reducing sugars [72]. Such reactions were well known to occur in cooked products and were first discovered in 1912 [73] and was originally of interest due to the loss of nutritive amino acids in processed foods. The possibility of the formation of a genotoxic carcinogen was not considered. Subsequent investigations indicated that acrylamide formation was a consequence of 'browning' reactions of certain foods however, the use of acrylamide as a biomarker for the consumption of certain foods is not possible due to the wide range of foods in which it is formed and the variable levels of exposure.

Biomarkers of susceptibility are one of the three classes of biomarker (alongside exposure and effect) defined by the World Health Organisation and mentioned above. Susceptibility can be a consequence of genetic polymorphisms or the epigenetic profile that results from prior exposure. In the case of the former, polymorphisms in xenobiotic metabolising enzymes (particularly cytochrome-P450 enzymes) [74] can affect the amount of active compound that is created. In addition, as noted previously, there are also differences in enzymes responsible for conjugation and excretion [75]. In addition to such variation occurring within individuals, there are also polymorphisms between species and this is an additional challenge to the extrapolation to humans of data derived from test animals. In addition to differences in metabolism, there are also genetic polymorphisms between target systems. One of the most significant of these is the presence of polymorphisms in the vitamin D receptor which affects the clinical consequences of dietary vitamin D [76]. A range of other receptors may be affected by polymorphisms and these can impact upon responses to secondary natural products. In the case of epigenetic profiles, a significant number of disease states have been found to have epigenetic biomarkers [77] and these affect disease susceptibility. The link between diet and previous environmental exposure (including diet) has been suggested [78] however further studies are needed to confirm the link and to use information in a predictive and mechanistic way [79].

Biomarkers of effect are the most challenging of the three classes of biomarker noted previously. At one extreme, biomarkers of effect can include predictive biomarkers that can reflect either limited or lifetime exposure to specific secondary natural products. In this category, they can also encompass biomarkers of susceptibility and changes in susceptibility as a result of exposure to bioactive agents. Biomarkers of effect can either be used to probe clinical endpoints or to investigate the effect of secondary natural products on certain biological mechanisms. Blood, serum and urine-based biomarkers are commonly used and these are generally indicative of a pharmacological effect [80]. Within the broad definition of biomarkers of effect, a number of types of biomarker are commonly defined. Diagnostic biomarkers and monitoring biomarkers are used to determine status and progression of a health or disease process. Response biomarkers are used to determine the effects of a specific intervention while predictive biomarkers can be used to determine the trajectory of a clinical situation. If a biomarker is measured in an easily accessible and relatively non-invasive way then it has to be validated as a surrogate for the clinical status of interest and it must also reflect, in a interpretable way, changes to that clinical status after an intervention has been carried out. In some cases, biomarkers can be developed in an in vitro model and then validated through a clinical trial by measuring an appropriate surrogate. For example, immune response biomarkers such as cytokines to health and resilience status can be measured systemically (in blood or serum) and changes to these can have clinical relevance [81]. In most cases the response to secondary natural products in foods is relatively weak and cohort sizes have to be large (and expensive) in order to achieve statistical significance. In addition, since pharmacological trial methodologies are often (mistakenly) used as a model for trials of bioactive natural products in foods, the measurement of deviations from a given homeostatic state is assumed to be important. Living systems do not normally exist in a state of homeostasis—they are, in the description of the balance of a clock used by German philosophers such as Leibniz, in a state of 'unruhe' or perpetual restlessness [82]. It is the measurement of that restlessness and what its normal state of being is, that is of greatest importance when trying to assess what is 'normal'. Pharmacological agents are designed to affect systems that have already been placed in an extreme situation (unwellness or illness) with a trajectory towards a clinical endpoint rather than being in a restless equilibrium with many options as to the progression to different healthy states. For this reason, the biomarker approach which was developed for pharmacology is inappropriate for measuring the effects of dietary secondary natural products on living systems. Unfortunately the approach taken by many regulatory authorities is predicated on the principle of a healthy homeostasis being a desirable state and food components are considered to be beneficial if they stabilise a given healthy state and protect it from perturbations. In reality, as noted above, living systems are continually in flux between different states so the concept of the existence of a state that can unequivocally be defined as 'healthy' to the exclusion of others is misguided. There have been some attempts to define health in terms of prevention of disease—for example the link between obesity and the later development of 'lifestyle' diseases such as type 2 diabetes—but these have rarely considered the possible benefits of secondary natural products. In addition many natural products have multiple biological activities (for example, in addition to being toxic to humans, the mycotoxin patulin was first isolated as part of a programme to find novel anti-microbials with a spectrum of activity akin to that of penicillin) [83] however it was soon found to be toxic to humans and to animals so research into its benefits was discontinued. The Dictionary of Natural Products database contains in excess of 200,000 compounds with a broad spectrum of activities [84]. Biomarkers are not simply used to assess beneficial secondary natural products. They can also be used to measure progression towards a toxic endpoint. One of the most commonly used biomarker approaches to assess toxic effects is the capacity of compounds to bind to nucleic acids (mutagenicity). This property—and the link to carcinogenicity—will be considered in more detail in Sect. 2.7 however it should be noted that binding to DNA is both a particular and a systemic consequence of exposure to certain classes of compounds (often compounds that are electrophilic, or able to be converted to electrophiles, and hence can bind to the nucleophilic base moieties of nucleic acids [85]—even if the clinical endpoints are not invariably predictable. In other words, mutagenicity does not invariably predict carcinogenicity and the importance of DNA damage as a clinical measure must include a consideration of mitigating factors such as DNA repair.

2.6 The 'Magic Bullet' and Pharmacology

The idea of a biological 'magic bullet' was expressed by Ehrlich as he sought to find an agent that would specifically kill an infective agent while simultaneously not harming the host. As a concept, it derives from the first great opera of the German Romantic period composed by Carl Maria von Weber, Der Freischütz. In the opera, which is based on a folk tale, a marksman makes a deal with the devil in order to win a shooting competition where the prize is the hand of Agathe, daughter of the head forester. The deal allows him to make magic bullets (Freikugeln) that will invariably hit their target and the target only. This aspect of the opera appealed to Ehrlich (who was active at about the time of its popularity) who developed the idea of a therapeutic agent that would only hit a pathogen and leave the host unharmed. While the first of these, Salvarsan, was not quite the completely harmless cure for syphilis that he had hoped for, but it was, to an extent, selectively toxic to the causative infective agent, the spirochete *Treponema pallidum* and therefore had some of the attributes of a 'magic bullet'.

The idea of selective toxicity is one which appeals to pharmacologists. There are two approaches—the first is to find a mechanism that only operates either in a target pathogen and not in the infected host or is effective against a specific disease state and does not impair the health of an unaffected individual so that, as the person is cured, they recover and grow in health status. The second is to find an agent that is detoxified by the host but remains effective against the pathogen. Antibiotics are often members of the first class. For example, penicillin and related compounds act by inhibiting bacterial cell wall synthesis. Erythromycin acts by inhibiting bacterial protein synthesis so it stops bacteria from growing rather than killing them. In addition, erythromycin and other macrolide antibiotics also show anti-inflammatory activity. In the area of non-infective disease, allopurinol is an effective agent for prevention of the build up of uric acid which can lead to gout and other conditions. It works by inhibiting the enzyme xanthine oxidase (the first enzyme in the metabolism of xanthine to uric acid. In the presence of allopurinol, the level of excretion of xanthine and hypoxanthine are increased and that of uric acid, decreased. This means that diseases associated with high levels of uric acid such as gout are prevented. The metabolism of drugs by the host is an important aspect of both efficacy and drug safety. Analysis of metabolites is a primary aspect of defining the therapeutic window of a drug. The therapeutic window is the ratio of the lethal dose to the effective dose and is a measure of the effectiveness of a therapeutic. An understanding of the metabolomics of a bioactive agent will help to define efficacy and safety [86].

As noted previously, the assessment of the effects of secondary natural products from food systems is problematic. This is a consequence of the following which have already been mentioned but are worth re-emphasising:

- The effects are generally small and chronic and therefore large cohorts are required and these must be examined over long periods of time.
- The effects show a high level of inter-individual variability.

- The presence of multiple effects and confounders makes it difficult to develop definitive conclusions as to the impact of individual secondary natural products.

Each of these is a consequence of the nature of food derived secondary natural products. There is also a key difference between pharmacological interventions and those which are a consequence of secondary natural product ingestion. Drugs are designed to deal with a specific malady and specificity is of prime importance. The system to which the drug is administered is generally out of balance—it is not in a state of healthy homeostasis (as was mentioned previously) and therefore one of the key ways of measuring success is whether the imbalance in the system is resolved. This is often determined by measuring clinical diagnostic biomarkers which are symptomatic of the disease. The absence of such biomarkers signifies that the drug has been effective and the disease is either cured or in remission. This does not, however, guarantee that the individual is healthy—simply that one specific cause of ill health has been resolved. In the case of secondary natural products from foods, they are not designed to cure a disease so diagnostic, clinical biomarkers are not a measure of efficacy and absence of disease is not proof of health (as noted above). This then raises the twin questions of what is the purpose of secondary natural products and how can this purpose be measured and quantified? Contrary to much general opinion, healthy systems are not in a state of homeostasis from which they can resist perturbations. Living systems are (as was recognised by Leibniz and noted previously) in a state of perpetual flux. The transition between states is stable and not chaotic except in the case where the system degenerates into a disease state as a consequence of some disturbance. The role of secondary natural products from foods can be to facilitate these ongoing, stable transitions. Measuring the effects of secondary natural products on this system is challenging since the features of the transitions are not able to be defined and therefore deviations from the norm cannot be detected. Epidemiological and observational studies suggest that there is an effect but quantification of the effect and definitive determination of the role of different bioactive agents is more problematic. This means that health claims associated with specific secondary natural products from foods are difficult to frame or substantiate. One area that is of interest and which may show potential is the effect on inflammatory-related diseases [87]. In this case, a relatively minor intervention may, by affecting an underlying pathway, have a significant effect on inflammation and diseases that are a consequence of inflammatory imbalance [88]. Another area that has attracted some attention is the impact of secondary natural products on the prevalence and development of cancer. This particular disease state is attractive for two reasons—first of all, it is a series of long term conditions and the pathway may be susceptible to different interventions at different stages, secondly, the disease pathway is, at least in part, for genotoxic carcinogens, defined from initial challenge and damage to DNA (mutagenesis) to cancer formation. This will be examined in more detail in Sect. 2.7.

2.7 Mutagenicity and Carcinogenicity of Secondary Natural Products

The pathway from DNA damage to cancer has been extensively studied. Damage to genes can lead to gain of function or loss of function. In addition, certain polymorphisms can increase the predisposition to the development of cancer. Mutations in certain genes can lead to the formation of a clonal population with a growth and proliferation advantage. The process can be divided into three sets of events—initiation, promotion and progression. Overlaid upon this is the repair of DNA which, if carried out successfully, can short circuit the cancer process. Natural products may operate at a number of events in the process of the development of disease and understanding these potential protective effects has been the subject of considerable research. One of the challenges in research into cancer development has been creating a better understanding of the significance of certain mutations compared to others. The creation of a list of those genes involved in driving the carcinogenesis process is an ongoing objective [89]. Such studies have enabled researchers to examine more closely the mutational landscape of tumours and provide clues as to which damaged sites have the greatest biological significance [90] and, through such studies, it has been revealed that certain combinations of mutations is more likely to be associated with survival than others.

More generally, the link between DNA damage, mutations and cancer has been extensively reviewed [91]. Perhaps the most significant finding is that, while not all mutations lead to cancer, all compounds that are capable of causing mutations should be considered to be potential carcinogens.

In addition to protective effects, some natural products are able to cause mutations and therefore are, potentially, carcinogens. While some classes of compounds have been considered to be beneficial, such classification can obscure harmful effects. For example, while flavonoids have been considered to have a positive health profile, some of them are mutagens and potential carcinogens however the data is somewhat contradictory. In addition to direct mutagenicity, some natural products have been found to be co-mutagens and this may be a consequence of the activation of Phase I enzymes responsible for activation of compounds. In very general terms, the more DNA adducts that are formed, the more significant are the consequences however other significant factors including the rate of DNA repair and the type of adduct formed also play a role. For example, the correlation of DNA adduct levels with degree of carcinogenic potential shows a range of about two orders of magnitude [92] and this is evidenced by the fact that the adducts formed by aflatoxin B1 appear to be 40 times more potent in causing hepatocellular carcinoma than those formed by dimethylnitrosamine. The reductionist approach of equating the level of DNA damage to cancer incidence is, therefore, inadequate. This approach, however, seems to underpin much of the thinking regarding natural products as cancer-preventive agents. In addition to natural products that are used directly in cancer therapy and which operate by a number of mechanisms including anti-proliferative effects [93], potential targets have included apoptopic pathways and pathways associated with

Nrf-2. The latter operates by activation of xenobiotic metabolising enzymes and reducing cellular oxidative stress [94]. The use of phytochemicals to prevent cancer has been broadly split into two mechanisms—impeding the consequences of DNA damage (including stimulation of repair processes or blocking the division of premalignant, damaged cells thereby preventing DNA damage from causing a permanent genotypic change. [95]. While such approaches have shown some promise, attempts to develop a fully functioning theory regarding particular natural products and cancer prevention remains elusive. Just as the correlation between firearm ownership and homicide rates due to firearms has been robustly demonstrated [96], it would seem reasonable to expect a correlation between DNA damage and rates of cancer occurrence however, as demonstrated above, different DNA adducts have different potencies for disease outcomes. Apart from the specific case of handgun ownership and suicide rate [97], there are relatively few studies on type of gun ownership and outcomes involving injury or death. In an analogous way, while the comet assay is a way of measuring DNA damage based on the principle that fragmented genomic DNA will produce an electrophoretic pattern with a 'tail' comprising DNA of different sizes and it has been used to assess DNA damage, there are a number of inconsistencies in comparing data [98] nonetheless it can provide a useful starting point for assessing the link between outcomes and more severe instances of DNA damage [99]. In general however, just as techniques such as the Ames test for mutagenicity are used to measure the mutation potential of particular compounds, the comet assay is a mutagenicity measure rather than a preventative potential assay [100]. The modification of the comet assay to detect different types of DNA damage and repair of lesions [101] suggests it may be of more use than a tool simply to measure the extent of DNA damage however problems still exist with inter-laboratory reproducibility and in the degree of invasiveness or the use of lymphocyte cells as surrogate markers and these will have to be resolved if the comet assay is to be more widely adopted as a measure of the degree of protection offered by secondary natural products and the extrapolation from in vitro to in vivo effects [102].

The use of epidemiological and observational studies to define agents implicated in cancer and related diseases has been successful in a number of cases. These have been supported by mechanistic studies designed to develop methods to prevent, in particular, the initiation and progression of the disease. This has led to the identification of a range of carcinogens, a greatly increased knowledge of the process of carcinogenesis and the discovery and development of therapies. There is substantial evidence of the role that increased consumption of certain foods (particularly fruits and vegetables) might play in the long term prevention of initiation and/or progression of cancer however application of reductionist methods to try to identify the precise nature of any protective agents and their mechanism of action has proven to be more challenging.

Since cancer cells are in an unbalanced state, the application of techniques developed to investigate pharmacological solutions may not be appropriate to the study of the healthy transitions of cells between different states. What may be needed is a paradigm shift away from regarding homeostasis as a desirable state (because it is the opposite of proliferative growth) and accepting the concept of transitions as being the more stable situation for living systems.

References

1. Dou Y, Mäkinen M, Jänis J. Analysis of volatile and nonvolatile constituents in Gin by direct infusion ultrahigh-resolution ESI/APPI FT-ICR mass spectrometry. J Agric Food Chem. 2023;71(18):7082–89. https://doi.org/10.1021/acs.jafc.3c00707. Epub 2023 Apr 27. PMID: 37103967; PMCID: PMC10176568.
2. Wu J et al. Recent advances in the understanding of off-flavors in alcoholic beverages: Generation, regulation, and challenges. J Food Compos Anal. 2021;103:104117.
3. Grant-Preece P, Barril C, Schmidtke LM, Scollary GR, Clark AC. Light-induced changes in bottled white wine and underlying photochemical mechanisms. Crit Rev Food Sci Nutr. 2017;57(4):743–54. https://doi.org/10.1080/10408398.2014.919246. PMID: 25879850.
4. Marques JG et al. Tonic water: A rare cause of exanthema. Allergologia et immunopathologia. 2012;40(1):60–1.
5. Fardet L, Nazareth I, Petersen I. Association between long-term quinine exposure and all cause mortality. JAMA. 2017;317(18):1907–9. https://doi.org/10.1001/jama.2017.2332.PMID:28492890;PMCID:PMC5815033.
6. Howard MA, Hibbard AB, Terrell DR, Medina PJ, Vesely SK, George JN. Quinine allergy causing acute severe systemic illness: report of 4 patients manifesting multiple hematologic, renal, and hepatic abnormalities. Proc (Bayl Univ Med Cent). 2003;16(1):21–6. https://doi.org/10.1080/08998280.2003.11927884. PMID:16278718;PMCID:PMC1200805.
7. Duncan ADS, Hapca S, De Souza N, Morales D, Bell S. Quinine exposure and the risk of acute kidney injury: a population-based observational study of older people. Age Ageing. 2020;49(6):1042–7. https://doi.org/10.1093/ageing/afaa079.
8. Makiso MU, Tola YB, Ogah O, Endale FL. Bioactive compounds in coffee and their role in lowering the risk of major public health consequences: A review. Food Sci Nutr. 2023;12(2):734–64. https://doi.org/10.1002/fsn3.3848.PMID:38370073;PMCID:PMC10867520.
9. Nuhu AA. Bioactive micronutrients in coffee: recent analytical approaches for characterization and quantification. ISRN Nutr. 2014;22(2014): 384230. https://doi.org/10.1155/2014/384230. PMID:24967266;PMCID:PMC4045301.
10. Jeszka-Skowron M, Zgoła-Grześkowiak A, Grześkowiak T. Analytical methods applied for the characterization and the determination of bioactive compounds in coffee. Eur Food Res Technol. 2015;240:19–31. https://doi.org/10.1007/s00217-014-2356-z.
11. Shofinita D, Lestari D, Aliwarga L, et al. Drying methods of coffee extracts and their effects on physicochemical properties: A review. Food Bioprocess Technol. 2024;17:47–72. https://doi.org/10.1007/s11947-023-03067-4.
12. Li Z, Zhao C, Cao C. Production and inhibition of acrylamide during coffee processing: A literature review. Molecules. 2023;28(8):3476. https://doi.org/10.3390/molecules28083476. PMID:37110710;PMCID:PMC10143638.
13. Kim H, Kang SH, Kim SH, et al. Drinking coffee enhances neurocognitive function by reorganizing brain functional connectivity. Sci Rep. 2021;11:14381. https://doi.org/10.1038/s41598-021-93849-7.
14. Willson C. The clinical toxicology of caffeine: A review and case study. Toxicol Rep. 2018;3(5):1140–52. https://doi.org/10.1016/j.toxrep.2018.11.002. PMID:30505695;PMCID:PMC6247400.
15. Nehlig A. Interindividual differences in caffeine metabolism and factors driving caffeine consumption. Pharmacol Rev. 2018;70(2):384–411. https://doi.org/10.1124/pr.117.014407. Epub 2018 Mar 7 PMID: 29514871.

16. Katz SN. Decaffeination of coffee. In: Clarke RJ, Macrae R, editors. Coffee. Dordrecht: Springer; 1987. https://doi.org/10.1007/978-94-009-3417-7_3.
17. Zou Y, Gaida M, Franchina FA, Stefanuto PH, Focant JF. Distinguishing between decaffeinated and regular coffee by HS-SPME-GC×GC-TOFMS, chemometrics, and machine learning. Molecules. 2022;27(6):1806. https://doi.org/10.3390/molecules27061806. PMID:35335174;PMCID:PMC8948847.
18. Córdoba N et al. Coffee extraction: A review of parameters and their influence on the physicochemical characteristics and flavour of coffee brews. Trends Food Sci Technol. 2020;96:45–60.
19. Niessen LW et al. Tackling socioeconomic inequalities and non-communicable diseases in low-income and middle-income countries under the sustainable development agenda. Lancet. 391(10134):2036–2046.
20. Singh AR, Singh SA. Diseases of poverty and lifestyle, well-being and human development. Mens Sana Monogr. 2008;6(1):187–225. https://doi.org/10.4103/0973-1229.40567. PMID:22013359;PMCID:PMC3190550.
21. Bui-Klimke TR, Wu F. Ochratoxin A and human health risk: A review of the evidence. Crit Rev Food Sci Nutr. 2015;55(13):1860–9. https://doi.org/10.1080/10408398.2012.724480. PMID:24874522;PMCID:PMC4247821.
22. Heussner AH, Bingle LE. Comparative ochratoxin toxicity: A review of the available data. Toxins (Basel). 2015;7(10):4253–82. https://doi.org/10.3390/toxins7104253. PMID:26506387;PMCID:PMC4626733.
23. Al-Khayri JM, Rashmi R, Toppo V, Chole PB, Banadka A, Sudheer WN, Nagella P, Shehata WF, Al-Mssallem MQ, Alessa FM, Almaghasla MI, Rezk AA. Plant secondary metabolites: The weapons for biotic stress management. Metabolites. 2023;13(6):716. https://doi.org/10.3390/metabo13060716. PMID:37367873;PMCID:PMC10302943.
24. Borgen BH, Thangstad OP, Ahuja I, Rossiter JT, Bones AM. Removing the mustard oil bomb from seeds: Transgenic ablation of myrosin cells in oilseed rape (Brassica napus) produces MINELESS seeds. J Exp Bot. 2010;61:1683–97. https://doi.org/10.1093/jxb/erq039.
25. Francis K, Smitherman C, Nishino SF, Spain JC, Gadda G. The biochemistry of the metabolic poison propionate 3-nitronate and its conjugate acid, 3-nitropropionate. IUBMB Life. 2013;65(9):759–68. https://doi.org/10.1002/iub.1195. Epub 2013 Jul 29 PMID: 23893873.
26. Bush K, Bradford PA. β-Lactams and β-Lactamase Inhibitors: An Overview. Cold Spring Harb Perspect Med. 2016;6(8): a025247. https://doi.org/10.1101/cshperspect.a025247. PMID:27329032;PMCID:PMC4968164.
27. Urugo MM, Tringo TT. Naturally occurring plant food toxicants and the role of food processing methods in their detoxification. Int J Food Sci. 2023;27(2023):9947841. https://doi.org/10.1155/2023/9947841. PMID:37153649;PMCID:PMC10159748.
28. Montagnac JA, Davis CR, Tanumihardjo SA. Processing techniques to reduce toxicity and antinutrients of cassava for use as a staple food. Compr Rev Food Sci Food Saf. 2009;8:17–27. https://doi.org/10.1111/j.1541-4337.2008.00064.x.
29. Meyerhof W, Batram C, Kuhn C, et al. The molecular receptive ranges of human TAS2R bitter taste receptors. Chem Senses. 2010;35:157–70.
30. Medapati MR, Singh N, Bhagirath AY, Duan K, Triggs-Raine B, Batista EL Jr, Chelikani P. Bitter taste receptor T2R14 detects quorum sensing molecules from cariogenic Streptococcus mutans and mediates innate immune responses in gingival epithelial cells. FASEB J. 2021;35(3): e21375. https://doi.org/10.1096/fj.202000208R. PMID: 33559200.
31. Kouakou YI, Lee RJ. Interkingdom detection of bacterial quorum-sensing molecules by mammalian taste receptors. Microorganisms. 2023;11(5):1295. 3390/microorganisms11051295. PMID: 37317269; PMCID: PMC10221136.
32. Kovács EG, Alatshan A, Budai MM, Czimmerer Z, Bíró E, Benkő S. Caffeine has different immunomodulatory effect on the cytokine expression and NLRP3 inflammasome function in various human macrophage subpopulations. Nutrients. 2021;13(7):2409. https://doi.org/10.3390/nu13072409. PMID:34371919;PMCID:PMC8308523.

33. Cornelis MC. The impact of caffeine and coffee on human health. Nutrients. 2019;11(2):416. https://doi.org/10.3390/nu11020416. PMID:30781466;PMCID:PMC6413001.
34. Andreozzi P, Sarnelli G, Pesce M, Zito FP, Alessandro AD, Verlezza V, Palumbo I, Turco F, Esposito K, Cuomo R. The bitter taste receptor agonist quinine reduces calorie intake and increases the postprandial release of cholecystokinin in healthy subjects. J Neurogastroenterol Motil. 2015;21(4):511–9. https://doi.org/10.5056/jnm15028. PMID:26351252;PMCID:PMC4622133.
35. Desai AJ, Dong M, Harikumar KG, Miller LJ. Cholecystokinin-induced satiety, a key gut servomechanism that is affected by the membrane microenvironment of this receptor. Int J Obes Suppl. 2016;6(Suppl 1):S22–27. https://doi.org/10.1038/ijosup.2016.5. Epub 2016 Nov 16. PMID: 28685026; PMCID: PMC5485878.
36. Workman AD, Maina IW, Brooks SG, Kohanski MA, Cowart BJ, Mansfield C, Kennedy DW, Palmer JN, Adappa ND, Reed DR, Lee RJ, Cohen NA. The role of quinine-responsive taste receptor family 2 in airway immune defense and chronic rhinosinusitis. Front Immunol. 2018;28(9):624. https://doi.org/10.3389/fimmu.2018.00624.PMID:29643854;PMCID:PMC 5882797.
37. Yan CH, Hahn S, McMahon D, Bonislawski D, Kennedy DW, Adappa ND, Palmer JN, Jiang P, Lee RJ, Cohen NA. Nitric oxide production is stimulated by bitter taste receptors ubiquitously expressed in the sinonasal cavity. Am J Rhinol Allergy. 2017;31(2):85–92. https://doi.org/10. 2500/ajra.2017.31.4424. PMID:28452704;PMCID:PMC5356199.
38. Liu K, Jaggupilli A, Premnath D, Chelikani P. Plasticity of the ligand binding pocket in the bitter taste receptor T2R7. Biochim Biophys Acta Biomembr. 2018;1860(5):991–9. https:// doi.org/10.1016/j.bbamem.2018.01.014. Epub 2018 Feb 12 PMID: 29355483.
39. Wang Y, Zajac AL, Lei W, Christensen CM, Margolskee RF, Bouysset C, Golebiowski J, Zhao H, Fiorucci S, Jiang P. Metal ions activate the human taste receptor TAS2R7. Chem Senses. 2019;44(5):339–47. https://doi.org/10.1093/chemse/bjz024. PMID:31066447;PMCID:PMC6538953.
40. Mayoral O, Solbes J, Cantó J, Pina T. What has been thought and taught on the lunar influence on plants in agriculture? Perspective from physics and biology. Agronomy. 2020;10(7):955. https://doi.org/10.3390/agronomy10070955.
41. Modenesi P. Skull lichens: A curious chapter in the history of phytotherapy. Fitoterapia. 2009;80(3):145–8. https://doi.org/10.1016/j.fitote.2008.12.005. Epub 2009 Jan 13 PMID: 19171183.
42. Zhang QW, Lin LG, Ye WC. Techniques for extraction and isolation of natural products: a comprehensive review. Chin Med. 2018;17(13):20. https://doi.org/10.1186/s13020-018-0177-x. PMID:29692864;PMCID:PMC5905184.
43. Friedrich C, von Domarus C. Carl Friedrich Wilhelm Meissner (1792–1853)—Apotheker und Alkaloidforscher. In: Pharmazie. Band 53, Nr. 1, Januar 1998:S67–73.
44. Fowsiya J, Madhumitha G. Extraction of bio organics phase from various food sources. In: Roopan S, Madhumitha G, editors. Bioorganic phase in natural food: An overview. Cham: Springer; 2018. https://doi.org/10.1007/978-3-319-74210-6_3
45. Verpoorte R, Schripsema J. Isolation, identification, and structure elucidation of alkaloids a general overview. In: Linskens HF, Jackson JF, editors. Alkaloids. Modern methods of plant analysis, vol 15. Berlin, Heidelberg: Springer; 1994. https://doi.org/10.1007/9783-642-842 26-9_1
46. Pannequin A, Laurini E, Giordano L, Muselli A, Pricl S, Tintaru A. Caution: Chemical instability of natural biomolecules during routine analysis. Molecules. 2020;25(14):3292. https://doi.org/10.3390/molecules25143292. PMID:32698478;PMCID:PMC7397321.
47. Ma X, Meng Z, Qiu L, Chen J, Guo Y, Yi D, Ji T, Jia H, Xue M. Solanesol extraction from tobacco leaves by Flash chromatography based on molecularly imprinted polymers. J Chromatogr B. 2016;1020:1–5. https://doi.org/10.1016/j.jchromb.2016.03.007.
48. Hanley AB, Heaney RK, Fenwick GR. Improved isolation of glucobrassicin and other glucosinolates. J Sci Food Agric. 1983;34:869.
49. Gadamer J. Ueber das Sinalbin. Ber Dtsch Chem Ges. 1897;30(2):2327–8.

50. Bussy, A. Sur la formation de l'huile essentielle de moutarde. J Pharmacol. 1840;27:464–71.
51. Ettlinger MG, Lundeen AJ. The structures of Sinigrin and Sinalbin; an enzymatic rearrangement. J Am Chem Soc. 1956;78:4172–73.
52. Benn MH, Ettlinger MG. The synthesis of sinigrin. Chem Commun (Lond.). 1965:445–47. https://doi.org/10.1039/c19650000445
53. Li JJ. Lossen rearrangement. In: Name reactions. Berlin, Heidelberg: Springer; 2009. https://doi.org/10.1007/978-3-642-01053-8_147
54. Steven D. A bitter tale of sweet synergy Munger. Cell Chem Biol. 24(10):1191–2.
55. Rakariyatham K, Yang X, Gao Z, Song M, Han Y, Chen X, Xiao H. Synergistic chemopreventive effect of allyl isothiocyanate and sulforaphane on non-small cell lung carcinoma cells. Food Funct. 2019;10(2):893–902. https://doi.org/10.1039/c8fo01914b. PMID:30694275;PMCID:PMC6611553.
56. Plaszkó T, Szűcs Z, Vasas G, Gonda S. Effects of glucosinolate-derived isothiocyanates on fungi: A comprehensive review on direct effects, mechanisms, structure-activity relationship data and possible agricultural applications. J Fungi (Basel). 2021;7(7):539. https://doi.org/10.3390/jof7070539. PMID:34356918;PMCID:PMC8305656.
57. Pyun M-S, Shin S. Antifungal effects of the volatile oils from Allium plants against Trichophyton species and synergism of the oils with ketoconazole. Phytomedicine: Int J Phytother Phytopharmacol. 2006;13(6):394–400.
58. Tepper BJ, Banni S, Melis M, Crnjar R, Tomassini BI. Genetic sensitivity to the bitter taste of 6-n-propylthiouracil (PROP) and its association with physiological mechanisms controlling body mass index (BMI). Nutrients. 2014;6(9):3363–81. https://doi.org/10.3390/nu6093363. PMID:25166026;PMCID:PMC4179166.
59. Hayes JE, Bartoshuk LM, Kidd JR, Duffy VB. Supertasting and PROP bitterness depends on more than the TAS2R38 gene. Chem Senses. 2008;33(3):255–65. https://doi.org/10.1093/chemse/bjm084.
60. Francisco M, Velasco P, Romero Á, et al. Sensory quality of turnip greens and turnip tops grown in northwestern Spain. Eur Food Res Technol. 2009;230:281–90. https://doi.org/10.1007/s00217-009-1163-4.
61. Fenwick G, Roger et al. Bitterness in brussels sprouts (*Brassica oleracea* L. var. gemmifera): The role of glucosinolates and their breakdown products. J Sci Food Agric. 1983;34:73–80.
62. Engel A, Lamm SH. Goitrogens in the environment. In: Braverman, L.E. (eds) Diseases of the thyroid. Contemporary endocrinology. Totowa, NJ: Humana Press; 2003. https://doi.org/10.1007/978-1-59259-352-1_15.
63. Galanty A, Grudzińska M, Paździora W, Służały P, Paśko P. Do brassica vegetables affect thyroid function?-A comprehensive systematic review. Int J Mol Sci. 2024;25(7):3988. https://doi.org/10.3390/ijms25073988. PMID:38612798;PMCID:PMC11012840.
64. Chripkova M, Zigo F, Mojzis J. Antiproliferative effect of indole phytoalexins. Molecules. 2016;21(12):1626. https://doi.org/10.3390/molecules21121626. PMID:27898039;PMCID:PMC6274154.
65. Sellam A, Iacomi-Vasilescu B, Hudhomme P, Simoneau P. In vitro antifungal activity of brassinin, camalexin and two isothiocyanates against the crucifer pathogens Alternaria brassicicola and Alternaria brassicae. Plant Pathol. 2007;56:296–301. https://doi.org/10.1111/j.1365-3059.2006.01497.x.
66. Nafisi M, Sønderby IE, Hansen BG, et al. Cytochromes P450 in the biosynthesis of glucosinolates and indole alkaloids. Phytochem Rev. 2006;5:331–46. https://doi.org/10.1007/s11101-006-9004-6.
67. Beretta HV, Torres-Palazzolo C, Ramírez DA, Camargo AB. Brassica vegetables: Rich sources of neuroprotective compounds. In: Gargiulo PÁ, Mesones Arroyo HL, editors. Psychiatry and neuroscience update. Cham: Springer; 2021. https://doi.org/10.1007/978-3-03061721-9_24.
68. Patel A, Shah D, Patel Y, Patel S, Mehta M, Bambharoliya T. A review on recent development of novel heterocycles as acetylcholinesterase inhibitor for the treatment of Alzheimer's disease. Curr Drug Targets. 2023;24(3):225–46. https://doi.org/10.2174/1389450124666221213114500. PMID: 36515018.

69. Rowland I, Faughnan M, Hoey L, Wähälä K, Williamson G, Cassidy A. Bioavailability of phytooestrogens. Br J Nutr. 2003;89(S1):S45–58. https://doi.org/10.1079/BJN2002796.
70. Landberg R, Karra P, Hoobler R, Loftfield E, Huybrechts I, Rattner JI, Noerman S, Claeys L, Neveu V, Vidkjaer NH, Savolainen O, Playdon MC, Scalbert A. Dietary biomarkers—An update on their validity and applicability in epidemiological studies. Nutr Rev. 2024;82(9):1260–80. https://doi.org/10.1093/nutrit/nuad119.
71. Tareke E, Rydberg P, Karlsson P, Eriksson S, Tornqvist M. Acrylamide: A cooking carcinogen? Chem Res Toxicol. 2000;13(6):517–22.
72. Stadler RH, Imre Blank N, Varga FR, Jörg Hau PA, Guy M-CR, Riediker S. Acrylamide from Maillard reaction products. Nature, Oct. 2002;419:449.
73. Finot PA. Historical perspective of the Maillard reaction in food science. Ann N Y Acad Sci. 2005;1043:1–8. https://doi.org/10.1196/annals.1333.001. PMID: 16037216.
74. Zhou SF, Liu JP, Chowbay B. Polymorphism of human cytochrome P450 enzymes and its clinical impact. Drug Metab Rev. 2009;41(2):89–295. https://doi.org/10.1080/03602530902843483. PMID: 19514967.
75. Chen CH. Metabolic enzymes: Polymorphisms and species differences. In: Activation and detoxification enzymes. Cham: Springer; 2024. https://doi.org/10.1007/978-3-031-552878_12
76. Usategui-Martín R, De Luis-Román DA, Fernández-Gómez JM, Ruiz-Mambrilla M, Pérez-Castrillón JL. Vitamin D receptor (VDR) gene polymorphisms modify the response to vitamin D supplementation: A systematic review and meta-analysis. Nutrients. 2022;14(2):360. https://doi.org/10.3390/nu14020360. PMID:35057541;PMCID:PMC8780067.
77. Skinner MK. Epigenetic biomarkers for disease susceptibility and preventative medicine. Cell Metab. 2024;36(2):263–77.
78. Zhang Y, Kutateladze TG. Diet and the epigenome. Nat Commun. 2018;9:3375. https://doi.org/10.1038/s41467-018-05778-1.
79. Choi SW, Friso S. Epigenetics: A new bridge between nutrition and health. Adv Nutr. 2010;1(1):8–16. https://doi.org/10.3945/an.110.1004. Epub 2010 Nov 16. PMID: 22043447; PMCID: PMC3042783.
80. Califf RM. Biomarker definitions and their applications. Exp Biol Med (Maywood). 2018;243(3):213–21. https://doi.org/10.1177/1535370217750088. PMID:29405771;PMCID:PMC5813875.
81. Llibre A, Duffy D. Immune response biomarkers in human and veterinary research. Comp Immunol Microbiol Infect Dis. 2018;59:57–62. https://doi.org/10.1016/j.cimid.2018.09.008. Epub 2018 Sep 24. PMID: 30290889; PMCID: PMC7172169.
82. Donougho M. 'Individuality' before Hegel: Leibniz, Herder, Goethe, Fichte, Schleiermacher, Humboldt, Schlegel. In: Hegel's 'individuality'. Cham: Palgrave Macmillan; 2023. https://doi.org/10.1007/978-3-031-21369-4_2
83. Birkinshaw JH, Michael SE, Bracken A, Raistrick H. Patulin in the common cold collaborative research on a derivative of Penicillium patulum Bainier. II. Biochemistry and chemistry. Lancet. 1943;245:625.
84. Rodrigues T, Reker D, Schneider P, Schneider G. Counting on natural products for drug design. Nat Chem. 2016;8(6):531–41. https://doi.org/10.1038/nchem.2479. Epub 2016 Apr 25 PMID: 27219696.
85. Chatterjee N, Walker GC. Mechanisms of DNA damage, repair, and mutagenesis. Environ Mol Mutagen. 2017;58(5):235–63. https://doi.org/10.1002/em.22087. Epub 2017 May 9. PMID: 28485537; PMCID: PMC5474181.
86. Wang P, Shehu AI, Ma X. The opportunities of metabolomics in drug safety evaluation. Curr Pharmacol Rep. 2017;3(1):10–15. https://doi.org/10.1007/s40495-016-0079-5. Epub 2017 Jan 3. PMID: 28758057; PMCID: PMC5526643.
87. Sai Priya T, Ramalingam V, Suresh Babu K. Natural products: A potential immunomodulators against inflammatory-related diseases. Inflammopharmacol. 2024;32:2821–59. https://doi.org/10.1007/s10787-024-01562-4

88. Gouda NA, Alshammari SO, Abourehab MAS, et al. Therapeutic potential of natural products in inflammation: Underlying molecular mechanisms, clinical outcomes, technological advances, and future perspectives. Inflammopharmacol. 2023;31:2857–83. https://doi.org/10.1007/s10787-023-01366-y.
89. Martínez-Jiménez F, Muiños F, Sentís I, et al. A compendium of mutational cancer driver genes. Nat Rev Cancer. 2020;20:555–72. https://doi.org/10.1038/s41568-020-0290-x.
90. Sinkala M. Mutational landscape of cancer-driver genes across human cancers. Sci Rep. 2023;13:12742. https://doi.org/10.1038/s41598-023-39608-2).
91. Basu AK. DNA damage, mutagenesis and cancer. Int J Mol Sci. 2018;19(4):970. https://doi.org/10.3390/ijms19040970. PMID:29570697;PMCID:PMC5979367.
92. Otteneder M, Lutz WK. Correlation of DNA adduct levels with tumor incidence: Carcinogenic potency of DNA adducts. Mutat Res. 1999;424:237–47. https://doi.org/10.1016/S00275107(99)00022-6.
93. Ali Abdalla YO, Subramaniam B, Nyamathulla S, Shamsuddin N, Arshad NM, Mun KS, Awang K, Nagoor NH. Natural products for cancer therapy: A review of their mechanism of actions and toxicity in the past decade. J Trop Med. 2022;11(2022):5794350. https://doi.org/10.1155/2022/5794350. PMID:35309872;PMCID:PMC89330790.
94. Li W, Kong AN. Molecular mechanisms of Nrf2-mediated antioxidant response. Mol Carcinog. 2009;48(2):91–104. https://doi.org/10.1002/mc.20465.
95. Ranjan A, Ramachandran S, Gupta N, Kaushik I, Wright S, Srivastava S, Das H, Srivastava S, Prasad S, Srivastava SK. Role of phytochemicals in cancer prevention. Int J Mol Sci. 2019;20(20):4981. https://doi.org/10.3390/ijms20204981. PMID:31600949;PMCID:PMC6834187.
96. Siegel M, Ross CS, King C. 3rd. The relationship between gun ownership and firearm homicide rates in the United States, 1981–2010. Am J Public Health. 2013;103(11):2098–105. https://doi.org/10.2105/AJPH.2013.301409. Epub 2013 Sep 12. PMID: 24028252; PMCID: PMC3828709.
97. Laqueur HS, Smirniotis C, McCort C, Wintemute GJ. Machine learning analysis of handgun transactions to predict firearm suicide risk. JAMA Netw Open. 2022;5(7): e2221041. https://doi.org/10.1001/jamanetworkopen.2022.21041. PMID:35816302;PMCID:PMC9274320.
98. Zare Sakhvidi MJ, Hajaghazadeh M, Mostaghaci M, Mehrparvar AH, Zare Sakhvidi F, Naghshineh E. Applicability of the comet assay in evaluation of DNA damage in healthcare providers' working with antineoplastic drugs: a systematic review and meta-analysis. Int J Occup Environ Health. 2016;22(1):52–67. https://doi.org/10.1080/10773525.2015.1123380. Epub 2016 Apr 25. PMID: 27110842; PMCID: PMC48942750.
99. Bonassi S, Ceppi M, Møller P, Azqueta A, Milić M, Neri M, Brunborg G, Godschalk R, Koppen G, Langie SAS, Teixeira JP, Bruzzone M, Da Silva J, Benedetti D, Cavallo D, Ursini CL, Giovannelli L, Moretti S, Riso P, Del Bo' C, Russo P, Dobrzyńska M, Goroshinskaya IA, Surikova EI, Staruchova M, Barančokova M, Volkovova K, Kažimirova A, Smolkova B, Laffon B, Valdiglesias V, Pastor S, Marcos R, Hernández A, Gajski G, Spremo-Potparević B, Živković L, Boutet-Robinet E, Perdry H, Lebailly P, Perez CL, Basaran N, Nemeth Z, Safar A, Dusinska M, Collins A; hCOMET project. DNA damage in circulating leukocytes measured with the comet assay may predict the risk of death. Sci Rep. 2021;11(1):16793. https://doi.org/10.1038/s41598-021-95976-7. Erratum in: Sci Rep. 2021 Sep 20;11(1):19043. PMID: 34408182; PMCID: PMC8373872.
100. Møller P, Stopper H, Collins AR. Measurement of DNA damage with the comet assay in high-prevalence diseases: Current status and future directions. Mutagenesis. 2020;35(1):5–18. https://doi.org/10.1093/mutage/gez018.
101. Møller P. The comet assay: Ready for 30 more years. Mutagenesis. 2018;33(1):1–7. https://doi.org/10.1093/mutage/gex046.

102. Praphasawat R, Munkong N. Anti-genotoxicity evaluation of cratoxylum formosum dyer leaves by comet assay and micronucleus test. Asian Pac J Cancer Prev. 2017;18(6):1475–8. https://doi.org/10.22034/APJCP.2017.18.6.1475. PMID:28669154;PMCID:PMC6373808.

Chapter 3
Analytical and Bioanalytical Methods

Contents

3.1 Recipe—Spices and Curry 69
3.2 Analytical Methods for Secondary Natural Products 72
3.3 Bioanalytical Methods for Secondary Natural Products 84
3.4 Chromatographic and Spectroscopic Methods—An Overview 90
3.5 Adulteration and Authenticity 95
3.6 Food Processing and Natural Products 96
References 97

Abstract The analysis and bioanalysis of natural products presents particular challenges. In some cases, the product found in nature can be decomposed during the isolative process and this may impact upon the choice of the best analytical procedure that is subsequently adopted. In principle a bioassay can be more sensitive but it may not discriminate between different compounds which may be structurally diverse but may have a similar bioactivity profile. Bioassays can take on different forms depending upon the objective of the investigator and can be linked to analytical methods in order to probe levels of efficacy and structure/activity relationships. The development of adjacent technologies (for example, spectroscopy) has facilitated the development of tools to increase the power of analysis and the measurement of structure/function relationships and authenticity/adulteration. Processing is an important aspect of food systems and can give rise to the formation of process-specific components and examples of these are available.

3.1 Recipe—Spices and Curry

Spices are an integral part of cooking and impart flavour as well as, in some instances, acting as an aid to preservation. Many of the organoleptic properties of spices are associated with secondary natural products. In Europe, the accepted implication is that curry will be hot and this is a guiding principle for many. It is, however, possible to create recipes in which the use of capsaicin (Fig. 3.1) is minimised or omitted completely.

B. Hanley, *Secondary Natural Products in Foods and Food Systems*,
https://doi.org/10.1007/978-3-032-00292-1_3

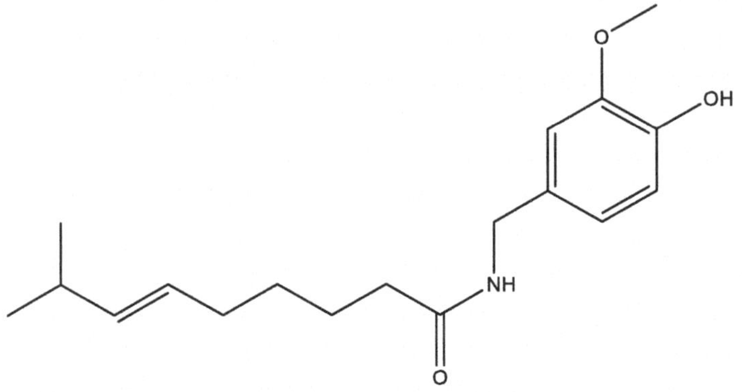

Fig. 3.1 Structure of capsaicin

Capsaicin (methyl-n-vanillyl nonenamide) is a secondary natural product and is lipophilic. It is also responsible for the hot, burning sensation associated with eating chilli fruits which is mediated by binding to a receptor (see below). It is one of a range of alkaloids that contain the same core structure (capsaicinoids) that are found in the fruit of the chilli pepper (*Capsaicin annuum*) which belongs to the family *Solanaceae*. Capsaicin biosynthesis in plants is a combination of two different pathways—the fatty acid pathway and the phenyl propanoid pathway. Capsaicin production in plants can be modulated by addition of precursors or by increased water stress. The analysis of capsaicin can be carried out by chromatography with ultraviolet spectroscopic detection or with various types of mass spectrometric detection [1]. Capsaicin binds to a specific receptor (transient receptor potential cation channel subfamily V member 1—TRPV-1) and much of its biological activity is mediated through this, although some alternatives modes of action also seem to operate. As shown in Fig. 3.1, the capsaicin molecule consists of two regions—an aromatic ring and a hydrophobic side chain connected to the aromatic ring by an amide bond. The molecular characteristics and the impact of structural variations on the biological activity of capsaicinoids has been reviewed [2]. TRPV1 is a member of the family of transient release potential ion channels that produce the sensation of heat and capsaicin is an agonist. It is used to control neuropathic pain as a dermal patch [3]. As noted in Chap. 2, a number of secondary natural products have been proposed to have anti-cancer properties. In the case of capsaicin, it is postulated to arrest the cell cycle in tumour cells and induce apoptosis [4] although other, less beneficial effects have also been noted. In addition, capsaicin is postulated to act on the prevention of tau protein hyperphosphorylation and the accumulation of beta amyloid plaques [5], both of which are thought to be important factors in the pathology of Alzheimer's disease [6] and this may provide a mechanistic rationale for the observation of a link between chilli consumption and the preservation of cognitive function [7]. The relatively short halflife in vivo [8] and the irritant side effects limit the pharmacological usefulness of capsaicin [9], particularly for oral rather than topical applications.

Curry can contain a number of other spices and other active constituents and these will be considered later. The protein source for a curry can be animal or vegetarian but the preparation method remains broadly equivalent. In most cases, a curry will be improved by pre-marinading of ingredients to develop flavours (for example both onion and garlic contain flavour precursors which are only released upon crushing or grating) or to facilitate the uptake and interactions of flavour molecules.

Assembly of Marinade

The protein of choice (500 g or so) is cut into bite sized pieces. The marinade is prepared in a sealable plastic bag. The marinade consists of:

Garlic (4 cloves crushed and chopped)

Cayenne pepper (1 teaspoon)

Ground coriander (1 teaspoon)

Ground turmeric (1 teaspoon)

Garam masala (2 teaspoons)

Lime juice (2 freshly squeezed limes)

Grated ginger (1/2 teaspoon)

Water (3 tablespoons)

Leave to marinade overnight in the plastic bag in a refrigerator.

Take a heavy bottomed saucepan or skillet pan. Add two tablespoons of cooking oil of choice (ground nut oil is good so long as nobody you are cooking for has a nut allergy). Heat the oil over a medium flame then add a piece of cinnamon stick and a teaspoon of cumin seeds. When the seeds start popping, add 4 cardamom pods. Skin and chop an onion into cubes and add to the oil. Chopping onions and tomatoes is a good test of how sharp your knife is—one way to cheat is to use a serrated knife [10]. ***Move the onion around with a spatula to stop it burning. After about 10 min the onion should be translucent. Add the protein and marinade and stir and cook. Add three chopped fresh tomatoes (or a can of chopped tomatoes) and cook on a low heat, covered, for 15 min. Add half a bag of chopped spinach or chard. Season with salt and pepper. Decide how hungry you are and, if necessary add a can of chick peas. Cook slowly (covered) for a further 30 min, adding water if it dries out. Add chopped coriander leaf (a good handful) then stir and serve either with rice (cooked with saffron) or a naan bread.***

Yoghurt (about 2 tablespoons) can be added alongside the tomatoes but can also be left out.

Marinating of meat and other proteins has been extensively investigated [11]. It affects both protein profile and the sensory properties. In addition, marinading with spices can enhance the formation and uptake of flavours. The addition of lime juice lowers the pH of the marinade and will also begin the process of breaking down proteins so they can better absorb flavours and become more tender. When garlic is crushed, a range of S compounds are formed catalysed by the enzyme, alliinase.The spectrum of compounds is affected by pH and by other factors [12] including the nature and source of the alliinase enzyme [13]. In addition, the volatile S compounds initially formed when garlic is crushed gradually, over time, react with each other to form slightly less volatile compounds which are less likely to be lost during cooking so the flavour profile of the dish will change depending on the length of the marinading time. This is considered in more detail in Chap. 9. Finally, the interaction between salt and sour taste has been studied [14] so the addition of acid to a marinade will not only influence the enzymically-produced hydrolysis products, it will also affect salt perception. In addition, some compounds (e.g. alkaloids) are likely to be extracted more effectively into a low pH aqueous solution and this may also alter the sensory characteristics.

3.2 Analytical Methods for Secondary Natural Products

A number of features which make up the ingredients of foods including availability, processability and concentration of function are all important aspects of the value of the ingredient. For example, saffron is a spice derived from part of the flower of *Crocus sativus*. It has a reputation as being the most expensive spice in the world because it only occurs in small amounts in a single plant part and only at a specific time in the life cycle of the plant. The chemical composition of saffron varies depending on the growing and processing conditions. The main chemical responsible for the colour of saffron is a carotenoid called crocin (Fig. 3.2) whose chemical name is more properly rendered as bis[β-D-glucopyranosyl-(1 → 6)-β-D-glucopyranosyl] 8,8′-diapocarotene-8,8′-dioate.

Crocin is hydrolysed in the gut after ingestion to give crocetin (Fig. 3.3) which has been proposed to have a range of biological activities [15].

The main aroma compound of saffron is safranal (Fig. 3.4).

Chemically it is a monoterpene aldehyde and is produced from picrocrocin (the major flavour component of saffron) by hydrolysis with a co-occurring glucosidase (Fig. 3.5).

All three of the major components of saffron, together with flavonols, are important quality parameters for saffron and contribute to its biological activity profile. They are analysed after extraction with ethanol/water (70:30) by HPLC coupled to a diode array detector [16]. The use of authentic standards is (as will become apparent

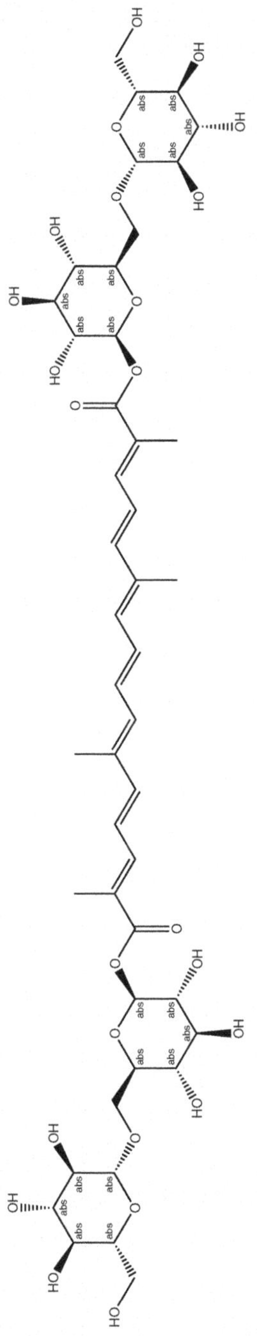

Fig. 3.2 Structure of crocin

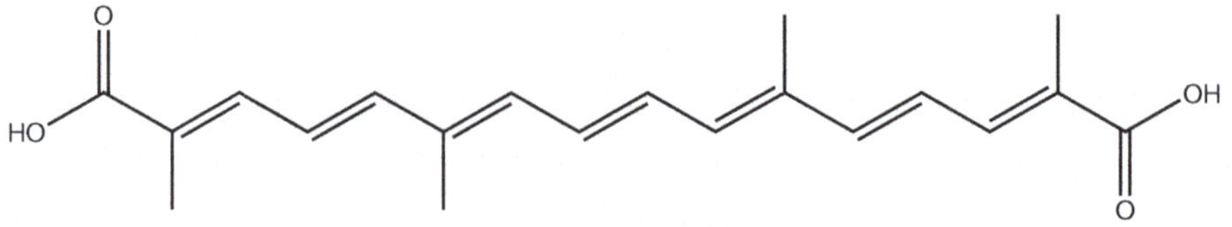

Fig. 3.3 Structure of crocetin

Fig. 3.4 Structure of Safranal

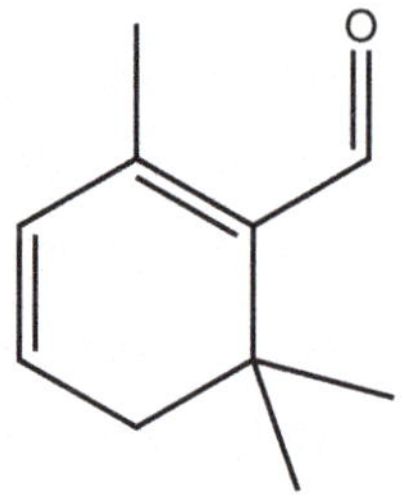

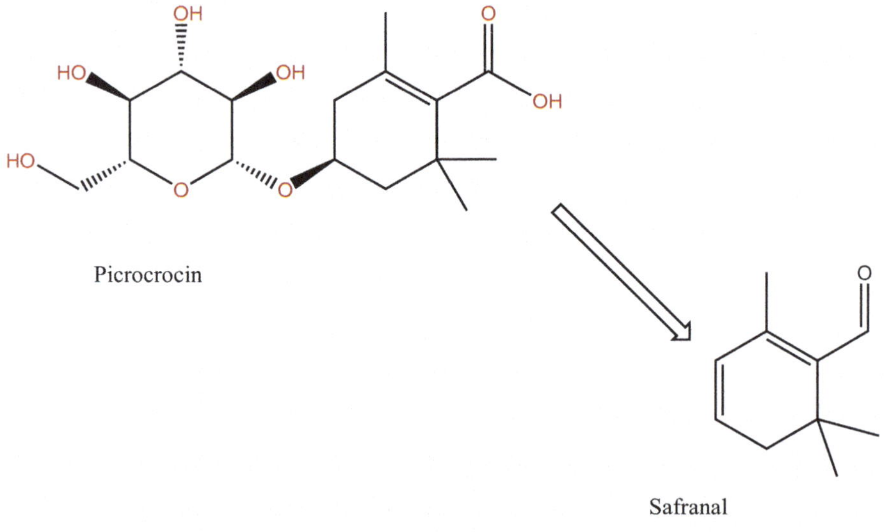

Fig. 3.5 Formation of safranal from picrocrocin

later) an essential aspect of quantitative analysis. In addition, correlation of quality parameters (particularly taste and colour) with chemical composition is important since it ensures the authenticity of a given sample and, in some cases, confirms its provenance. For example, vanillin can be derived from pods, biosynthetically or synthetically [17]. Authentication of the source has a significant effect on the value of the product and on the use since natural vanillin is favoured from a regulatory standpoint and the price of the natural compound is at least 200 times the price of the synthetic flavouring [18]. The use of adulterated or synthetic vanillin in place of the natural material has a significant financial incentive. Food authenticity confirmation [19] often relies on secondary natural product profiles and even isotope ratios of products (because of the nature of biosynthetic production and the isotope effect observed in some enzymically catalysed reactions) [20], products from different sources often have different ratios of the isotopes of H, C, S and O and these can be quantified [21].

The analysis of secondary natural products is closely linked to the amounts and complexity of the material and the form the material takes. In the nineteenth century, the analysis of natural products was linked to their isolation and characterisation. The latter was achieved by the reductionist approach of breaking a substance down until it formed recognisable units then combining units to synthesise the material again. This tradition of lysis followed by synthesis was the accepted route for structural confirmation in pre-spectroscopic times [22] and was derived from the philosophical tradition of chemistry [23]. Indeed, from alchemical beginnings, lysis followed by synthesis was the common method used in the eighteenth century. In practical terms, the process of analysis starts with the whole molecule and then fragments it into its constituents. Synthesis begins with fragments and reconstructs them to give the whole molecule. The complementarity of these approaches is clearly evident. This tradition predates even the eighteenth century. Paracelsus (1493–1541) envisioned the *tria prima* (salt, sulphur, and mercury) as being the constituent parts of all substances. These could be separated, purified then recombined to either give the original material or a different substance. It is important to realise that analysis and synthesis (decomposition and recomposition) were not simply philosophical tools but were underpinned by the reductionist and empirical tradition of chemistry. They were practical techniques used to understand specific aspects of particular materials. In addition, this approach relied, in part, on the assumption that the logic applied by a chemist would be guided by (or be at least sufficiently similar to) the logic that applies in living systems. An example which illustrates this is that of the precursors of mustard oils—the glucosinolates.

The pungent essence derived from mustard seeds crushed in water has a long and significant history. The study of the chemistry of the pungent principle of mustard can be traced to Bussy who isolated mustard oil in 1840 [24]. Further work suggested that the essential oil derived from a glucoside which was present in the seeds of black mustard. The structure of the precursor was proposed by Gadamer in 1897 but was later found to be incorrect using the analysis/synthesis approach by Ettlinger and Lundeen [25, 26]. The structures of the glucosides themselves and that of the isothiocyanate breakdown products suggested two approaches to analysis. The first of these was the analysis of the involatile glucosides as intact molecules. The problem with this approach is that, while in some seeds, there was only one principal glucosinolate present and this could be isolated and the amount determined, in other plant parts and even in many seeds, there were several glucosinolates present and separation of these molecularly and structurally similar compounds, containing a glucose moiety and a sulphonated oxime and only differing in the presence of different side chains (Fig. 3.6), was challenging.

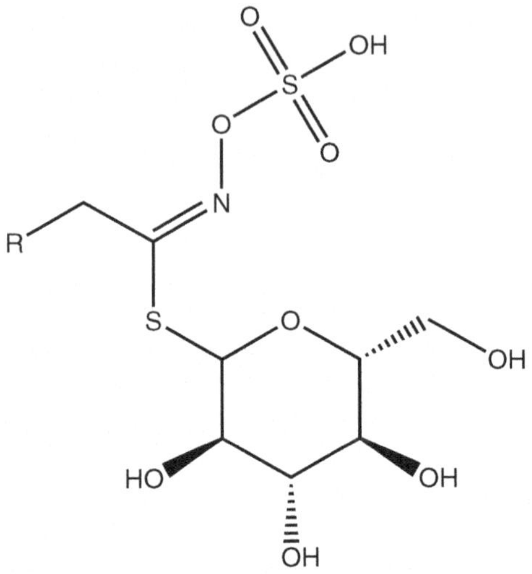

Fig. 3.6 Representative glucosinolate structures

Direct analysis of the involatile glucosinolates did not become a practical proposition until liquid chromatographic methods were developed that had sufficient resolution (and were fast enough) to be practically viable as an analytical tool [27]. Isothiocyanates, on the other hand, are relatively volatile and therefore analysis based on volatility, which was in a more advanced state of development, became the method of choice. The isothiocyanates were produced when the plant material was crushed and the glucoside came into contact with a co-occurring (but differentially located) thioglucosidase enzyme, myrosinase (Fig. 3.7).

The analysis of isothiocyanates had been described [28]. The potential of gas chromatography (gc) for the analysis of volatile compounds had been realised prior

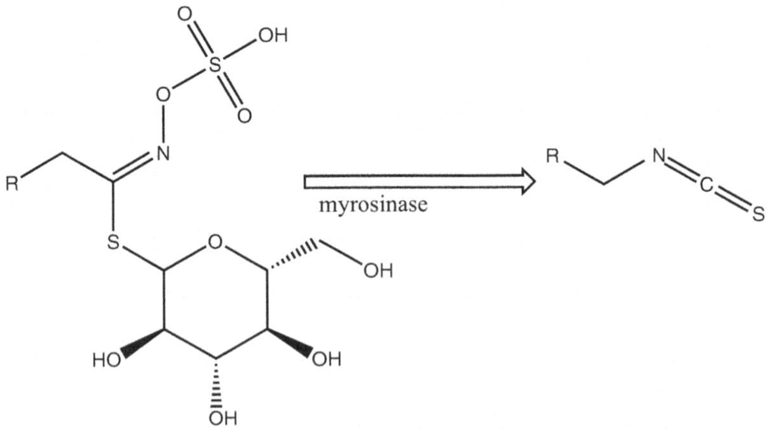

Fig. 3.7 Enzymic hydrolysis of glucosinolates

to 1952 [29] so the analysis of the glucosinolate content of plants could be carried out by inference [30]. In addition, methods for the synthesis of authentic isothiocyanates were known [31]. While most glucosinolates are able to be analysed by gc, some of them either do not produce volatile hydrolysis products (e.g. indole glucosinolates) or the initially formed products undergo further reactions to give involatile products (e.g. oxazolidinethiones). As can be seen from the structure of glucosinolates, hydrolysis of the thioglucose moiety, in addition to releasing a functionalised side chain, also produces glucose and (after rearrangement) a sulphate molecule. The direct assay of released glucose has been used as a method for the analysis of total glucosinolates [32] and this has been used for screening purposes. It relies upon the specificity of the myrosinase enzyme and that it will not act as a more general glucosidase and hydrolyse other glucosides.

A second method developed for the analysis of glucosinolates makes use of the sulphonated oxime moiety. In this case, a purified sulphatase from the snail *Helix pomatia* is used to selectively hydrolyse the sulphate moiety of glucosinolates previously adsorbed onto an anion exchange resin and the desulphated glucosinolate is eluted with water. In essence, this is an in situ transformation of a characteristic of a molecule that enables it to bind to a particular matrix (a negatively charged sulphate group) specifically into an uncharged molecule that can no longer bind to the matrix. The unpurified, commercially available sulphatase also contains beta-glucuronidase activity and this can prevent quantitative production of desulpho glucosinolates. A microcolumn method has been developed and the individual desulpho glucosinolates can then be analysed by HPLC [33] or gc (after derivatisation with trimethylsilyl groups). In addition, intact glucosinolates can also be measured directly [34]. The method for gc analysis was developed by Thies [35] and the method, which makes use of trimethylsilyl derivatisation to volatilise the desulphoglucosinolates, has been modified to extend its range and usefulness [36]. The two principal analytical methods outlined above—analysis of the desulpho glucosinolate sidechain and analysis of isothiocyanates—represent two different approaches to the lysis of the parent natural product. The desulphation approach might be considered to be closer to being biomimetic since sulphation is the final step in the biosynthesis of glucosinolates [37] and specific reversal of this step is an example of a retro-biosynthesis.

Conceptually, therefore, sulphation of a desulphoglucosinolate to give the complete structure ought to be a logical step in the proof of structure by synthesis mentioned previously. Unfortunately, sulphation (often carried out using pyridine—sulphur trioxide) requires that the OH groups of the glucose moiety be protected so a direct synthesis of a glucosinolate from a desulphoglucosinolate has not been achieved. It is likely that a biomimetic (enzymic) route will offer the best possibility to achieve this philosophically satisfying synthesis. The isothiocyanate cannot be easily conceptually envisioned as being converted into a glucosinolate since it is the product of a rearrangement reaction (the Lossen Rearrangement reaction mentioned previously). Retrosynthesis from chemical principles has been examined and a route involving an electrophilic glucosyl donor and a thiohydroxamic acceptor group was developed by Ettlinger and Lundeen [26] for the synthesis of benzyl glucosinolate (Glucotropaeolin) (Fig. 3.8). This has been termed the ‘anomeric’ route.

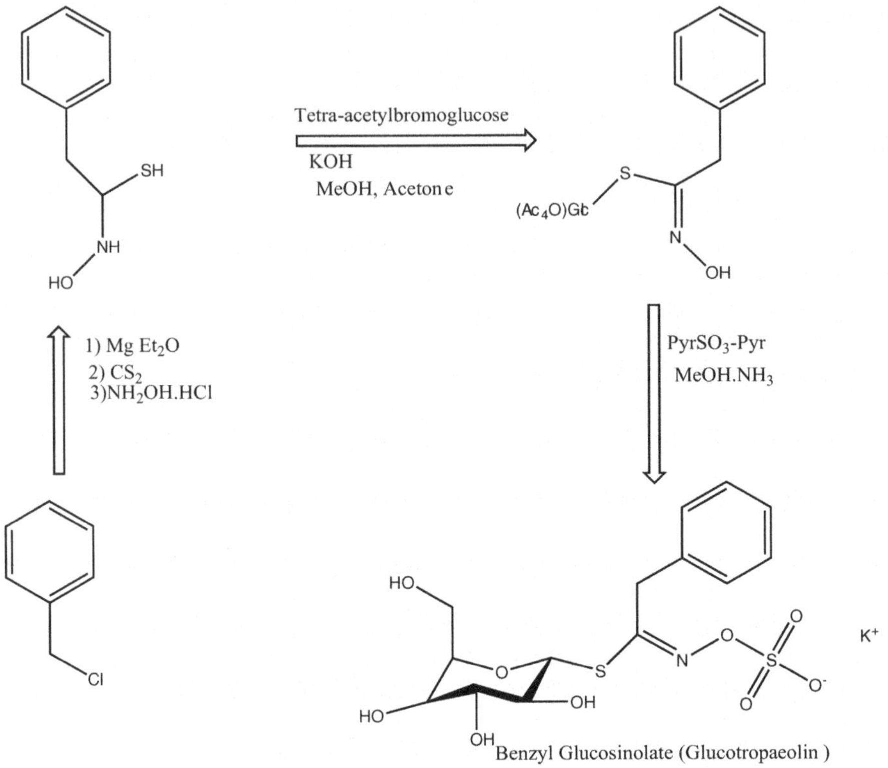

Fig. 3.8 The anomeric approach to glucosinolate synthesis

A nitronate pathway has also been developed (Fig. 3.9) and was used in the synthesis of sinigrin however a biomimetic pathway has not been achieved and, as noted above, the relative configurations of the N, C and S atoms in the rearrangement product, the isothiocyanate means that a biomimetic synthesis directly from the corresponding isothiocyanate is not possible.

In the case of another class of natural products, phytoestrogens, some of the same issues arise as with glucosinolates. In this case the primary classification is by function however structural diversity presents an even more significant challenge in the development of analysis (or bioanalysis). Common structural features that might aid the development of rapid analytical tests do not exist. Endogenous estrogen (estradiol) (Fig. 3.10) binds to estrogen receptors (ERα and Erβ) and is implicated in a wide range of physiological effects.

The two estrogen receptors mediate different effects and this is reflected in differences in the clinical outcomes of natural, exogenous estrogens with different selective estrogen receptor modulating (SERM) activities. This activity is, in part, modulated by selective estrogen receptor sub-type binding. For example, one natural phytoestrogen class, the isoflavones have a five-fold greater affinity for βER than

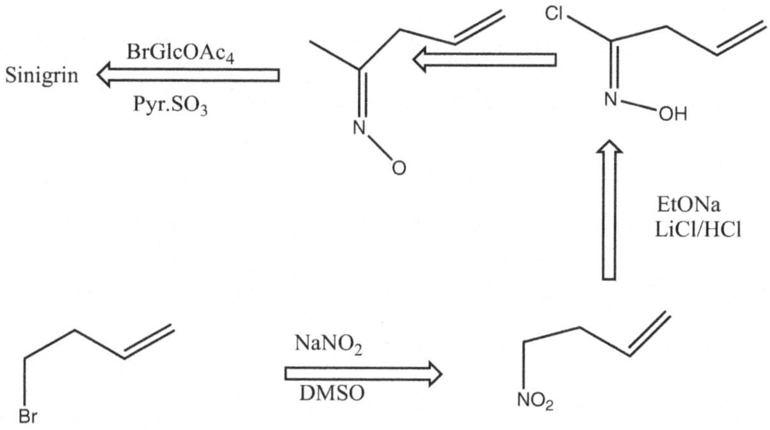

Fig. 3.9 The nitronate route to glucosinolate synthesis

Fig. 3.10 The structure of estradiol

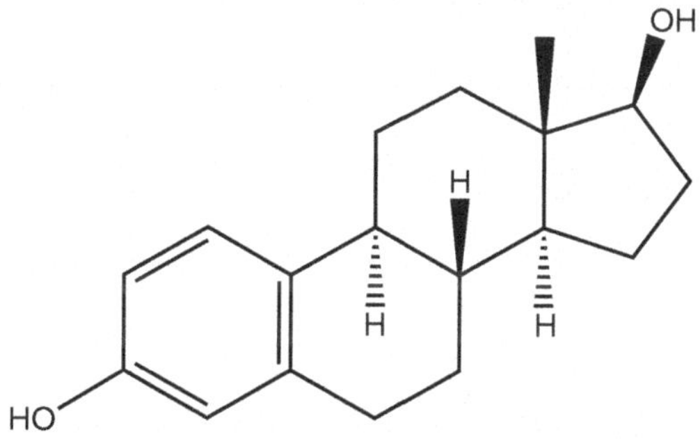

α-ER [38]. They may also operate selectively as estrogen receptor agonists or antagonists. Estrogen receptors have some functional similarity to other nuclear receptors including the androgen receptor, progesterone receptor, glucocorticoid receptor, vitamin D receptor, retinoid acid receptor and thyroid receptor. The transcription factor activity of estrogen receptors has been described in detail [39]. Exogenous estrogens derive from plants (phytoestrogens) or from fungi (mycoestrogens). Phytoestrogens are generally classified by structural characteristics into isoflavones, coumestans and lignans (Fig. 3.11) however these do not contain specific functionalities that can be used to facilitate analysis.

Mycoestrogens include macrocyclic compounds like zearalenone (Fig. 3.12).

Other exogenous natural estrogens include stilbenes and pterocarpans and also the division of flavonoid like compounds including flavones, flavanones, isoflavones (noted above) and isoflavanes. One of the most significant members of this final class is the metabolite equol which is produced from daidzein by gut bacteria [40] but only in a sub-group of the population. Although there are no moieties that could help in the developement of analytical protocols, given the degree of structural similarity

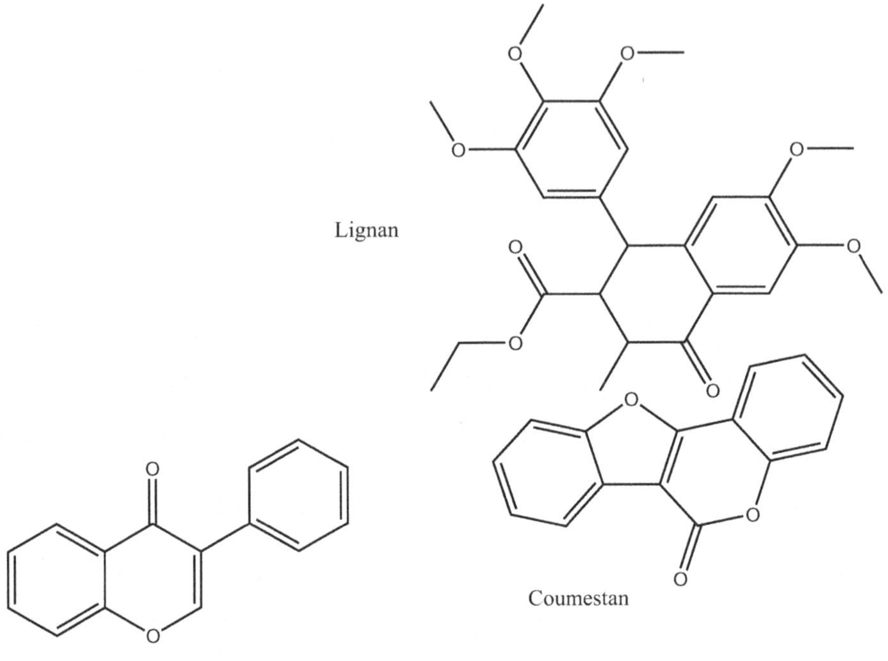

Fig. 3.11 Structures of isoflavones, coumestans and lignans

Fig. 3.12 Structure of zearalenone

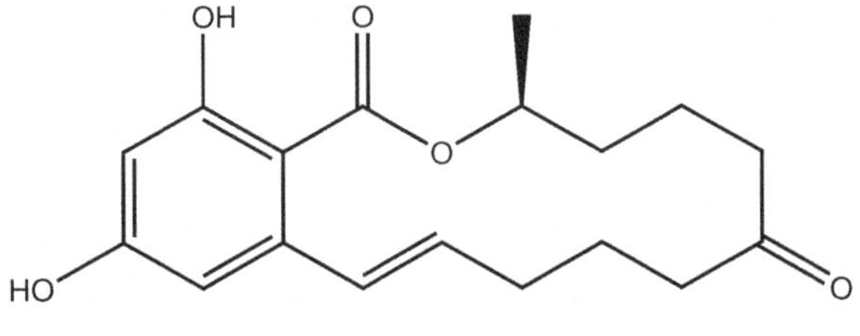

between flavonoids, stilbenes and pterocarpans, there are some biosynthetic links between them and these have been reviewed [41] along with their biological activity. The analysis of the class of phytoestrogens has been extensively investigated and reviewed [42]. The classification of phytoestrogens includes compounds that interfere with the pathways initiated by the natural hormone, estradiol, and, from a plant protection standpoint, production of phytoestrogens can affect herbivore fertility and hence may lower the threat to the plants. In this way, phytoestrogens can be considered to be a version of the 'magic bullet' that favours plants rather than animals. Phytoestrogens act specifically on mammalian metabolic pathways and not on plants. As with glucosinolates, isoflavones are often found as glycosides however in this case, hydrolysis to give the aglycone (which occurs readily but non-specifically) is not the initiation of further reactions. Analysis of phytoestrogens should take into consideration the plant source since this will, in part, direct the analytical methods to be used. Isoflavones are found primarily as glycosides in common plants such as *Glycine max, Glycrrhiza glabra, Medicago sativa, Phaseolus vulgaris, P. lobata,* and *Trifolium. pratense.* Coumestans are found mainly in *Medicago sativa* (alfalfa) while lignans occur in flaxseed and in some other plants (pumpkin seeds in particular). Since some phytoestrogens occur as acetylated or malonated glycosides, in order to obtain a complete spectrum of the compounds that are present, some care must be taken during extraction to prevent hydrolysis of the acetyl or malonyl groups or, indeed, the glycosidic residues. The primary methods for analysis of flavonoid phytoestrogens is liquid chromatography linked to a suitable detector. Because of the structural variation and functional similarity of different types of phytoestrogens, biological assays and bioanalytical methods are sometimes considered to be of more use. These will be considered in more detail in Sect. 3.3. In the case of lignans the situation is more complex because of the presence of a range of glycosides and higher molecular weight derivatives. For example secoisolariciresinol cannot be directly determined in flaxseed extract because it is present as a glucoside, which is further ester-linked with 3-hydroxy-3-methyl-glutaric acid to form high molecular weight (4000) oligomers. Methods have however been developed to analyse free lignan, glycosides and oligomers [43]. The analysis of coumestrol in plant material is also complicated by the occurrence of glycosides. The analysis of isoflavones has also been carried out by enzyme-linked immunosorbent assay (ELISA) [44]. Since this method makes use of the recognition of a molecule by a biological system (an antibody) albeit in vitro, it can be considered, in some ways, to be a biological assay. While the sulphatase activity from *Helix pomatia* was used in the analysis of glucosinolates, the same biological source (but a different enzyme activity) was used to cleave the glycosides of isoflavones prior to analysis of the aglycones [45]. While glucosinolates have both a range of structures (and biological activities), the members of the class have structural features in common (the presence of a glucose moiety and a sulphonated oxime) that facilitate analysis of the natural products as a class. Phytoestrogens are linked more firmly by a common functional biological activity and while this activity (a consequence of binding to estrogen receptors) means that there will likely be some structural similarities between members of the class, these are not of the form (or, as is the case with glycosylation, do not occur to an equal extent in all members) that will facilitate analysis of the complete class of molecules simultaneously. This means that many analytical procedures have been developed for individual (or small groups) of compounds from a single biological source. For example, the fungally produced toxin zearalenone has been analysed by a variety of methods [46]. As a fungal toxin, it is particularly important as a potential contaminant of foods and feeds and it has a range of toxic activities including mutagenicity, teratogenicity, carcinogenicity, nephrotoxicity, immunotoxicity, and genotoxicity. Zearalenone, like aflatoxin B1 and patulin, also has some antibacterial activity [47]. These different functional profiles mean that direct comparison of secondary natural products on the basis of a core biological activity is not feasible.

The characterisation and isolation of alkaloids was a substantial step for early natural product chemists. Before the development of chemical sub-structure awareness, secondary natural products were often defined as extracts containing mixtures of compounds with particular biological properties. The extract was given a name that alluded either to the biological properties or to the characteristic features of the plant or other species. The concept of 'signatures' was a combination of these two ideas. This postulated that the physical appearance of a plant gave a clue as to its potential medicinal properties. For example, walnuts resembled brains and therefore would be expected to have a particular medicinal use in treating headaches. The concept, which was known to the ancient Greeks, was popularised by the German shoemaker Jacob Boehme in a book produced in 1621 in which he claimed to have a vision that illustrated the link between God and humans through the 'signatures' with which God, through divine design, showed how to use certain plants for specific uses. This has a more modern manifestation in the concept of 'intelligent design' in which some external agency can be recruited as the instigator of some aspects of the universe. Trying to see something with agency that connects apparently disparate entities has, therefore, a long history. The concept of signatures was, therefore, in some ways, an attempt to characterise the world in divine terms. Naming plants and extracts by their apparent medicinal function was another aspect of this and would seem to be a reasonable reductionist step that would be recognisable to current day natural product chemists. For example, the willow tree grows in damp conditions and the bark of the tree was thought to be effective against rheumatic conditions. It is commonly mythologised that antirheumatic ancient remedies included the bark of willow (which contains salicylic acid) but there is little evidence for this, and it was not until the nineteenth century that a connection between willow bark and rheumatism was suggested [48]. The use of functional terms such as 'knitbone' or 'devil's breath' to describe certain organisms and to allude, thereby, to their medicinal effects has a long tradition. This has continued—in the alkaloid and other areas of secondary natural product chemistry—with the classification of compounds by biological function (for example anxiolytics, diuretics etc.) rather than by chemical formula however this has only been of limited value since, as noted previously, alkaloids have a wide range of functional properties and classification solely by biological type is of limited value. In the case of alkaloids, the development of methods for their isolation was facilitated by one particular physicochemical aspect of their collective properties—their capacity to neutralise acids—rather than a common biological property. The concept of acids and bases developed over a number of years culminating in the concept that Arrhenius stated that acids are simply substances that add hydrogen cations to a solution and bases could combine with acids to form neutral salts. This led to the idea that if the strength of an acid was known, the number of units of an unknown base that were required to neutralise the acid then allowed the basicity to be measured relative to the acid. The idea that plants could contain 'alkali-like' substances led to the description of these as alkaloids. While the plant-based origin for alkaloids is commonplace, alkaloids are also synthesised and secreted by other organisms. Fungi and lichens, in particular have been known to contain bioactive compounds that are alkaloidal in nature [49] and subclasses of fungi such as mushrooms contain a number of alkaloids including, in particular, β-carbolines, pyrroloquinolines, pyrroles, and indoles [50]. In the case of lichen, secondary natural products only occur when a suitable fungal partner and an appropriate phycobiont are present. A number of alkaloids exhibit antimicrobial (and specifically antibacterial) activity and bacterially derived alkaloids are relatively rare however indoles, pyrroles, pyridines, indolizidines, amides, and diketopiperazines have been found and, as the application of omics technology increases, the discovery of biosynthetic potential is likely to be found to be more widespread. While the isolation of classes of alkaloids was based upon their chemical properties as bases, most of the interest in them arose from their wide-ranging biological properties.

One of the challenges that arose in the use of alkaloids was determining how much of an extract to administer to obtain optimal beneficial medicinal effects. This required the application of physicochemical techniques driven by a biological impetus. The first alkaloid to be isolate and purified so an effective dose could be calculated and given was morphine which was purified from poppy seed by Serturner in (1804) [51]. Other alkaloids including quinine, atropine, caffeine, strychnine and nicotine were isolated during the first half of the nineteenth century. The first successful synthesis of an alkaloid was achieved in 1886 by Ladenburg [52]. This was, then, the first verifiable proof of the analysis/synthesis cycle for alkaloids. The development of, particularly, chromatographic and spectroscopic methods greatly facilitated the isolation and identification of alkaloids and other secondary natural products. The defining characteristic of alkaloids (their solubility in acid and then neutralisation and back extraction into an organic solvent) was a significant feature in their popularity as a class of natural products. Structural determination by analogy and by decomposition to give core molecules was also of great assistance in increasing both the knowledge concerning and the usefulness of alkaloids. The analysis of the different classes of alkaloids can be challenging. The development of a nitrogen-specific detector has increased the applicability of gc in the analysis of alkaloids [53] however many alkaloids are involatile so analysis by liquid chromatography is more widely applied [54]. A range of sample preparation techniques have been developed to remove interfering molecules such as tannins and organic acids. Alkaloids are usually isolated as free bases prior to chromatographic analysis. Ion exchange cartridges are also commonly used in sample cleanup and optimisation of chromatographic conditions have been studied [55]. Since alkaloids have some chemical similarity which will, to an extent influence their physicochemical (and chromatographic) properties but very different biological profiles, analysis by chemical similarity (as with glucosinolates) may be more successful than using a single biological assay.

Terpenes (isoprenoids) are a class of molecules with the formula $(C_5H_8)_n$ while terpenoids (i.e. terpene-like molecules) contain the same core but have other additional functional groups attached. The most well known members of the class are those found in cannabis plants however they also contribute significantly to the aromatic smell of a number of other species. Terpenes have a range of biological properties—from repelling insect and herbivore attack to anti-inflammatory effects [56]. Plants containing terpenes are among the most popular for 'alternative' medicine however the lack of convincing clinical data is a drawback to their more extensive use. In addition, the use of crude extracts containing a range of bioactive molecules means that traditional pharmacological investigations are comparatively rare and there are a large number of anecdotal reports of the effectiveness of crude extracts. The development of analytical methods is a crucial step in the understanding of the effects of different terpenes and in establishing dose/response relationships. Terpenen/terpenoids are built up from hydrocarbon units containing a prenyl (isopropyl, 3methylbut-2-en-1-yl) unit. This leads to a diverse number of cyclic and linear molecules which can combine with sugars and fatty acids and also contain a diverse number of functional groups (see below). All terpenes are formed from two, 5-carbon units isopentenyl diphosphate (IPP) and dimethylallyl diphosphate. Addition of further 5-carbon units (isopentyl diphosphate) gives rise to a range of core units including monoterpenes (comprised of 10 carbon atoms, C10), sesquiterpenes (C15), diterpenes (C20), triterpenes (C30), carotenoids (C40) and polyprenols (> 45). These units can have further functionalisation including hydroxylation, glycosidation, methylation, acylation etc. to give rise to a diverse range of molecules. Depending on the degree of polymerisation and the types and numbers of functional groups added, extraction and analysis is based around polarity (choice of extraction solvent) and volatility (type of chromatography favoured). A number of protocols for extraction of different types of terpenes/terpenoids have been summarised [57] however some prior knowledge of the expected size, volatility and polarity of the target molecules is required for the choice of an optimal method. The use and development of specific methodologies such as gc coupled to different detectors has increased the numbers of target molecules that can be quantified at once [58] and the use of both gc and liquid chromatography [59] has greatly increased the level of analytical capability.

The development of methods that enable better and more complete structural determination of secondary natural products has enabled chemists to develop rapid and appropriate analytical methods. These can make use of particular functional groups (as is the case with glucosinolates and alkaloids), gross molecular properties such as volatility (as occurs for terpenoids and isothiocyanates) or a combination of analytical and bioanalytical techniques (as is the case for naturally-occurring estrogens). Since much of the interest in secondary natural products, in food and elsewhere, is a consequence of their biological properties, bioanalytical methods that are more focused on functional properties rather than using chemical reductionism to define groups of compounds based on physicochemical characteristics, is of particular value.

3.3 Bioanalytical Methods for Secondary Natural Products

Much of the interest in natural products arose from an appreciation of their biological properties. This led to a growth in the development of rapid and sensitive bioanalytical techniques. We have seen (Sect. 3.2) how some groups of compounds are more amenable to a biologically-based assay which relies upon function rather than purely on structural characteristics. Some of these will be summarised. It should be noted that, in some cases, bioanalytical (bioresponse) techniques are most conveniently used as a screening method prior to more accurate quantitation. In this case sensitivity, reproducibility and limit of detection are key attributes of any test method. Bioanalysis is not solely the measurement of the response of a biological system to a chemical stimulant—it may also refer to the measurement of biological systems themselves or of chemicals measured in a biological system. The definition of bioanalysis will be briefly considered.

There are three different concepts that are commonly covered by the term 'bioanalysis'. It can mean **the analysis of substances in biological systems**—this can also be referred to as biomarkers of exposure since it refers directly to the measurement of specific entities (and their metabolites) in biological systems. The principles that underpin the detection of secondary natural products in biological systems lies outwith the scope of this chapter and is mentioned only for the sake of completeness. The second meaning is the **measurement of biological entities themselves**. This can include, for example, the detection and quantitation of viruses, bacteria, fungi and other microorganisms in systems. Methods developed for such quantitation include the detection of microbially-produced ATP—often using bioluminescence as a detection method [60]. While ATP detection can provide evidence for viable (actively metabolising) biological systems, there are a number of challenges to such an assay (particularly when considering microbial species. The principal challenges associated with the detection of ATP are that some bacterial cells (particularly Gram – ve cells) are less easily detected, viruses and spores are not detected and some species are viable but non-culturable (vbnc) and may not give an ATP signal. Simply measuring microbial ATP production therefore gives an incomplete picture of the presence of microbes and other living cells. It is, however, a very rapid and effective screening method and has significant advantages over traditional culture methods. A common feature of most living systems (and the basis of an alternative way of measuring the presence of biological species) is that they contain DNA and RNA and this can be used as the basis of another method of detection. The particular advantages of nucleic acid-based testing includes sensitivity, specificity and breadth of application [61]. The two principal types of methods are unamplified and amplified testing. In the case of the former the target organism must be present in high enough concentrations to enable its DNA to be detected directly. The alternative is to amplify a small specific sequence from a sample using the polymerase chain reaction (PCR) and to then detect a specific sequence using a probe that hybridises to a specific sequence [62] which is unique to the organism being detected. Descriptions of how PCR works are available. The development of PCR is one of the most significant advances in the detection of biological material. The obvious problem with this approach is that the amplification step relies on having some knowledge of the identity of the target organism and its DNA sequence. A second problem is that detection of DNA does not confirm that the source organism is alive or even present. It simply records that a 'signature' of a species is detectable. An additional use of DNA manipulation which is of direct relevance to secondary natural product investigations is the automatic reconstruction of secondary natural product metabolic pathways from biosynthetic gene clusters [63] however that is not the primary focus of this chapter. **The third definition of bioanalytical methods has already been mentioned—the principle of bioresponse to specific chemical substances**—and is probably the most commonly understood definition of the term. This involves using biological systems for the detection and analysis of secondary natural products. Such methods can range from the analysis of biologically defined classes of compounds to specific chemical entities. Examples from both groups will be given.

As noted previously, natural products can be classified in a number of ways—chemical/structural characteristics (e.g. alkaloids), biosynthetic derivation (e.g. terpenoids), biological activity (e.g. phytoestrogens) or a combination of all three. The classification of secondary natural products has been driven in the first instance by the usefulness and the importance of their biological properties.

Biological properties (such as smell or taste or some types of healing properties) can be measured at a number of levels and can be available to an untrained population group. Understanding and making use of either chemical structure or biosynthetic derivation require a relatively high level of specialist knowledge. The different levels of biological classification can be demonstrated by the case of phytoestrogens. At a whole organism level, phytoestrogens can affect fertility while, on a molecular level, their importance derives from their capacity to bind to estrogen receptors. This also can provide the basis for their detection and quantification. Phytoestrogens are structurally diverse and a functionally based assay might be favoured however chemists tend to feel more comfortable with chemical methods of analysis and large numbers of phytoestrogens are able to be analysed using chromatography combined with mass spectrometry [64], however chemical analysis can only be used to analyse known chemical entities and may not, therefore, provide a full assessment of the total estrogenic potential of a given sample. An alternative is to use a biologically-based system to assess the overall endocrine-disrupting activity of a sample. Such a system can be specific and receptor-based [65], whole cells using, for example, a transgenic yeast-based screen [66] or can incorporate a whole animal system [67]. The last of these can distinguish estrogen agonists and antagonists and provide a more comprehensive view but it is both expensive and a poor use of a valuable resource when a high degree of granularity is not required and cheaper methods are available. Estrogenic activity is primarily associated with a very specific set of molecular events—binding to different forms of the estrogen receptor. Similar assays, based on binding to receptors or to enzymes can also be used to identify a range of different beneficial and deleterious properties. For example, the presence of anti-thyroid compounds from a range of sources can be assayed simultaneously and the goitre prevention assay was developed in the 60 s [68]. This assay relied upon measuring the excretion rate of thyroxine. A more specific assay which measured the sodium iodide symporter (NIS) (an essential component for thyroid hormone synthesis) in a transfected cell line [69]. More generic biologically based assays have also been used in the analysis of secondary natural products. Two examples of these are the measurement of mutagenicity and modification of inflammation. In the latter case, in particular, multiple biological mechanisms may be impacted so a range of secondary natural products may elicit an effect. In both cases, biological relevance must be carefully considered—particularly when the relative importance of multiple molecules or potential synergies is being considered. The use of bioassays can, as has been previously indicated, be particularly valuable for screening purposes. The added value of a bioanalytical approach is in its biological relevance and how it may measure the overall (and both additive and synergistic) effects of molecules that are structurally disparate. One of the most studied areas of investigation using bioanalytical procedures at various levels of sophistication is the investigation of mutagens and genotoxic carcinogens.

Mutagens are defined as compounds or processes that cause changes in DNA. Food mutagens are considered to be implicated in diet-derived carcinogenesis [70]. It is worth noting that not all mutations are equal. Some occur in DNA that is less biologically significant and some types of lesions are easier to repair than others. If damage to DNA is repaired before the function of the DNA is affected then the mutation is biologically non-existent. If the mutation is in a part of DNA that is not biologically significant (sometimes called a silent mutation) then the mutation has no effect on phenotype and is irrelevant. Mutagens are often, although not invariably, carcinogenic. There are a limited number of mechanisms by which mutagens operate. These include:

- Intercalation—interferes with the normal structure of the DNA and can lead to mutations.
- Alkylation—can alter the structure of the DNA and interfere with its function.
- Oxidation—creation of highly reactive free radicals can damage DNA by adding or removing electrons from the molecule.
- Hydrolysis—breaking the bonds that hold the DNA molecule together can lead to leading to mutations.
- Deamination—removal of the amino group from the bases of the DNA molecule can lead to mutations.

Deciding on how to measure mutagens is dependent upon the level of biological relevance that is considered to be important. If the capacity to damage DNA (mutagenicity) is the prime criteria then the test system will be simpler than that employed to measure carcinogenic potential. The simplest method for measuring mutagenesis is the Ames test [71]. The principle underlying the test is simple. The test uses several strains of bacteria (*Salmonella, E. coli*) that carry a particular mutation which means that the bacteria cannot grow unless a particular amino acid (generally histidine or tryptophan) is supplied. If the test compound causes a mutation that restores the amino acid biosynthesis capability and the bacteria can then grow in the absence of exogenous amino acid, then that compound is a mutagen. While the most common form of the Ames test uses bacteria [72], it has also been adapted for use with eukaryotic cells and other complementary tests have been evaluated [73]. Testing different concentrations, allows mutagens to be ranked. An alternative approach is to use structure:activity relationships based on known mutagens to predict if a given molecule is likely to be mutagenic. Mutagenicity prediction based upon electrophilicity has also been investigated [74]. In addition, in silico models for the prediction of mutagenicity have been reviewed [75]. As with many model systems, the interpretation of the data is of crucial importance [76] and the development of advanced techniques such as neural networks will be important [77].

While techniques for the investigation of food-borne mutagens and carcinogens have been developed and are widely used, the importance of inflammation in chronic disease [78] and its manipulation by secondary natural products has become realised as being significant. The rise in interest in compounds that can attenuate inflammation has grown as the importance of inflammation as a factor in many diseases has risen and techniques to measure inflammation have become more sophisticated. This is in part a consequence of the awareness of the primary inflammation pathways—NF-κB, MAPK, and JAK-STAT. This has increased the numbers of targets that can be investigated. The process by which inflammation cascades occur is broadly similar—no matter the cause of the initial inflammatory provocation. These can be summarised as:

- Provocation of inflammatory stimuli
- Activation of inflammation
- Release of inflammatory biomarkers
- Stimulation of production of pro-inflammatory cells.

Unfortunately, the codification of the inflammatory response has not necessarily been accompanied by an increase in the development of validated methods of ensuring the biological relevance of measurements and techniques. In many cases, the data generated has also suffered from a lack of reproducibility or accuracy. In addition, inflammation is also associated with chronic disease states (see above) therefore an underlying condition may be responsible for a given inflammatory response. Ensuring that there is a mechanistic basis for correlation between a particular exposure event, the appearance of specific inflammatory biomarkers and a given clinical endpoint can be challenging. Nonetheless a number of models of organ-specific inflammation and its consequences have been developed—primarily for pharmacological use [79]. The use of animal models is often preferred over an in vitro approach however cost, biological relevance and ethical issues have contributed to the development of non-animal alternatives. These will be discussed in more detail in Chap. 4 when new Approach Methodologies (NAMs) will be introduced. Animal models of different types of inflammation have been reviewed [80] however these are less useful for the development of methods for the bioanalysis of secondary natural products and have more use in a drug discovery type process. The important role of chronic inflammation in disease has been widely recognised. While many of the usual outcomes are known and the impact of inflammation has been recognised, the range of disease outcomes that appear to be affected by inflammation appears to be increasing. For example, there is considerable support for the 'inflammation hypothesis' of Alzheimer's disease [81]. This seems to be a consequence of activated microglia and the release of cytokines. The possible role of inflammation and therapies to treat this will be discussed in more detail below but first of all, the possible involvement of quantum effects will be considered. An interesting part of the beta amyloid/tau theory of Alzheimer's disease is that phosphorylation of the tau protein is necessary in order that it can be removed from microtubules. It can then be dephosphorylated and return to stabilise the microtubules. Hyperphosphorylation of tau leads to the formation of tau tangles and tau cannot then be reincorporated into microtubules. This results in the collapse of unstabilised microtubules. One theory about consciousness suggests a key role for microtubules in the modulation of quantum effects and consciousness [82]. This theory, which was originally introduced by Penrose and Hameroff [83], suggested a key role for microtubules, in what they referred to as the orchestrated objective reduction (Orch-OR) effect. Tau is one of the most physiologically significant microtubule-associated proteins (MAPs) and microtubule stabilisation has been proposed as a theory to combat Alzheimer's disease [84]. Since it has been postulated that a minimum number of microtubules are required for consciousness then the collapse of these due to a loss of tau integrity could lead to a loss of consciousness. Although is issue is far from being resolved, the idea that Alzheimer's disease may be a type of 'quantum abnormality' has some coherence. Indeed a link between the immune system, inflammation and consciousness has been suggested however this remains a controversial area [85]. A meta analysis gave some support to the hypothesis of a link between Alzheimer's disease and inflammation since it suggested that exposure to nonsteroidal anti-inflammatory drugs (NSAIDs) may be associated with a reduced risk of Alzheimer's disease [86], however placebo controlled studies with both selective and non-selective cyclooxygenase -2 (COX-2) inhibitors failed to demonstrate an effect except, perhaps, in those at a very early stage of the disease [87]. Overall, observational studies seem to support the use of NSAIDs to prevent the cognitive decline associated with Alzheimer's disease but a randomised controlled trial (RCT) did not verify these findings [88]. The increase in interest in 'anti-inflammatory' effects of secondary natural products derived from foods and other sources has been stimulated by the ease of use of various methods to measure the proposed effects in clinical samples. This can be misleading, however, particularly because many studies are designed to monitor short term effects. There are a range of inflammatory transcription factors however measurement of these can be challenging [89] and, as a result, the simpler to measure, anti-inflammatory response or production of cytokines are used as surrogate end points. Some protocols look at the effects of specific components on inhibition of pro-inflammatory species such as lipopolysaccharide on cells in culture. As with all such in vitro studies, care must be taken when interpreting the results. The two major drawbacks are probably the difficulty in carrying out long term studies in any in vitro cell culture-based model and the development of a dose response relationship. In addition, provocation of an inflammatory response purely by addition of an exogenous agent is, as mentioned previously, both a philosophically and scientifically flawed approach. Finally, the relationship between any model system and the in vivo situation must be validated otherwise any results will show nothing more significant than general trends.

3.4 Chromatographic and Spectroscopic Methods—An Overview

Many of the advances in natural product chemistry have arisen as a consequence of advances in adjacent fields or from the application of techniques in unexpected or unanticipated ways. The development of isolative and purification techniques (with the exception of chromatographic separation) can be traced from the alchemical principles that pre-dated atomism and molecular theories. Indeed the seven stages of the alchemical process can be seen to have analogues in the principles of chemistry.

The first stage of alchemy is referred to as calcination. This is the process of heating and breaking down material into its constituent parts. Such a process forms the basis of the analysis/synthesis process used in eighteenth century chemistry to define substances. One area of confusion was the observation that calcination often led to an increase in weight of the material. This is, of course, reasonable from our perspective since calcination led to the formation of oxides. To the early chemical way of thinking, the idea of a gain in weight was counter-intuitive since calcination was thought to drive off a component of the material (phlogiston) and so would result in a loss of weight. This could occur quickly—in which case the presence of a flame was a signifier of the rapid loss of phlogiston—or slowly as was the case with a metal.

The second stage, after calcination, is dissolution of the post-calcination residue in water and the third stage is filtration to separate the soluble from the insoluble material. After separation comes conjunction in which the materials are joined. This is analogous to synthesis in the analysis/synthesis cycle used by chemists to prove a structure. Fermentation is then used to break down the formed material using a biological system. This is clearly a similar process to that which is carried out during incubation but it may also be analogous to the use of synthetic biology to create something that is new from a biological source. The final two stages (distillation and coagulation/crystallisation) are directly analogous to operations used by chemists. One missing element from the alchemical stages is the chromatographic separation of similar (but not identical) compounds. While adoption of the alchemical approach enabled the extraction and isolation of some compounds (e.g. morphine by crystallisation and mustard oils by distillation), a very different approach was required for the separation of other, closely related materials or complex mixtures and this developed from nonalchemical roots.

The essence of chromatography is to separate components in a mixture by using two immiscible components. These can be two liquids, a liquid and a solid or a gas and a solid. The history of chromatography has been extensively reported and, while it is of some historical interest, from a secondary natural product standpoint, it is only really of value in cases where it gives examples of specific principles. There have been a number of suggestions of the use of solid and liquid phases to test substances in antiquity however these are invariably spot tests where a test mixture is absorbed onto a solid surface that has been pretreated with something that reacts with a specific entity in the mixture. Such approaches do not involve separation and so are not chromatography. The 'father' of chromatography is a title most often given to Mikhail Tsvet. Tsvet took an extract of plant leaves and filtered this through a glass column filled with powdered chalk. The constituent pigments were separated, and this was the first reported example of adsorption chromatography. While Tsvet carried out his first experiments in 1903, it was not until 1942 that Martin and Synge developed the theory and practice of chromatography which led to them receiving the Nobel prize [90]. The idea that compounds can be separated as a consequence of relatively small molecular differences is an important conceptual idea. What is more, rather than studying structures by decomposition, chromatography enables intact molecules to be isolated and studied—this is a significant advance on the traditional analytical methods used in the nineteenth century. The other aspect of chromatography that is important and greatly increases its applicability is that, unlike, for example, distillation that employs just one aspect of a mixture (volatility) to effect a separation, by varying the liquid and stationery phases, molecules can be separated according to polarity, molecular size and shape and, in the case of some bioactive molecules, biological properties. This then creates a significant number of criteria that can be used (sometimes successively) to fractionate mixtures. For example, the preparative fractionation of glucosinolates was carried out using molecular charge (acidic alumina), charge/size ratio (DEAE Sephadex), size (G-10 Sephadex) and finally changes to the counter ion to facilitate crystallisation [91]. Similarly, the analysis of secondary natural products can make use of liquid and gas chromatography with a range of detectors. While the objective selection of the most appropriate technology should be a major criterium, there is a tendency for people to use the 'I have a hammer so every problem is a nail' mode of thinking [92], attributed to Maslow—although, as a concept, it probably predates this example. One possible exhibition of this is the use of supercritical fluid chromatography (SFC) for the analysis of alkaloids (not a common target for SFC because of their polarity) and other natural products. Although investigators have claimed superior analysis times, the need to add alkali to mobile phases and the lack of validation data suggests that other options may be more suitable. SFC nonetheless remains an increasingly applicable analytical tool in the investigation of secondary natural products and its use is likely to increase despite high initial capital costs [93]. The development of new chromatographic methods within the basic paradigm of the use of two (or more) immiscible phases will continue with drivers including the need to increase sensitivity and selectivity, the analysis of analytes from complex and different matrices, better, noninvasive and non-destructive methods and better predictive tools to help in the development of new methods.

Different forms of spectroscopy have been developed and then applied to the analysis of molecules. The essential beginnings of spectroscopy lie in the work on optics carried out by Newton. Chemists were aware that certain substances, if burned, would produce specific colours in the flame. Sulphur burned with a pink flame while compounds containing sodium had a typically yellow flame. When this was combined with analysis of light, dark lines appeared that were characteristic of the metal that was present—even if other compounds were also present. This allowed chemists to identify components (elements) in mixtures based upon their spectral characteristics even if the elements themselves could not be separated and characterised. Previously

Herschel and Ritter had discovered that radiation from the sun extended into the infra-red and ultraviolet parts of the spectrum respectively. Classical spectroscopy enables scientists to study objects at a distance without requiring that they isolate and purify the object of their enquiry. In order for spectroscopy to be of value to the study of natural products, it had to be capable of probing the functional characteristics of molecules and not simply their elemental composition. This is analogous to the reductionist analytical approach developed in the eighteenth and nineteenth centuries. To be fully effective, spectroscopy needed to probe the constituent parts of a molecule and how these parts were related to each other. One of the challenges to the use of spectroscopy was alluded to earlier. While some forms of spectroscopy are very powerful probes of molecular function, when it becomes a tool looking for an application, it becomes objectively less useful. Some of the different types of spectroscopy and their characteristics are shown in Table 3.1. Spectroscopy is the interaction of light (or other radiation) and matter. Although it is often grouped with spectroscopic techniques, mass spectrometry is, in fact, not a spectroscopic technique. It will therefore be described briefly but the criteria used to discuss it are different from those of spectroscopic techniques. The other important difference is that, on a molecular level, spectroscopic methods are non-destructive of the sample while mass spectrometry relies upon the breaking down of a molecule into its constituent parts. While most spectroscopic (and mass spectrometric) techniques will help to add to the evidence of a natural product structure, the importance of selective degradation and identification of constituent parts (and how they are formed biosynthetically) is still an important aspect of the study of secondary natural products. Despite the use of libraries of spectra and automation, structural determination of natural products still requires chemical intuition and awareness.

The most common methods used for the study of natural products are uv/vis, IR and NMR spectroscopy and mass spectrometry. Each method has particular characteristics and, when used in conjunction with knowledge of the chemical plausibility of a given structure, can be very powerful. In particular, the use of stable or radioisotopes isotopes to follow the incorporation of specifically labelled biosynthetic precursors can also provide considerable insights. These can help to build up an idea of the biological plausibility of structures and of biosynthetic methods and mechanisms used in the biological synthesis of molecules. Techniques that are less commonly used but which have particular uses in biological systems include Raman spectroscopy [94] and circular dichroism spectroscopy [95]. As noted previously, the hammer/nail hypothesis should be guarded against and the most appropriate technique for a required outcome should always be considered. Single techniques, in isolation, are considerably less powerful than the use of multiple approaches, each of which may provide particular insights.

Table 3.1 A non-exhaustive summary of some spectroscopic methods

Technique	Principles and measurement	Uses of method	Limit of detection
Absorption spectroscopy	Different molecules absorb radiation at different frequencies	Determination of the amount of a particular substance present	Can be as low as a single atom
Atomic absorption spectroscopy	Detection of free electrons in the gaseous state	Quantitative and qualitative analysis of metals in complex systems	Variable depending on signal/noise ratio of an individual instrument
Circular dichroism spectroscopy	Difference in the absorption of left and right polarised light	Measurement of optically active chiral molecules—particularly protein conformation	Of the order of micrograms per ml of sample
Electrochemical impedance (EI) spectroscopy	Measures electrical impedance of a substance by applying different AC frequencies	Populations of microorganisms, screening for cancer tissue, changes in electron transfer capabilities	Depends on the nature of the sample but typically nano molar
Electron spin resonance (esr) spectroscopy	Atoms, molecules or ions with unpaired electrons have magnetic properties because of electron spin. Can measure the electronic structure of a sample	Measurement of metalloproteins and catalysed reaction rates. Protein degradation and reactive N and O species	Can be nano scale but depends on application
Emission spectroscopy	When compounds are heated they emit energy in discrete amounts. Each element emits a discrete wavelength	Elemental composition of a sample. Detection of metals in alloys, metals in liquids etc.	Inductively coupled plasma–optical emission spectrometry (ICP-OES) has a detection limit of ppb for most metals
Energy dispersive spectroscopy (electron dispersive Xray spectroscopy)	Electron excitation by x-rays leads to release of energy and x-ray beams that are characteristic of a given atomic structure	Chemical composition of subcellular structures. Deposition of chemicals on living surfaces (e.g., pesticides on a plant leaf)	Qualitative rather than quantitative but can detect to about 0.1%
Fluorescence spectroscopy	Excited molecules can emit fluorescence	Fluorescence properties depend on the structure of the molecule so it can be used to detect specific molecules also detection of metals	Depends on the sample but typically pico molar

(continued)

Table 3.1 (continued)

Technique	Principles and measurement	Uses of method	Limit of detection
Fourier transform infrared (FTIR) Spectroscopy and infrared spectroscopy	A multiple frequency beam irradiates the sample and the frequencies that are adsorbed determined. FTIR measures the vibrations of atoms, and based on this it is possible to determine the functional groups present	FTIR is used to determine the functional groups of molecules since different functional groups will create different spectra	The limit of detection is of the order of 0.1 mg/ml
Molecular spectroscopy	This involves the interaction of materials with electromagnetic radiation. A range of different interactions can take place and these provide different types of information about the sample	The analysis of the atomic and molecular structures of various compounds can be carried out	Depends on the technique and the information required
Mossbauer spectroscopy	Gamma rays emitted by one nucleus can be completely absorbed by a sample containing nuclei of the same isotope	Used to measure chemical, magnetic and structural properties of materials (e.g. ironcontaining proteins)	
Nuclear magnetic resonance (NMR) spectroscopy	Nuclei have spin. When a magnetic field is applied the nuclei can align in one of two energy states and move between them by absorbing or emitting photons of energy. These can be measured and reflect the environment the atom or molecule is in	NMR can measure purity, structure and different types of atom, as well as the environment of the atom	For C-13, 5 mM of material. For H-1, 0.05 mM of material
Photoelectron Spectroscopy	Photoionisation (uvinduced) gives the binding energy of electrons	Can measure oxidation states and relative ratios of different elements and isotopes	

(continued)

Table 3.1 (continued)

Technique	Principles and measurement	Uses of method	Limit of detection
Raman spectroscopy	Measures the difference in vibrational energy between two states in the atoms in a molecule	Can determine the chemical composition and structure in solution	Varies depending on analyte and technique/instrumentation used
Ultra violet/visible (Uv/Vis) spectroscopy	Radiation is absorbed by materials at different frequencies. The absorption spectrum depends on the atomic and molecular composition	Can be used to identify classes of materials. Can be used to 'fingerprint' extracts and mixtures	

3.5 Adulteration and Authenticity

One of the principal uses of analytical methodology is in the identification of authentic materials and prevention of adulteration. While techniques to confirm the chemical authenticity (and/or the adulteration) of a sample has been established for a number of years (including isotope analysis), the use of biological authenticity has increased greatly with the development of DNA-based methods including polymerase chain reaction (PCR) and sequencing technologies. While more and more advanced techniques have been employed, these are increasingly being supported by AI traceability methodologies such as Blockchain [96]. Such solutions, while being increasingly popular by the developers, are not the panacea that ardent users of AI would have you believe. As with any system, technology is only as foolproof as its weakest link. One of the major problems with blockchain-type approaches is the scalability and the generation of data—huge amounts of which must be handled effectively. This can increase costs and, if Blockchain network members take shortcuts, then the integrity of the whole system is at risk. Another issue with any type of mutual validation technology is the input (accidently or on purpose) of false data. In addition, Blockchain is a complex technology and slow adoption is likely to limit its widespread implementation. In addition, if different blockchain methodologies are adopted in different parts of the food chain then a system-wide method will not be possible. The history of food security and adulteration is replete with examples of technologies that were developed to generate and to then interpret complex datasets but which had crucial issues that could either inadvertently or due to purposeful exploitation, create a false sense of security.

The mass production of food led to an increase in the scale of adulteration [97] and a concomitant increase in the pace of development of analytical methods to find chemical contaminants. The first food adulteration Act in the UK was passed in 1860. A House of Commons report from 1856 noted the flagrant adulteration of food with, among other materials, plaster of Paris, arsenic, nux vomica (a mixture of strychnine and brucine), alum, copper, red lead, lime and powdered glass. It was also recognised that the problem was more acute in poorer districts [98]. The development of analytical techniques has resulted in an overall improvement in quality of chemical constituents in food however some large-scale instances of adulteration have still been observed. For example, there was a significant product recall in 2003 when Sudan Red contamination of chilli powder was uncovered and led to significant action [99]. There are two issues that tend to drive the development of food authenticity. The first of these is food safety and quality. If an adulterant poses a risk to health or adversely affects food quality then there is a public protection argument. If the adulterant is primarily used because it is a cheap alternative. For example, there are economic reasons for unscrupulous manufacturers and suppliers to use a cheaper alternative to saffron and chemical methods have been developed to enable quality markers for saffron to be confirmed [100]. A number of methods have been developed to authenticate saffron including biological markers of adulterants [101] however it should be noted that a priori identification of unknown adulterants can still be challenging.

The use of DNA-based technology for assessing the adulteration of foods has been reviewed [102]. A range of targets including species authenticity, microbial contamination and the presence of genetically modified material have been detected and methods validated. Analogously to the antimicrobial resistance (AMR) ‘arms race’, regulatory authorities have to be aware of advances in fraud and adulteration and to work with analytical scientists to develop new methods. The EU provides scientific support for policy and their resource can be searched (https://knowledge4policy.ec.europa.eu/home_en). These include a monthly food fraud awareness newsletter.

3.6 Food Processing and Natural Products

Food processing has been developed over long periods to improve the safety and quality of food and to enable consumption to be more convenient. In some cases, processing is required to make food edible and, in some cases, it improves the bioavailability of nutrients and is, therefore, beneficial, however a number of factors may be implicated in the optimisation of food nutrient bioavailability [103]. There has also been a growth in the concept of ‘smart processing’ to purposely improve the quality and health characteristics of food [104].

There have also been instances, however, where food processing has had unforeseen deleterious consequences. One of the most significant was the formation of the toxin, acrylamide, as a result of a Maillard reaction taking place between asparagine and a reducing sugar during food processing [105]. Considerable efforts were made to try to understand the reaction and to minimise acrylamide formation. Other compounds formed by Maillard type reactions include heterocyclic aromatic amines [106] which are potent carcinogens (Fig. 3.13). A representative member of the class is 2-amino-3,4-dimethylimidazo[4,5-f]quinoline (MeIQ)Other compounds formed

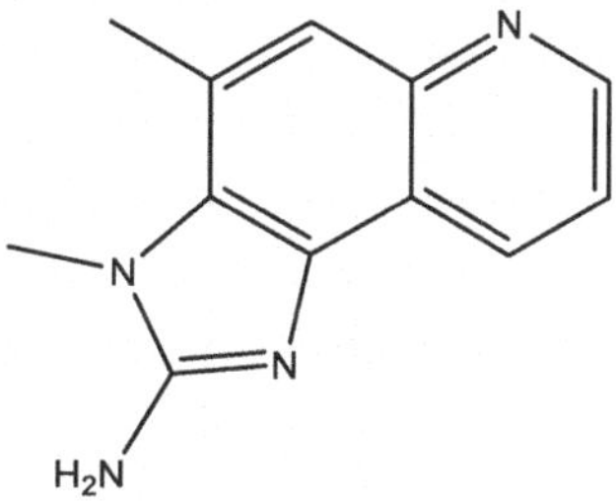

Fig. 3.13 Structure of a representative heterocyclic aromatic amine (MeIQ)

during processing/cooking which may pose a small risk to some consumers include 3- monochloro propanediol (3-MCPD), furan and glycidyl fatty acid esters.

There has also been considerable discussion regarding certain types of processing as a risk factor in disease promotion. Of particular interest has been the growth in the definition of 'ultra processed foods' (UPFs). These were defined by researchers in Brazil as one of four categories. While three of the categories are relatively simple to define in terms of processing (unprocessed, processed ingredients such as salt sugar and oils, processed foods which are altered in a way that home cooks could do themselves), the definition of UPFs relies heavily on examples and on processing that uses ingredients not normally found in a home kitchen. This is a problematic definition since it is defining a process on the basis of ingredients and products rather than by a process itself. In essence, nutritionists have used the UPF definition to demonise their own particular betes noir. Overall the use of the term UPF is profoundly and disturbingly unscientific. It also implies that chemicals are 'bad' which is an evidence-free value judgement.

References

1. Srinivas NR. LC/MS/MS analysis of capsaicin using multiple transition pairs-some view points to ponder. Biomed Chromatogr. 2009;23:1129–30. https://doi.org/10.1002/bmc.1227.
2. Zhang W, Zhang Y, Fan J, Feng Z, Song X. Pharmacological activity of capsaicin: Mechanisms and controversies (Review). Mol Med Rep. 2024;29(3):38. https://doi.org/10.3892/mmr.2024.13162. Epub 2024 Jan 19. PMID: 38240083; PMCID: PMC10828990.
3. Peppin JF, Pappagallo M. Capsaicinoids in the treatment of neuropathic pain: a review. Ther Adv Neurol Disord. 2014;7(1):22–32. https://doi.org/10.1177/1756285613501576. PMID:24409200;PMCID:PMC3886382.
4. Luján-Méndez F, Roldán-Padrón O, Castro-Ruíz JE, López-Martínez J, García-Gasca T. Capsaicinoids and their effects on cancer: The "double-edged sword" postulate from the molecular scale. Cells. 2023;12(21):2573. https://doi.org/10.3390/cells12212573. PMID:37947651;PMCID:PMC10650825.
5. Wang J, Sun BL, Xiang Y, Tian DY, Zhu C, Li WW, Liu YH, Bu XL, Shen LL, Jin WS, et al. Capsaicin consumption reduces brain amyloid-beta generation and attenuates Alzheimer's diseasetype pathology and cognitive deficits in APP/PS1 mice. Transl Psychiatry. 2020;10:230. https://doi.org/10.1038/s41398-020-00918-y.
6. Kent SA, Spires-Jones TL, Durrant CS. The physiological roles of tau and Aβ: implications For Alzheimer's disease pathology and therapeutics. Acta Neuropathol. 2020;140(4):417447.

https://doi.org/10.1007/s00401-020-02196-w. Epub 2020 Jul 29. PMID: 32728795; PMCID: PMC7498448.

7. Shi Z, El-Obeid T, Riley M, Li M, Page A, Liu J. High chili intake and cognitive function among 4582 adults: An open cohort study over 15 years. Nutrients. 2019;11:1183. https://doi.org/10.3390/nu11051183.
8. Rollyson WD, Stover CA, Brown KC, Perry HE, Stevenson CD, McNees CA, Ball JG, Valentovic MA, Dasgupta P. Bioavailability of capsaicin and its implications for drug delivery. J Control Release. 2014 Dec 28;196:96–105. https://doi.org/10.1016/j.jconrel.2014.09.027. Epub 2014 Oct 12. PMID: 25307998; PMCID: PMC4267963.
9. Reyes-Escogido Mde L, Gonzalez-Mondragon EG, Vazquez-Tzompantzi E. Chemical and pharmacological aspects of capsaicin. Molecules. 2011;16(2):1253–70. https://doi.org/10.3390/molecules16021253. PMID:21278678;PMCID:PMC6259610.
10. Atkins AG, Xu X, Jeronimidis G. Cutting, by 'pressing and slicing', of thin floppy slices of materials illustrated by experiments on cheddar cheese and salami. J Mater Sci. 2004;39:2761–6. https://doi.org/10.1023/B:JMSC.0000021451.17182.86.
11. Latoch A, Moczkowska-Wyrwisz M, Sałek P, Czarniecka-Skubina E. Effect of marinating in dairy-fermented products and sous-vide cooking on the protein profile and sensory quality of pork longissimus muscle. Foods. 2023;12(17):3257. https://doi.org/10.3390/foods12173257. PMID:37685190;PMCID:PMC10486606.
12. Lawson LD, Hughes BG. Characterization of the formation of allicin and other thiosulfinates from garlic. Planta Med. 1992;58(4):345–50. https://doi.org/10.1055/s-2006-961482. PMID: 17226483.
13. Yutani M, Taniguchi H, Borjihan H, et al. Alliinase from *Ensifer adhaerens* and its use for generation of fungicidal activity. AMB Expr. 2011;1:2. https://doi.org/10.1186/2191-08551-2.
14. Simon SA. Interactions between salt and acid stimuli: a lesson in gustation from simultaneous epithelial and neural recordings. J Gen Physiol. 2002;120(6):787–91. https://doi.org/10.1085/jgp.20028735. PMID:12451049;PMCID:PMC2229563.
15. Hashemi M, Hosseinzadeh H. A comprehensive review on biological activities and toxicology of crocetin. Food Chem Toxicol. 2019;130:44–60. https://doi.org/10.1016/j.fct.2019.05.017. Epub 2019 May 14 PMID: 31100302.
16. Predieri S, Magli M, Gatti E, Camilli F, Vignolini P, Romani A. Chemical composition and sensory evaluation of saffron. Foods. 2021;10(11):2604. https://doi.org/10.3390/foods10112604. PMID:34828885;PMCID:PMC8618029.
17. Amr K, Rasheed DM, Khachila M, Farag MA. Production, extraction, and authentication of natural and non-natural vanillin. A comprehensive review and economic future biotechnology perspectives. Food Chem. 2025;466:142249. https://doi.org/10.1016/j.foodchem.2024.142249. Epub 2024 Nov 26. PMID: 39612858.
18. Khoyratty S, Verpoorte R, Kodja H. Vanillin: Biosynthesis, biotechnology, and bioproduction. In: Mérillon JM, Kodja H, editors. Orchids phytochemistry, biology and horticulture. Reference series in phytochemistry. Cham: Springer; 2022. https://doi.org/10.1007/978-3-030-38392-3_14
19. Cichna-Markl M, Mafra I. Techniques for food authentication: Trends and emerging approaches. Foods. 2023;12(6):1134. https://doi.org/10.3390/foods12061134. PMID:36981061;PMCID:PMC10048066.
20. Silverman SN, Wijker RS, Sessions AL. Biosynthetic and catabolic pathways control amino acid δ2H values in aerobic heterotrophs. Front Microbiol. 2024;5(15):1338486. https://doi.org/10.3389/fmicb.2024.1338486. PMID:38646628;PMCID:PMC11026604.
21. Wüst M. Authenticity control of natural products by stable isotope ratio analysis. In: Schwab W, Lange B, Wüst M, editors. Biotechnology of natural products. Cham: Springer; 2018. https://doi.org/10.1007/978-3-319-67903-7_10.
22. Bensaude-Vincent B. Questioning the symmetry between analysis and synthesis in chemical practices. In: Newman WR, Schickore J, editors. Traditions of analysis and synthesis. Archimedes, vol. 73. Cham: Springer; 2025. https://doi.org/10.1007/978-3-031-76398-4_14.

23. Weisberg M, Needham P. Matter, structure, and change: Aspects of the philosophy of chemistry. Philos Compass. 2010;5:927–37. https://doi.org/10.1111/j.17479991.2010.00335.x.
24. Bussy A. Sur la formation de l'huile essentielle de moutarde. J Pharmac Chim. 1840;26:39.
25. Ettlinger MG, Lundeen AJ. The structures of sinigrin and sinalbin; an enzymatic rearrangement. J Amer Chem Soc. 1956;78:4172.
26. Ettlinger MG, Lundeen AJ. First synthesis of a mustard oil glucoside; the enzymatic lossen rearrangement. J Amer Chem Soc. 1957;79:1764.
27. Touchstone JC. History of chromatography. J Liq Chromatogr. 1993;16(8):1647–65. https://doi.org/10.1080/10826079308021679.
28. Siggia S, Hanna JG. Determination of organic isocyanates of isothiocyanates. Anal Chem. 1948;20:1084–1084.
29. Kolomnikov IG, Efremov AM, Tikhomirova TI, Sorokina NM, Zolotov YA. Early stages in the history of gas chromatography. J Chromatogr A. 2018;16(1537):109–17. https://doi.org/10.1016/j.chroma.2018.01.006. Epub 2018 Jan 3 PMID: 29373128.
30. Kjær A, Conti J, Larsen I. isoThiocyanates. IV. A systematic investigation of the occurrence and chemical nature of volatile isoThiocyanates in seeds of various plants. Acta Chem Scand. 1953;7:1276.
31. Eschliman K, Bossmann SH. Synthesis of isothiocyanates: An update. Synthesis (Stuttg). 2019;51(8):1746–52. https://doi.org/10.1055/s-0037-1612303. Epub 2019 Mar 14. PMID: 39193180; PMCID: PMC11349037.
32. VanEtten CH, McGrew CE, Dazenbichler ME. Glucosinolate determination in cruciferous seeds and meals by measurement of enzymically released glucose. J Agric Food Chem. 1974;22:483–7.
33. Grosser K, van Dam NM. A straightforward method for glucosinolate extraction and analysis with high-pressure liquid chromatography (HPLC). J Vis Exp. 2017;121:e55425. https://doi.org/10.3791/55425
34. Bjerg B, Sørensen H. Quantitative analysis of glucosinolates in oilseed rape based on HPLC of desulfoglucosinolates and HPLC of intact glucosinolates. In: Glucosinolates in rapeseeds: Analytical aspects. Netherlands: Springer; 1987. pp. 125–50.
35. Thies W. Detection and utilization of a glucosinolate sulfohydrolase in the edible snail, *Helix pomatia*. Naturwissenschaften. 1979;66:364–5.
36. Heaney RK, Fenwick GR. Glucosinolates in Brassica vegetables. Analysis of 22 varieties of Brussels sprout (*Brassica oleracea* var. gemmifera). J Sci Food Agric. 1980;31:785–93.
37. Piotrowski M, Schemenewitz A, Lopukhina A, Müller A, Janowitz T, Weiler EW, Oecking C. Desulfoglucosinolate sulfotransferases from *Arabidopsis thaliana* catalyze the final step in the biosynthesis of the glucosinolate core structure. J Biol Chem. 2004;279:50717–25.
38. Křížová L, Dadáková K, Kašparovská J, Kašparovský T. Isoflavones. Molecules. 2019;24:1076. https://doi.org/10.3390/molecules24061076.
39. Fuentes N, Silveyra P. Estrogen receptor signaling mechanisms. Adv Protein Chem Struct Biol. 2019;116:135–70. https://doi.org/10.1016/bs.apcsb.2019.01.001. Epub 2019 Feb 4. PMID: 31036290; PMCID: PMC6533072.
40. Mayo B, Vázquez L, Flórez AB. Equol: A bacterial metabolite from The Daidzein Isoflavone and its presumed beneficial health effects. Nutrients. 2019;11(9):2231. https://doi.org/10.3390/nu11092231. PMID:31527435;PMCID:PMC6770660.
41. Lecomte S, Demay F, Ferrière F, Pakdel F. Phytochemicals targeting estrogen receptors: Beneficial rather than adverse effects? Int J Mol Sci. 2017;18(7):1381. https://doi.org/10.3390/ijms18071381. PMID:28657580;PMCID:PMC5535874.
42. Adam-Dima I, Olteanu AA, Olaru OT, Popa DE, Purdel C. Methods of analysis of phytoestrogenic compounds: An up-to-date of the present state. Separations. 2024;11:205. https://doi.org/10.3390/separations11070205.
43. Popova IE, Hall C, Kubátová A. Determination of lignans in flaxseed using liquid chromatography with time-of-flight mass spectrometry. J Chromatogr A. 2009;1216:217–29.

44. Sakamoto S, Yusakul G, Pongkitwitoon B, Paudel MK, Tanaka H, Morimoto S. Simultaneous determination of soy isoflavone glycosides, daidzin and genistin by monoclonal antibody-based highly sensitive indirect competitive enzyme-linked immunosorbent assay. Food Chem. 2015;169:127–33.
45. Fiechter G, Opacak I, Raba B, Mayer HK. A new ultra-high pressure liquid chromatography method for the determination of total isoflavone aglycones after enzymatic hydrolysis: Application to analyze isoflavone levels in soybean cultivars. Food Res Int. 2013;50:586–92.
46. Caglayan MO, Şahin S, Üstündağ Z. Detection Strategies of Zearalenone for food safety: A review. Crit Rev Anal Chem. 2022;52(2):294–313. https://doi.org/10.1080/10408347.2020.1797468. Epub 2020 Jul 26 PMID: 32715728.
47. Boutibonnes P. A propos de l'activité antibactérienne de la zéaralénone [Antibacterial activity of zearalenone]. Can J Microbiol. 1979;25(3):421–2. French. PMID: 455155.
48. MacLagan T. The treatment of acute rheumatism by salicin. Lancet. 1876;108(2774):383.
49. Silva J et al. Alkaloids from fungi. In: Carocho M, Heleno SA, Barros L, editors. Natural secondary metabolites. Cham: Springer; 2023. https://doi.org/10.1007/978-3-031-185878_16
50. Zorrilla JG, Evidente A. Structures and biological activities of alkaloids produced by mushrooms, a fungal subgroup. Biomolecules. 2022;12:1025. https://doi.org/10.3390/biom12081025.
51. Krishnamurti C, Rao SC. The isolation of morphine by Serturner. Indian J Anaesth. 2016;60(11):861–2. https://doi.org/10.4103/0019-5049.193696. PMID:27942064;PMCID:PMC5125194.
52. Ladenburg A. Synthese der activen Coniine. Berichte der deutschen chemischen Gesellschaft. 1886;2578.
53. Dagnino D, Verpoorte R. Gas chromatography in the analysis of alkaloids. In: Linskens HF, Jackson JF, editors. Alkaloids. Modern methods of plant analysis, vol. 15. Berlin, Heidelberg: Springer; 1994. https://doi.org/10.1007/978-3-642-84226-9_4
54. Petruczynik A. Analysis of alkaloids from different chemical groups by different liquid chromatography methods. Open Chem. 2012;10(3):802–35. https://doi.org/10.2478/s11532012-0037-y.
55. Tuzimski T, Petruczynik A. Influence of cations and anions of ionic liquids and salts as additives in RP HPLC systems to the analysis of alkaloids on C18 and Polar RP columns. J Liq Chromatogr Relat Technol. 2024. https://doi.org/10.1080/10826076.2024.2435648.
56. Cox-Georgian D, Ramadoss N, Dona C, Basu C. Therapeutic and medicinal uses of terpenes. Med Plants. 2019;12:333–59. https://doi.org/10.1007/978-3-030-31269-5_15. PMCID:PMC7120914.
57. Jiang Z, Kempinski C, Chappell J. Extraction and analysis of terpenes/terpenoids. Curr Protoc Plant Biol. 2016;1:345–58. https://doi.org/10.1002/cppb.20024. Epub 2016 Jun 10. PMID: 27868090; PMCID: PMC5113832.
58. Nguyen TD, Riordan-Short S, Dang TT, O'Brien R, Noestheden M. Quantitation of select terpenes/terpenoids and nicotine using gas chromatography-mass spectrometry with high temperature headspace sampling. ACS Omega. 2020;5(10):5565–73. https://doi.org/10.1021/acsomega.0c00384. PMID:32201850;PMCID:PMC7081649.
59. Lü H, Lee RP, Huang J, et al. A new HPLC–UV method for the quantification of terpenoids and antioxidant activity of commercial loquat leaf tea and preparation. Food Measur. 2020;14:1085–91. https://doi.org/10.1007/s11694-019-00358-3.
60. Turner DE, Daugherity EK, Altier C, Maurer KJ. Efficacy and limitations of an ATP-based monitoring system. J Am Assoc Lab Anim Sci. 2010;49(2):190–5. PMID: 20353694; PMCID: PMC2846007.
61. Banerjee P, Kumar P. Microbial DNA testing. In: Puri A, Mahalakshmi N, Chauhan T, Mishra A, Bhatnagar P, editors. Fundamentals of forensic biology. Singapore: Springer; 2024. https://doi.org/10.1007/978-981-99-3161-3_24
62. Gannon F. DNA probes for the identification of microorganisms. J Ind Microbiol. 1994;13(2):71–6. https://doi.org/10.1007/BF01584101.

63. Sulheim S, Fossheim FA, Wentzel A, et al. Automatic reconstruction of metabolic pathways from identified biosynthetic gene clusters. BMC Bioinf. 2021;22:81. https://doi.org/10.1186/s12859-021-03985-0.
64. Myrtsi ED. Food Biosci. 2023;56:103357.
65. Mueller SO. Overview of in vitro tools to assess the estrogenic and antiestrogenic activity of phytoestrogens. J Chromatogr B Analyt Technol Biomed Life Sci. 2002;777(1–2):15565. https://doi.org/10.1016/s1570-0232(02)00282-9. PMID: 12270209.
66. Beck V, Pfitscher A, Jungbauer A. GFP-reporter for a high throughput assay to monitor estrogenic compounds. J Biochem Biophys Methods. 2005;64(1):19–37. https://doi.org/10.1016/j.jbbm.2005.05.001. PMID: 15992933.
67. Diel P, Schmidt S, Vollmer G. In vivo test systems for the quantitative and qualitative analysis of the biological activity of phytoestrogens. J Chromatogr B Analyt Technol Biomed Life Sci. 2002;777(1–2):191–202. https://doi.org/10.1016/s1570-0232(02)00494-4. PMID: 12270212.
68. Wiberg GS, Carter JR, Stephenson NR. The effects of various goitrogens on the determination of the relative potency of thyroid by the goitre prevention assay. Acta Endocrinol (Copenh). 1964;45:370–80. https://doi.org/10.1530/acta.0.0450370. PMID: 14131769.
69. Dong H, Atlas E, Wade MG. Development of a non-radioactive screening assay to detect chemicals disrupting the human sodium iodide symporter activity. Toxicol In Vitro. 2019;57:39–47. https://doi.org/10.1016/j.tiv.2019.01.021. Epub 2019 Feb 7 PMID: 30738889.
70. Goldman R, Shields PG. Food mutagens. J Nutr. 133(3):965S–73S.
71. Zeiger E. The test that changed the world: The Ames test and the regulation of chemicals. Mutat Res Genet Toxicol Environ Mutagen. 2019;841:43–8. https://doi.org/10.1016/j.mrgentox.2019.05.007. Epub 2019 May 15 PMID: 31138410.
72. Vijay U, Gupta S, Mathur P, Suravajhala P, Bhatnagar P. Microbial mutagenicity assay: Ames test. Bio Protoc. 2018;8(6): e2763. https://doi.org/10.21769/BioProtoc.2763. PMID:34179285;PMCID:PMC8203972.
73. Kirkland D, Zeiger E, Madia F, Gooderham N, Kasper P, Lynch A, Morita T, Ouedraogo G, Parra Morte JM, Pfuhler S, Rogiers V, Schulz M, Thybaud V, van Benthem J, Vanparys P, Worth A, Corvi R. Can in vitro mammalian cell genotoxicity test results be used to complement positive results in the Ames test and help predict carcinogenic or in vivo genotoxic activity? I. Reports of individual databases presented at an EURL ECVAM Workshop. Mutat Res Genet Toxicol Environ Mutagen. 2014;775–776:55–68. https://doi.org/10.1016/j.mrgentox.2014.10.005. Epub 2014 Oct 23. PMID: 25435356.
74. Zheng M, Liu Z, Xue C, Zhu W, Chen K, Luo X, Jiang H. Mutagenic probability estimation of chemical compounds by a novel molecular electrophilicity vector and support vector machine. Bioinformatics. 2006;22(17):2099–106. https://doi.org/10.1093/bioinformatics/btl352. Epub 2006 Jul 12 PMID: 16837526.
75. Bakhtyari NG, Raitano G, Benfenati E, Martin T, Young D. Comparison of in silico models for prediction of mutagenicity. J Environ Sci Health C Environ Carcinog Ecotoxicol Rev. 2013;31(1):45–66. https://doi.org/10.1080/10590501.2013.763576. PMID: 23534394.
76. Mombelli E, Raitano G, Benfenati E. In silico prediction of chemically induced mutagenicity: How to use QSAR models and interpret their results. Methods Mol Biol. 2016;1425:87–105. https://doi.org/10.1007/978-1-4939-3609-0_5. PMID: 27311463.
77. Kumar R, Khan FU, Sharma A, et al. A deep neural network–based approach for prediction of mutagenicity of compounds. Environ Sci Pollut Res. 2021;28:47641–50. https://doi.org/10.1007/s11356-021-14028-9.
78. Chavda VP, Feehan J, Apostolopoulos V. Inflammation: The cause of all diseases. Cells. 2024;13(22):1906. https://doi.org/10.3390/cells13221906. PMID:39594654;PMCID:PMC11592557.
79. Peng Y, Chu S, Yang Y, Zhang Z, Pang Z, Chen N. Neuroinflammatory in vitro cell culture models and the potential applications for neurological disorders. Front Pharmacol. 2021;23(12): 671734. https://doi.org/10.3389/fphar.2021.671734. PMID:33967814;PMCID:PMC8103160.

80. Rafiyan M, Sadeghmousavi S, Akbarzadeh M, Rezaei N. Experimental animal models of chronic inflammation. Curr Res Immunol. 2023;11(4): 100063. https://doi.org/10.1016/j.crimmu.2023.100063. PMID:37334102;PMCID:PMC10276141.
81. Kinney JW, Bemiller SM, Murtishaw AS, Leisgang AM, Salazar AM, Lamb BT. Inflammation as a central mechanism in Alzheimer's disease. Alzheimers Dement (N Y). 2018;6(4):575–90. https://doi.org/10.1016/j.trci.2018.06.014. PMID:30406177;PMCID:PMC6214864.
82. Li T, Tang H, Zhu J, Zhang JH. The finer scale of consciousness: quantum theory. Ann Transl Med. 2019;7(20):585. https://doi.org/10.21037/atm.2019.09.09. PMID:31807566;PMCID:PMC6861790.
83. Hameroff S. 'Orch OR' is the most complete, and most easily falsifiable theory of consciousness. Cogn Neurosci. 2020;12(2):74–6. https://doi.org/10.1080/17588928.2020.1839037.
84. Brunden KR, Yao Y, Potuzak JS, et al. The characterization of microtubule-stabilizing drugs as possible therapeutic agents for Alzheimer's disease and related tauopathies. Pharmacol Res. 2011;63:341–51. https://doi.org/10.1016/j.phrs.2010.12.002.
85. Ruggiero M, Hiratsuka A. On the natural intelligence and consciousness of the immune system and its relationship with the brain. Am J Immunol. 2024;19(1):11–23. https://doi.org/10.3844/ajisp.2024.11.23.
86. Zhang C, Wang Y, Wang D, Zhang J, Zhang F. NSAID exposure and risk of Alzheimer's disease: an updated meta-analysis from cohort studies. Front Aging Neurosci. 2018;28(10):83. https://doi.org/10.3389/fnagi.2018.00083. PMID:29643804;PMCID:PMC5882872.
87. Imbimbo BP, Solfrizzi V, Panza F. Are NSAIDs useful to treat Alzheimer's disease or mild cognitive impairment? Front Aging Neurosci. 2010;21(2):19. https://doi.org/10.3389/fnagi.2010.00019. PMID:20725517;PMCID:PMC2912027.
88. Wang J, Tan L, Wang HF, Tan CC, Meng XF, Wang C, Tang SW, Yu JT. Anti-inflammatory drugs and risk of Alzheimer's disease: an updated systematic review and meta-analysis. J Alzheimers Dis. 2015;44(2):385–96. https://doi.org/10.3233/JAD-141506. PMID: 25227314.
89. Lee JO, Yang Y, Tao Y, Yi YS, Cho JY. Korean red ginseng saponin fraction exerts anti-inflammatory effects by targeting the NF-κB and AP-1 pathways. J Ginseng Res. 2022;46(3):489–95.
90. Martin AJP, Synge RLM. A new form of chromatogram employing two liquid phases A theory of chromatography. 2. Application to the micro-determination of the higher monoamino-acids in proteins. Biochem J. 1941;35(12):1358–68.
91. Hanley AB, Heaney RK, Fenwick GR. Improved isolation of glucobrassicin and other glucosinolates. J Sci Food Agric. 1983;34:869.
92. Maslow AH. The psychology of science: A reconnaissance. Harper & Row; 1966. pp. 15–6. ISBN 978-0-8092-6130-7.
93. Ganzera M, Zwerger MJ. Analysis of natural products by SFC—Applications from 2015 to 2021. TrAC trends in analytical chemistry. 2021:116463.
94. Allakhverdiev ES, Khabatova VV, Kossalbayev BD, Zadneprovskaya EV, Rodnenkov OV, Martynyuk TV, Maksimov GV, Alwasel S, Tomo T, Allakhverdiev SI. Raman spectroscopy and its modifications applied to biological and medical research. Cells. 2022;11(3):386. https://doi.org/10.3390/cells11030386. PMID:35159196;PMCID:PMC8834270.
95. Hussain R, Jávorfi T, Hughes CS, Siligardi G. Circular Dichroism and synchrotron radiation circular dichroism applications to biomaterials. In: Pereira A, Tavares P, LimãoVieira P, editors. Radiation in bioanalysis. Bioanalysis, vol. 8. Cham: Springer; 2019. https://doi.org/10.1007/978-3-030-28247-9_5
96. Ikram A, Mehmood H, Arshad MT, Rasheed A, Noreen S, Gnedeka KT. Applications of artificial intelligence (AI) in managing food quality and ensuring global food security. CyTA—J Food. 2024;22(1). https://doi.org/10.1080/19476337.2024.2393287
97. Collins EJT. Food adulteration and food safety in Britain in the 19th and early 20th centuries. Food Policy. 1993;18:95–109.
98. Hansard, House of Commons debate 29 February 1860, vol. 156 cc2025–42.

99. Commission E. Commission decision of 23 May 2005 on emergency measures regarding chilli, chilli products, curcuma and palm oil. Off J Eur Union. 2005;135:34–6.
100. Alonso GL, Salinas MR, Sánchez-Fernández MA, Garijo J. Note. Physical parameters in controlling saffron quality/Nota. Algunos parámetros físicos del control de calidad del azafrán. Food Sci Technol Int. 2000;6(1):59–65. https://doi.org/10.1177/108201320000600109
101. Ryparova Kvirencova J, Navratilova K, Hrbek V, Hajslova J. Detection of botanical adulterants in saffron powder. Anal Bioanal Chem. 2023;415(23):5723–34. https://doi.org/10.1007/s00216-023-04853-x. Epub 2023 Aug 17. PMID: 37587313; PMCID: PMC10474180.
102. Fanelli V, Mascio I, Marilena Miazzi M, Savoia MA, De Giovanni C, Montemurro C. Molecular approaches to agri-food traceability and authentication: An updated review. Foods. 2021;10(07):1644.
103. Rein MJ, Renouf M, Cruz-Hernandez C, Actis-Goretta L, Thakkar SK, da Silva PM. Bioavailability of bioactive food compounds: a challenging journey to bioefficacy. Br J Clin Pharmacol. 2013;75(3):588–602. https://doi.org/10.1111/j.1365-2125.2012.04425.x. PMID:22897361;PMCID:PMC3575927.
104. Bioavailability U. Elevating food quality through smart processing. Food Sci Technol. 2024;38(4):44–7. https://doi.org/10.1002/fsat.3804_11.x.
105. Zhang Y, Zhang Y. Formation and reduction of acrylamide in Maillard reaction: a review based on the current state of knowledge. Crit Rev Food Sci Nutr. 2007;47(5):521–42. https://doi.org/10.1080/10408390600920070. PMID: 17558658.
106. Nadeem HR, Akhtar S, Ismail T, Sestili P, Lorenzo JM, Ranjha MMAN, Jooste L, Hano C, Aadil RM. Heterocyclic aromatic amines in meat: Formation, isolation, risk assessment, and inhibitory effect of plant extracts. Foods. 2021;10(7):1466. https://doi.org/10.3390/foods10071466. PMID:34202792;PMCID:PMC8307633.

Chapter 4
Bioassays

Contents

4.1 Recipe—Lamb, Ginger, Cumin and Spinach 105
4.2 Bioassays—Introduction and Validation 108
4.3 Enzyme Bioassays 110
4.4 Receptor Bioassays 113
4.5 Whole Organ and Whole Organism (Prokaryote, Animal and Human) Bioassays 114
4.6 New Approach Methodologies (NAMs) for Assessing Biological Activity 119
4.7 Assessing Complex Mixtures of Secondary Natural Products 127
References 128

Abstract A bioassay (in the context of natural product chemistry) is a testable system for determining the effects of a substance (or group of substances) on a living system. They can vary in complexity from simple binding assays (receptors or enzymes) to measurement of the effects on a complete organism. The natural product chemistry tradition has focused on the isolation of individual components and the assessment of their potency followed by the synthesis of the active ingredient. This reductionist approach brings a degree of control to the study of the usefulness of secondary natural products however in food systems where multiple, chronic effects are more usual, this paradigm may not operate effectively. Bioassays (particularly those which involve the use of whole organisms) can be costly and difficult to interpret so new approach methodologies (NAMs) are growing in importance. These can involve the use of cells in culture in a controlled environment. These approaches have advantages (particularly as a screening tool) and are becoming more sophisticated and relevant. The challenge of assessing complex mixtures and long-term, chronic exposure remains prevalent.

4.1 Recipe—Lamb, Ginger, Cumin and Spinach

The use of different spices at different stages in a recipe is an important aspect of cooking. Some flavours clash while others are either obviously complementary or are surprisingly compatible. Ginger, in particular, produces warmth without the heat derived from capsaicin and may, therefore be complementary to a number of other

B. Hanley, *Secondary Natural Products in Foods and Food Systems*,
https://doi.org/10.1007/978-3-032-00292-1_4

flavours. The clearest example of a complementary flavour is umami which, together with monosodium glutamate (MSG) provides an enhanced savoury experience in a number of foods [1]. This also relates to philosophical studies regarding substances and their chemical constituents. The knowledge of taste has been considered too idiosyncratic and subjective to be examined from a purely philosophical standpoint but this view has been challenged [2]. When the philosophy of taste (including, in particular, the aesthetics of the sensation) is combined with the reductionism of chemistry then what results is the correlation of sensation with the chemical composition of food. This can be seen in the history of the science of the sensation of heat. In purely physicochemical, scientific terms, the heat capacity of a substance is a measure of the amount of heat an object contains while the temperature of an object is a measure of the intensity of the heat at a point in time. Capsaicin is the equivalent of an intense temperature. Its heat is measured in Scoville Heat Units (SHUs). Originally SHUs were measured organoleptically by trained tasters hence there was a degree of subjectivity to the measurement. More recently, since the degree of heat is related directly to the amount of capsaicin present, chromatographic measurement of capsaicin gives a direct and quantifiable measure of the SHU of a chilli. Returning to a biological equivalence, capsaicin binds to transient receptor potential cation channel vanilloid subfamily member 1(TRPV-1) and induces pain [3]. It also seems to have a neuroprotective function [4]. Polymorphisms in the receptor are responsible for different sensitivities to capsaicin. Indeed the role of an individual amino acid in determining species sensitivity to capsaicin has been investigated [5].

Ginger (*Zingiber officinale* Roscoe) produces an irregular shaped rhizome which intensively used in Asia but is also common in Western cooking. It is related to cardamom and to turmeric. Ginger contains a mixture of volatile and involatile compounds and, unlike chilli which contains one principal organoleptically important secondary natural product, capsaicin, the flavour of ginger is associated with at least three classes of involatile compounds (gingerol, shogaol, and zingerone—Fig. 4.1) and several volatile components.

Ginger oil is composed of volatile secondary natural products—particularly monoterpenes and sesquiterpenes. Drying and processing of ginger changes the chemical composition and this has consequences both for its flavour and biological properties [6]. In particular, the conversion of gingerols to shogaols is thought to change the biological properties of dried compared to fresh ginger [7]. By using a mixture of fresh and dried ginger, a broader spectrum of flavour compounds can be obtained and some of the volatile constituents are incorporated. Like capsaicin, compounds in ginger act on TRPV-1 and this activity has been modelled [8] however compounds in ginger also act as antagonists of cholinergic and serotonergic receptors [9]. The biological properties of the plethora of compounds found in ginger have been reviewed [10] and both the agonist and antagonist properties of some of the compounds have been noted. While the interaction of some ginger compounds with TRPV-1 has been noted and this may explain, in part, the heat sensation experienced, the potency is considerably reduced compared to capsaicin and this can, at least in part, be explained by structural differences [11]. Compared to capsaicin, ginger

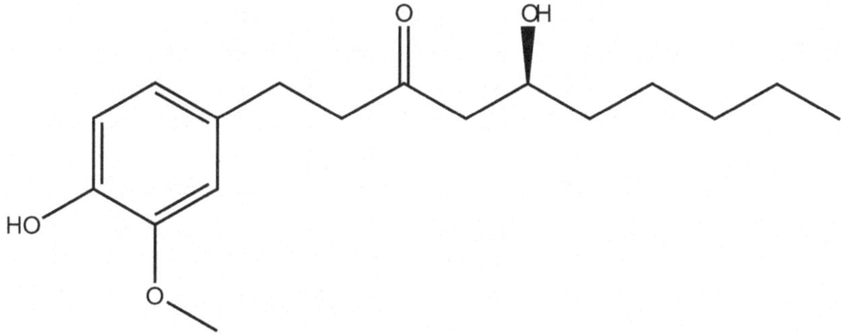

Gingerol

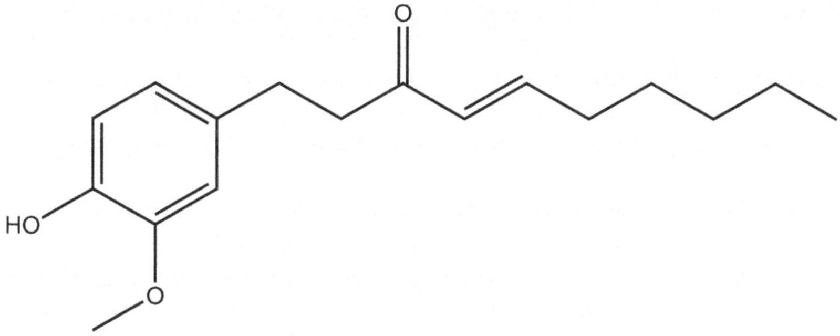

Shogaol

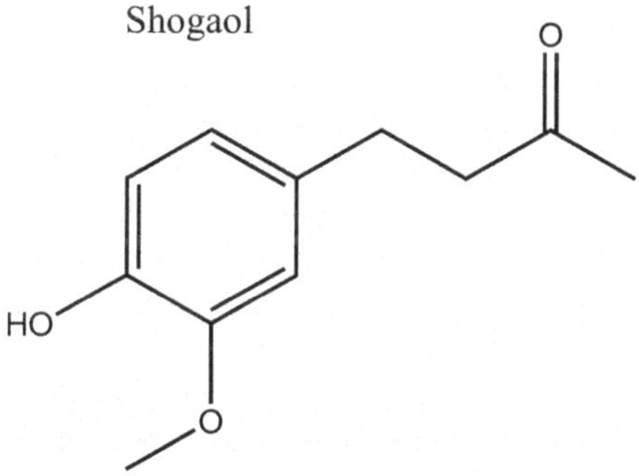

Zingerone

Fig. 4.1 Structure of gingerol, shogaol and zingerone

compounds may be acting as partial antagonists of TRPV-1 (see below). The activity of ginger is more complex than that of chilli due to the range of compounds produced and their interactions with several receptors. It is not unreasonable to suppose that the addition of ginger may reduce the heat experienced by consumption of chilli as a consequence of competition for the active site of the TRPV-1 receptor and the reduced potency of the compounds from ginger. In this way, ginger may reduce the intensity (temperature) experienced on consumption of capsaicin while the overall heat potential may be more varied due to the presence of different compounds.

The understanding of the biological activity of secondary natural products can be assembled at various levels. The interaction of individual compounds can be determined from a consideration of molecular level binding to receptors or enzymes (or other biological macromolecules). This is the most reductionist approach and is the basis for the development of pharmaceuticals. A further level of complexity is introduced if mixtures of compounds or compounds and their metabolites are considered. These can still be assessed in terms of a single biological target. Living systems are, however, composed of cells containing multiple biological characteristics and individual cells interact at a cellular and organotypic fashion to give a complete response. For example, interaction with receptors can lead to signal transduction and an effect experienced in the brain, for example. Binding of agonists such as capsaicin to the TRPV-1 receptor causes the activation of multiple signal transduction pathways [12]. Activation via nociceptive impulses are transmitted to the brain and the individual has the experience of pain. As noted previously, the activation pathway can be interrupted by the binding of less potent secondary natural products to the TRPV-1 receptor. Such modulation of TRPV1 activity by natural products has been reviewed and may form the basis for pharmacological interventions [13].

Place a tablespoon of olive oil in a thick-bottomed pan and apply medium heat. Add two chopped onions (1–2 cm dice) and stir to prevent burning. Once the onion is translucent (5–10 min) add some chopped, cooked lamb shoulder or leg (1–2 cm dice). It is also possible to use fresh lamb. Cook for 5 min until the lamb takes on a little colour (Maillard reaction) then add 2 teaspoons of grated root ginger and half a teaspoon of cayenne pepper. Grating the ginger leads to greater release of flavour compounds. Unlike the case of garlic, disruption of the cell structure to release a flavour precursor that is hydrolysed by a co-occurring hydrolytic enzyme, is not necessary for ginger. Grating the ginger simply leads to greater release of flavour compounds from the root matrix. Stir and add about a tablespoon of tomato puree. Cook for 5 min on a low heat. Add a tin of chopped tomatoes and a tin of chick peas. Half fill each can with water and add to the mixture along with freshly ground black pepper and a teaspoon of salt. Cover and cook for 40 min. Add 150 g of washed chopped spinach (or chard). Stir and cook for 10 min, uncovered. Add 1 teaspoon of dried ground ginger and 1 teaspoon of ground cumin. Cook for 5 min. Stir and serve with a warmed pita bread.

4.2 Bioassays—Introduction and Validation

Binding to a receptor (as capsaicin and the components of ginger are able to do) is one of the simplest representations of a bioassay. A bioassay can be defined as a testable system for determining the effects of a substance (or group of substances) on a living system. They may operate on a molecular, cell or whole organism scale however no matter what biological scale is used, the bioassay must be specific and demonstrate a dose response relationship. Predictive biomarkers derived from the bioassay will provide measurements that will reflect the likelihood of later clinical outcomes. For example, the link between serum cholesterol and cardiovascular disease has been extensively studied and is established [15] and the mechanistic basis for this link has been investigated [16]. A bioassay can be used to measure serum cholesterol and the relationship between this and the risk of developing cardiovascular disease can be assessed and appropriate remedial action taken. In the context of secondary natural products, bioassays can be used to detect (and optimise) the biological attributes of a ligand or as a detection method for the deleterious effects of a harmful chemical or group of chemicals. Among the key characteristics of a bioassay method are the following:

- Specificity
- Robustness and reproducibility
- Interpretability
- Sensitivity and accuracy over the required range and dose responsive
- Type [direct, indirect, qualitative or quantal (yes/no type of assay)]
- Cost effectiveness and appropriateness.

The specificity of a bioassay is the primary essential characteristic. The test system must be specific and fit for purpose and should not generate false positive (or false negative) results. It should be clear precisely what the bioassay results mean (and what they do not mean) together with any caveats. For example, in studies of statin drug screening, the type of animal model used is important [14], with large differences in the effect size depending on which animal model is used (− 30% in rabbits, − 20% in mice and − 10% in rats). In food systems, the mycotoxin aflatoxin B1 can be formed as a consequence of contamination by the fungus *Aspergillus flavus* or some related species. The toxin is bioactivated *in* vivo to form the corresponding epoxide which is the main toxic entity [17]. The sensitivity of different animal models to the toxin has been found to vary [18] and this is considered to be due to the biotransfomative potential of different species. It is important that other, non-target substances should not provoke an unspecific reaction in the bioassay. The assay should be able to detect both positive (agonist) and negative (antagonist) type activity. It should respond without producing a biologically false positive or false negative result. As part of the validation process for a bioassay, the chosen system should be able to be inhibited with a specific inhibitor (preferably with a known mechanism of action). In addition, bioassays must be validated with known ligands and these should be of known potency so the bioassay can be shown to be effective and test samples can be measured relative to a known standard response. It is not necessary for a bioassay to be completely linear over the whole concentration range (see below). Biological test systems can show a degree of inherent variability and therefore absolute potency can be a misleading measure and relative potency is generally preferable. An unknown test sample should behave in a similar way to a dilution series from a sample of known potency in the bioassay system and its activity should lie within the response range of the dilution series. Bioassays should be repeated at least 5 times at a range of concentrations to demonstrate reproducibility. The assay and the test substance should also be stable under the conditions employed. This will define the working concentration range for the bioassay and both the limit of detection and the limit of saturation (the latter is the concentration of analyte beyond which the bioassay does not respond in a predictable fashion due to saturation of the bioassay by the analyte). Bioassays are a valuable tool in the determination of the biological potency of an analyte however they must be used with care and both optimised and validated for the analyte in question under the conditions in which it is formed. Frequent revalidation should be carried out to ensure the bioassay remains responsive and has not degraded over time. The limitations of bioassays must be recognised and it should be realised that they only reflect certain aspects of a biological system and not the whole, complete system—they are generally indicative of a biological outcome, not absolutely confirmatory [19].

4.3 Enzyme Bioassays

Enzyme bioassays can give some indication of biological potency but may also be simply a detection system that responds to a particular biological interaction rather than a measure of chemical identity. Enzyme-based bioassays can be either direct or indirect and can be quantitative or quantal. In addition, they can be used to determine both substrate and inhibitors and, to an extent, the mode of action of each. In addition, the reverse reaction, the bioassay of enzyme activity in a biological entity, may also indicate the presence of a particular enzyme in a system and enable its production and activity to be optimised. While the use of enzyme-based bioassays has developed significantly over time, the use of rapid throughput methodology has greatly increased the utility of enzyme bioassays. The use of microtiter plates remains the core of many methods however advances in microfluidics [20] has enabled advances in technologies including microfluidic arrays and droplet microfluidics [21].

Rapid methods that employ enzymes to measure specific analytes include assays for glucose [22] that can operate through the action of hexokinase and glucose-6phosphate dehydrogenase or glucose oxidase which produces hydrogen peroxide which then reacts with a dye in the presence of peroxidase to give a product that can be quantitated spectrophotometrically. Such coupled assays are commonly employed and are used for a range of analytes including proteins and nucleic acids. These routine assays rely on a supply of stable enzymes with an appropriate level of specificity and sensitivity. While assays for common analytes comprise a large part of the use of enzymes in bioassays, there are three particular methods that have been particularly transformative. These are the specific amplification of DNA, the use of enzyme-based quantitation methods coupled to specific detection of a target using antibodies and the use of enzymes for high throughput screening. Each of these will be briefly described. In each case, more detailed descriptions of the methods and protocols are available.

One of the most significant (and specific) uses of enzymes in a bioassay is the use of the polymerase chain reaction (PCR) to amplify a specific DNA sequence which can then be quantitated or used in other applications. The essence of PCR has been described in many publications and technical notes [23]. The key to the reaction is the existence of a thermostable DNA polymerase isolated from the bacterium *Thermus aquaticus*. The *in* vitro reaction copies a single strand piece of DNA multiple times through a heating (denaturation of double stranded DNA) cooling (annealing) and replication cycle, repeated multiple times. The polymerase requires a double stranded piece of DNA (created using a primer sequence) to bind to in order to start the replication process and to then add complementary bases to the rest of the single strand until it reaches a second double stranded piece which is created by a second primer. The DNA is then denatured by heating and the process of annealing and polymerisation repeated. This enables the DNA sequence between the two primers to be specifically copied many times.

The principle of an enzyme-linked immunosorbent assay (ELISA) is simple. The binding and reporting of an analyte are decoupled. In a typical ELISA assay, a particular type of multiwell plate is coated with an antibody that is able to specifically bind the analyte of interest. The test solution is applied and the analyte captured by the antibody. The rest of the sample is removed by washing. A second antibody which can bind to a different epitope of the analyte is added. This antibody also contains an enzyme (often peroxidase or phosphatase) and a colourless substrate which is converted to a coloured product is added. The amount of analyte present in the test sample is obtained by measuring the colour development compared to that of a standard concentration curve of the authentic analyte. Interfering substances may inhibit initial binding but the basic principle of the ELISA method—using a quantifiable biological response (the use of specific antibodies) as a detection method for measuring the unknown concentration of an analyte in a range of types of sample is both robust and reproducible. Immunoassays (like other biologically-based assay methods) have a number of potential issues that may limit their effectiveness [24]. The use of other confirmatory methods may help to overcome some of these [25].

The use of an enzyme to carry out high throughput screening (HTS) is a conventional and very valuable approach which has met with considerable success. It can form the basis of therapies designed to target specific organisms or to counter a deleterious activity caused by a particular attribute of a result of disease. One of the most illustrative examples of the former is the action of penicillins and cephalosporins. Penicillin kills susceptible bacteria by specifically inhibiting the transpeptidase that catalyses the final step in cell wall biosynthesis, the cross-linking of peptidoglycan [26] by irreversibly acylating the transpeptidase. An example of the latter is the use of kinase inhibitors in anti-cancer strategies [27]. Inhibition of enzyme activity to counter a disease process is a common target for pharmacologically active species [28]. In essence an enzyme catalyses a reaction by binding to a substrate, facilitating the change of that substrate and formation of a product that is then released from the enzyme. An inhibitor of the enzyme can operate in different ways but, most commonly, competes with the substrate for the enzyme active site and prevents substrate binding and hence inhibits the reaction. When seeking competitive inhibitors of an enzyme, it is usual to investigate the reaction when $< 10\%$ of the substrate has been used (or $< 10\%$ of the product has been formed). There have been a number of studies that give a rationale for choosing these conditions and these can be investigated further however, the general rule applies and should be followed. The other key criteria for investigating enzyme inhibition as a high throughput screening tool include the following.

- The enzyme must be isolated in sufficient purity and without competing enzymes or inhibitors being present.
- The substrate (or a valid surrogate) must be available.
- Other requirements for the reaction (buffers, cofactors etc.) must be identified and be available.
- The enzyme must be stable under the assay conditions.
- The reaction should display linearity under the required conditions

Note there is no a priori requirement for the reaction conditions to mimic the physiological conditions under which the enzyme operates in vivo. The purpose of HTP is to eliminate molecules which have no biological relevance to the system under investigation. The success of a molecule in a HTS does not guarantee that it will be medicinally effective. In addition, the investigation of chronic or long term effects is not part of an HTS approach therefore it has less relevance to food natural products.

One of the key parameters is the km for the reaction. This is derived from the equation:

$$\text{V(reaction rate)} = [\text{S}].\text{Vmax}/[\text{S}] + \text{km}$$

The value of Km is a key parameter [29]. The other variables are the maximum rate of the reaction (Vmax) which occurs when all of the enzyme active sites are occupied by a substrate molecule and the substrate concentration ([S]). In this equation, Km is the Michaelis constant, representing the substrate concentration at which the reaction rate is half of Vmax.

In addition, there are a range of freely available sources of information on the quantification of enzyme inhibitors and the use of HTS methodologies. A detailed discussion of these lies outwith the scope of this chapter however some examples will illustrate some salient points.

The long term use of enzyme inhibitors is demonstrated in the use of cholinesterase and monoamine oxidase (MAO) inhibitors [30] to treat long term conditions such as Alzheimer's disease. In the case of cholinesterase inhibitors, their activity is designed to prevent the breakdown of the neurotransmitter, acetyl choline while MAO inhibitors act to prevent the breakdown of dopamine, serotonin and noradrenaline. Both are targets for the prevention of cognitive decline. In both cases, their therapeutic role is to ameliorate the consequences of Alzheimer's disease rather than preventing or curing the disease itself. In addition to the treatment of diseases of ageing, antimicrobial activity and cancer, noted above, one of the most widely prescribed class of drugs are statins. These are used to treat hypercholesteremia and act by inhibiting the first enzyme in the cholesterol biosynthesis pathway—hydroxymethylglutaryl-CoA reductase [31]. Other targeted for inhibition biosynthesis enzyme systems include riboflavin [32], nucleotides [33] and cysteine [34]. All of these, and others, are potential routes to the development of anti-disease strategies. One of the most significant areas of research are those that focus on anti-viral strategies and, in particular, the inhibition of viral replication [35]. The development of enzyme inhibitors was an important factor in the development of Covid-19 strategies [36] however better, more efficacious and potent (and more specific) molecules will be required.

With the increase in occurrence (and cost and importance) of non-infectious, chronic diseases, the development of new methods to facilitate the study of the interrelationships between the whole range of enzyme activities in living systems will be necessary. In addition to the development of specific enzyme inhibitors, future health will also depend on the development of multifunctional enzyme inhibitors which are able to act in concert. Many food secondary natural products are partial rather than pharmacologically relevant enzyme inhibitors and may also have a spectrum of activity including inhibition of more than one enzyme. This, combined with the presence of several chronically bioactive compounds, means that far more subtle ways of measuring the interactions of secondary natural products with enzymes (and its consequences) will need to be developed.

4.4 Receptor Bioassays

Receptor bioassays (along with other bioactive compound/biological macromolecular interactions) are the first step in the screening of a molecule and its target. In order that a receptor bioassay can demonstrate biological relevance, it is important that a bioactive molecule should both bind to a target receptor and elicit a biological effect. One of the advantages of receptor bioassays is that (as is the case with some biologically important enzymes) receptors can often be cloned and their use optimised in a suitable in vitro cell system. While this may reduce the extent to which biological context can be investigated, it is a cost and time effective way of screening large numbers of potential leads. In addition receptor bioassays can be combined with reporter genes to provide a sensitive way of measuring agonists and antagonists [37]. One area of interest, in the context of secondary natural products, is the development of methods to measure thyroid disrupting compounds [38]. While considerable information is available concerning the binding of agonists and antagonists to receptors and this has been supplemented by techniques like X-ray crystallography (sometimes in conjunction with functional assays [39], pathway analysis can be a more effective tool to examine the effect of specific ligands on the whole cell, phenotypic consequences of ligand binding [40]. One of the drawbacks to crystallographic techniques is that they provide only a snapshot of a single conformation of a receptor and its interactions with a ligand. In order to probe the dynamics of interactions, nuclear magnetic resonance (NMR) spectroscopy is a more valuable tool [41]. Significant advances have been made by examining the relaxation properties of receptors and this has shed light on side chain dynamics. The class of G protein coupled receptors (GPCRs) have been studied by using NMR in conjunction with specific stable isotope labelling of different domains [42]. The use of specific expression systems has facilitated such studies [43]. In most cases, the transgenic receptor has been isolated and membrane mimetic systems have been used. As techniques improve, the combination of real systems together with better membrane mimetics and the use of NMR together with static techniques (crystallography and electron microscopy) will yield information about receptors in greater detail and resolution. Ultimately the greatest relevance will accrue from systems in which NMR is used to study GPCRs and other receptors in vivo rather than using biomimetic systems. Ultimately the aim will be to investigate both small molecule/receptor interactions and large biomacromolecule/ligand interactions in a dynamic, real time environment.

4.5 Whole Organ and Whole Organism (Prokaryote, Animal and Human) Bioassays

As noted previously, while binding to biological macromolecules can be a useful means of differentiating bioactive molecules—particularly in screening assays to identify agonists and antagonists or substrates and inhibitors—information on the effectiveness of molecules is most completely obtained from whole cell/organism or phenotypic assays. There are a number of issues with whole organism (or even whole cell) assays.

- Application of a whole system approach to a reductionist chemistry solution using individual molecules and complete biological systems. This is scientifically and philosophically, a conflict.
- Extrapolation from pathways to cells and ultimately to whole organisms.
- Cross-species extrapolation (this will be examined in more detail in Sect. 4.6 on NAMs and the growth of alternative methods).
- The problem of measuring long term effects with finite experiments and predictive biomarkers.

The essence of natural product chemistry has been the analysis (and confirmation) of structures and the development of synthetic methods to confirm both structural and biological interrelationships. This decomposition/recomposition duopoly has enabled structures to be determined and biosynthetic hypotheses to be proposed. The derivation of both analytical and synthetic chemistry as applied to organic and natural product chemistry lies in the developments of inorganic (mineral) chemistry. One of the reasons for this is the concept of viable scale. Inorganic chemistry is concerned with a range of elements that can be combined and decomposed to give stable entities. Organic chemistry, if decomposed to the limit, comprises primarily C, H and O with a small number of other elements. This level of granularity provides relatively little information about the isolated substances. To study organic chemistry, a larger scale is required. Molecules have to be considered in the context of groups rather than irreducible elemental entities. The concept of acids and bases combining to give stable salts, the idea of 'affinities' and displacement reactions has its roots in the end of alchemy and the beginning of chemistry [44]. Indeed, Geoffroy, by reducing substances to their core components, hoped to develop a link between microscopic and macroscopic properties. The idea that atoms (or molecules) reflected macroscopic properties was one that appealed to the Ancient Greeks. They postulated that water 'atoms' must be smooth and slippery while iron atoms were hard and had hooks that held them together. There were other examples including taste and other macroscopic properties that could be explained on the basis of the properties of atoms. This is, in some way, analogous to the biological homunculus theory of the alchemists that proposed that germ cells contained a small but complete person which then grew into a fully formed person. In the case of both inorganic and organic chemistry, it became clear that synthesis might actually create completely new substances rather than the anticipated chemical proof of identity of structure. This enabled organic and inorganic chemists to begin from different starting points and to create new paradigms that would take them in different directions. Experimental physical chemistry, by contrast, was more concerned with the development of better and better tools to measure the manifestation of existing paradigms by devising laboratory experiments that substantiate the commonly accepted paradigm. One example of the fluidity in approach of organic chemistry starts with natural products. Application of the reductionist analytical methodology (combined with increasingly accurate means of measurement) allowed natural products to obtain the empirical formulas of isolated materials. By the nineteenth century, analysis of specific functional groups was being developed. This led to the isolation of, for example, classes of compounds such as alkaloids like quinine [45]. When Perkin attempted a synthesis of quinine in 1856, his starting point was the empirical formula which he postulated could be prepared by the addition of an oxidising agent (potassium dichromate) to aniline (Fig. 4.2).

This made empirical sense however the reaction proceeded through the reaction of an impurity in his aniline sample (o-toluidine) to give the colour mauveine (Fig. 4.3).

As it turned out, Perkin had actually synthesised a range of chemical structures that were related to the core mauveine compound and the structural enigma was not solved until 1994 [46]. Nonetheless, although Perkin failed completely in his attempt to prepare quinine, the whole area of aniline dyes became commercially important, particularly in Germany [47]. While the discovery of an aniline dye was somewhat serendipitous, Perkin had stumbled upon a colour that was difficult to obtain in nature. The classic example of a colour of this hue, Tyrian Purple was isolated in a laborious process from part of a particular sea snail [48]. As a result, not only did Perkin synthesise a new colour that would create a whole new field of chemistry, he also prepared a substance for which there was a significant unmet need. Prior to his synthesis of the first aniline dye, there was no economically viable way of isolating or preparing purple dyes so Perkin created a new area of chemistry and, simultaneously, met a market for the product he synthesised. Clearly, from a scientific standpoint, empirical formulas are insufficient to obtain a complete understanding of how organic natural products are constructed. The advent of methods to identify specific functional groups was a necessary prerequisite for a better understanding of how molecules existed. The understanding that some products occurred in classes

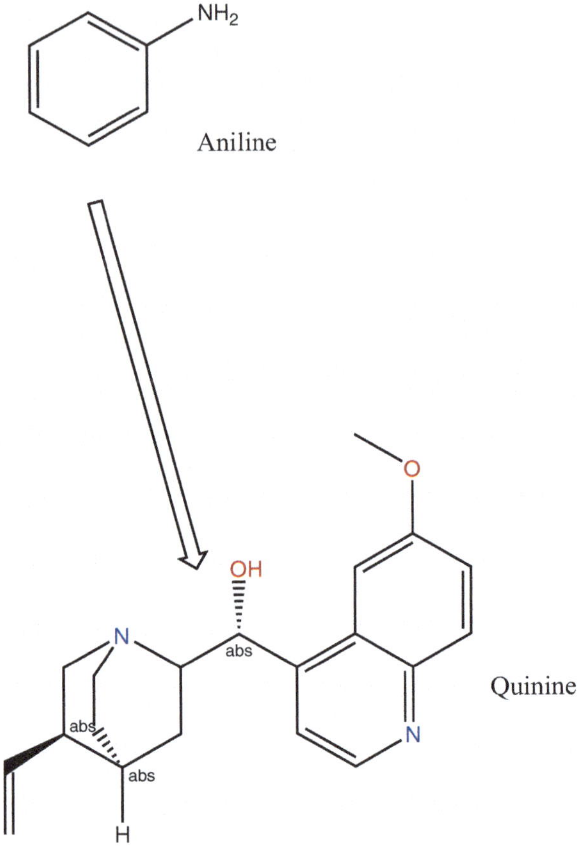

Fig. 4.2 A postulated synthesis of quinine from aniline

in natural product mixtures from specific sources was a significant advance. Once a particular natural product had been isolated, there was a good chance that other compounds with the same core structure would be found from the same source. In addition to providing new molecules for investigation, this approach also helped chemists to think in terms of biosynthetic pathways, their occurrence in nature and the types of modifications to functional groups that might be expected to occur. An example of this was the increased knowledge about alkaloids that built up in the nineteenth century. The first of the opioid alkaloids, morphine, was allegedly isolated from an extract of the opium poppy in 1805 by Serturner [49]. His isolation made use of the basic nature of the molecule and its crystallisability. This was in contrast to the accepted wisdom of the time which suggested that plant natural products were assumed to be primarily acidic in nature (for example, tartaric, benzoic, citric, oxalic, malic, glucuronic and gallic acids) so his discovery of an alkali-like (alkaloid) structure was a considerable breakthrough [50]. Numerous other alkaloids were then isolated including codeine, quinine, strychnine, veratrine and emetine (Fig. 4.4).

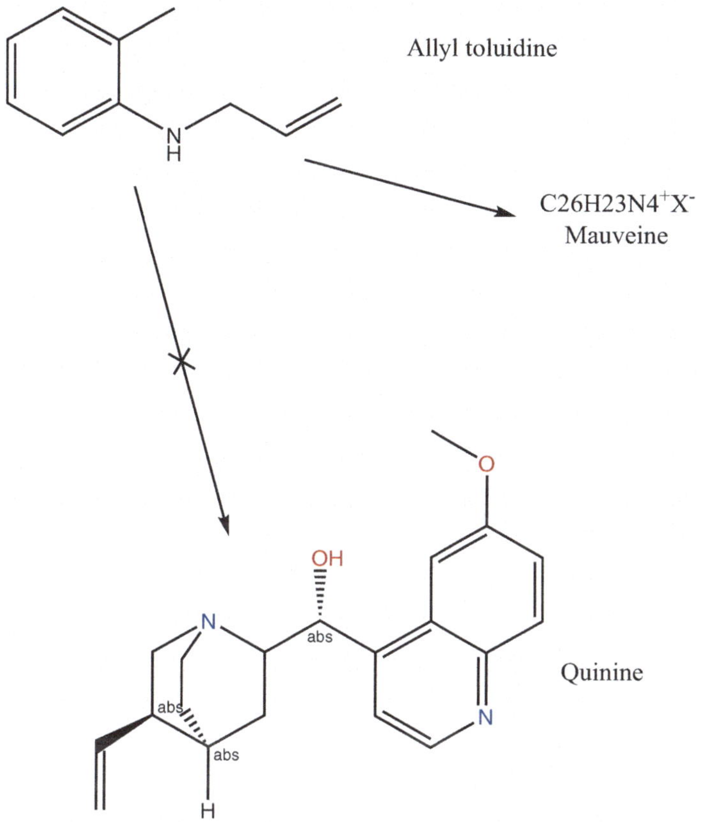

Fig. 4.3 The synthesis of mauveine from o-toluidine

Although they were not necessarily structurally similar (indeed in many cases the structures were not determined until a considerable time after their initial isolation), each had the typical basic N atom that facilitated isolation. It was not, however, a rapidly evolving area of scientific investigation. After its discovery by Pelletier and Caventou in 1818, it then took 140 years for the structure of strychnine to be confirmed.

Indeed, isolation was more advanced than structure elucidation or synthesis and coiine (the toxic principle of hemlock) was the first alkaloid to have its structure established in 1871 [51] although it was isolated in 1827 [52] and it was not synthesised until 1886 [53]. In part, the time taken was a combination of the lack of suitable structure determination techniques and, in part, because attention was focussed on the purification of plant principles (such as quinine) with distinctly useful biological properties rather than the determination of their structures. Biological activity driven isolation and purification was the precursor of methods such as combinatorial chemistry where screening for biological activity was the principal scientific motivation.

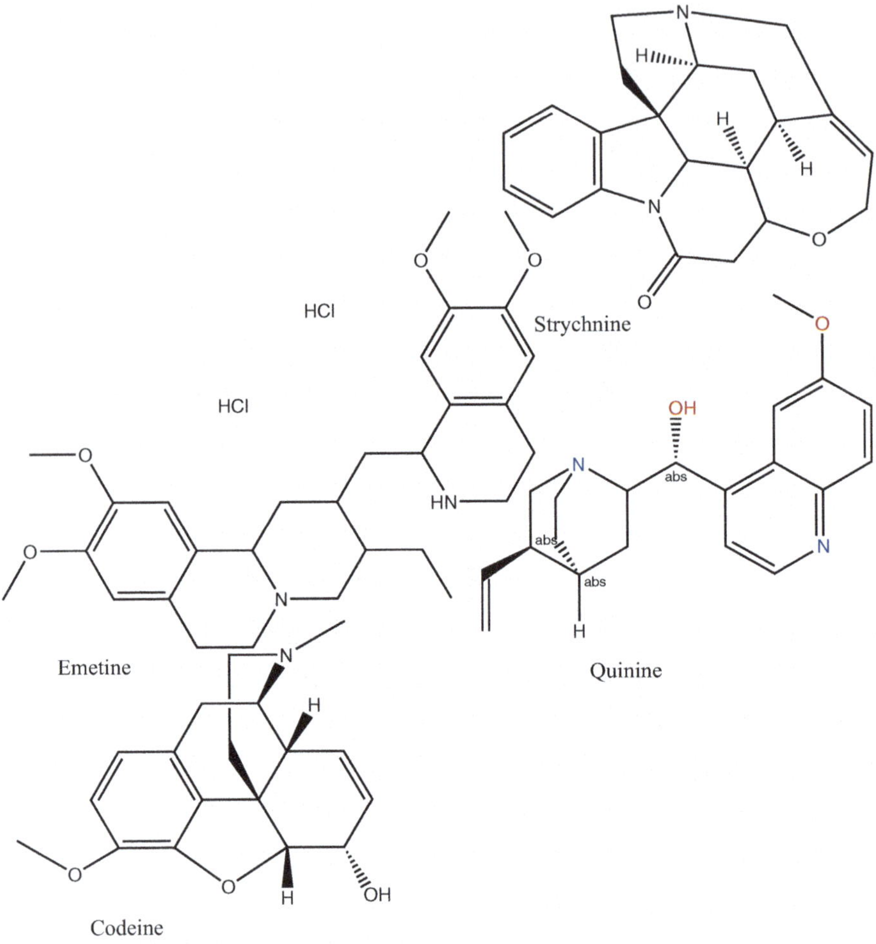

Fig. 4.4 Structures of representative alkaloids

Clearly the order of development was isolation and purification as a primary driver followed by the development of methods to identify particular functionalities (such as the Hoffman degradation and oxidation reactions to characterise certain features of alkaloids—Fig. 4.5) and, eventually, the determination of complete structures. Once the structures of secondary natural products had been determined, ways of combining functionalities and the development of new reagents capable of adding specific functionalities could be contemplated.

As more and more structures (and sub-structures) were identified, the knowledge base upon which the structures of novel compounds could be related, increased and familial structures became more apparent. This increased the opportunity for biosynthesis to be regarded as an intellectually worthy discipline and changed the mind-set of natural products from destructive analysis to developing methods to

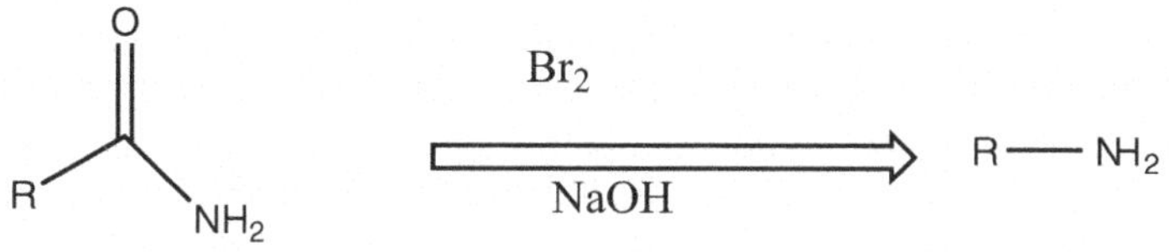

Fig. 4.5 The Hoffman degradation reaction

build structures. This marked a gradual paradigm shift in thought compared with the eighteenth and nineteenth centuries and paved the way for the 'Golden Age' of natural product chemistry. Much of the progress was driven by the commercial attractiveness of pharmaceuticals and dyestuffs. Research into the chemical components of foods was relatively slow to progress however the application of techniques developed in other areas helped to fuel the growth in understanding of secondary natural products in food systems.

4.6 New Approach Methodologies (NAMs) for Assessing Biological Activity

The concept of New Approach Methodologies builds upon the twin challenges of assessing both the biological activity of chemicals (both natural and synthetic) and doing this in the most meaningful and cost-effective fashion [54]. The idea of NAMs can be traced to the '3Rs' philosophy of 'reduce, refine and replace' animal models [55]. The main drivers, so far as the pharmaceutical industry is concerned, is the replacement of expensive animal models with quicker methodologies that are also more relevant to the human situation. There is also a growing belief that animal models are not the most suitable for pharmaceutical development or regulation and, as noted above, are expensive. In the case of food related secondary natural products, the effects are often much less (and more difficult to define) and this, combined with the presence of significant confounders, means that test cohort sizes have to be significant therefore rapid and cost-effective screening methods are required. The commonly adopted paradigm is that systems should be developed for pharmaceutical use and then extended to bioactive components from food (see below). While this approach may be flawed, it provides a useful starting point for the development of NAMs. It is important to recognise that economic, regulatory or even ethical issues should not be the most significant drivers. If the scientific challenges of NAMs are not met, then there is a danger that a system with a particular set of negative attributes (cost & ethics) will be replaced with a more cost-effective and more ethical model which is still not fit for purpose. There are clearly a number of challenges that will have to be overcome if a scientifically valid set of methodologies are to be incorporated. These include the complexity of living systems and the persistent regulatory requirements for safety testing. In addition, current models are complicit in the high wastage in developing new pharmaceuticals. Although considerable efforts have been made to optimise the pipeline for drug discovery (including optimisation of early stage structure:activity confirmation, for example), it remains the case that 90% of drugs discovery candidates fail to reach the clinical trial stage. The reasons for this have been extensively studied and a number of improvements to improve the success rate have been suggested [56]. It is important, therefore, to consider the basic principles of some of the currently available NAMs methods to examine if advances in these may achieve the goals of an effective NAMs approach.

In the case of secondary natural products from foods, the development of appropriate model systems is further complicated by the need to incorporate long term (often lifetime) exposures to bioactive constituents. In addition, as we shall examine in more detail in Sect. 4.7, exposure to food constituents requires the deconvolution of multiple effects from complex mixtures. The systems most commonly used to investigate secondary natural products from foods are:

- Testing using microbial species to determine anti-microbial and other activity or mutagenicity of secondary natural products.
- Secondary cell lines—short and long term systems.
- Primary cell lines and human-derived cells.
- Organoids.
- Animal models (these will not be considered in detail since they fall outside the scope of NAMs).
- Human nutritional and clinical studies.

The considerable effect of microbial cells on a human (or animal) host and the role of secondary natural products in this association will be considered in Chap. 9 in a consideration of probiotic organisms and prebiotics.

The importance of secondary natural products as anti-microbial agents was appreciated in folk medicine. One significant source was moulds (fungi) [57], however many fungal species, in addition to containing anti-microbial species, are also toxic to humans. Examples given previously include aflatoxin B1 and 3-nitropropanoic acid. In addition, a mould (*Claviceps purpurea*) that affects badly stored grain can give rise to ergot poisoning that has been implicated in a disease typified by uncoordinated movement of limbs. This was called called Sydenham's chorea (St Vitus' Dance, St Anthony's Fire) however it has been suggested that all the observed clinical symptoms are not consistent simply with ergot poisoning by ergot alkaloids and other components may contribute. The scientific and pharmacological literature is replete with examples of anti-bacterial toxins which started out as promising drugs (e.g. patulin) but were later abandoned due to toxicity concerns. A range of antimicrobial species have also been isolated from plants. These include the S compounds from *Allium* and *Brassica* species which will be considered in more detail in Chaps. 6 and 9. A range of other plants have also been sources of potent anti-microbial compounds [58] and other bioactive species [59]. These products can be characterised by structure (e.g. alkaloids, terpenes, polyketides and phenolics), derivation (plant species or environment (e.g. marine secondary metabolites)) or biological activity (e.g. antipsychotics, opioids). In many cases, however, classical clinical trials have not been carried out and claims are sometimes inconsistent or disjointed. The major use of microbial species to examine the potential biological activity of secondary natural products is the Ames test. The Ames test is a microbiological procedure to determine if a test compound is mutagenic [60]. While there are several variations on the test procedure, the basic principles remain standard. A microbial strain (*Salmonella typhimurium*) is engineered such that it contains a point mutation that prevents it from synthesising histidine. It therefore requires exogenous histidine in order to grow. If a test compound is able to induce a mutation to restore histidine biosynthesis capability (reversion) then the strain will be able to grow in the absence of histidine. The Ames test, then, can detect the mutagenicity potential of individual compounds. Mutating species are often activated into electrophilic species prior to binding to DNA [61] and liver microsomal preparations are generally included in an Ames test since these contain phase I enzymes that are capable of carrying out the activation reaction. While the Ames test is not 100% accurate at detecting genotoxic carcinogens, it is rapid, technically simple to carry out and is, therefore, a much-favoured primary screen for assessing genotoxic potential.

The convenience of secondary cell lines has greatly increased their utility for assessing secondary natural products. In addition, unlike the Ames test which makes use of a microbial reaction to a particular type of toxin, cell lines can be derived from a number of species (including humans). Secondary cell lines have been maintained for considerable periods of time and are generally phenotypically and genotypically stable. They are effectively 'immortalised'. The major drawback to the use of secondary cell lines is that they have, in general, lost any organotypic characteristics and are therefore of most value in the study of biological events that are common to most cells and are not typical of any particular organ. One of the most significant aspects of the use of secondary cell lines, from a secondary natural product standpoint, is that they have little xenobiotic metabolising capability. One exception to this is the Hep G2 cell line (derived from a hepatocellular carcinoma). This cell line retains some morphological distinctness and has some similarity to the organ that it is derived from [62], however the cell line is severely compromised in terms of expression of xenobiotic metabolising enzymes and care must be taken when extrapolating to any whole organ (or whole body) effects that rely on xenobiotic metabolism [63]. The other problem is that a proportion of secondary cell lines have, over the years, become contaminated with one particular cell line (called HeLa) [64]. This particular cell line (which was the first to be successfully cultured) was later found to be able to outgrow other cell lines and it has been estimated that up to 20% of secondary cell lines are, in fact composed of a HeLa contamination [65] rather than being derived from the original isolated cells. Despite these drawbacks, secondary cell lines are generally stable, well characterised and have been used in multiple studies. They are, therefore, the type of cell line of choice for many investigations. They are particularly valuable for studying primary metabolism, compounds that do not require activation (e.g. 3-nitropropanoic acid) and nonphenotypic pathways that are not organotypic.

Primary cell lines and human tissue-derived material is particularly valuable for the study of secondary metabolite pathways and more specific effects. Primary cells are derived directly from biological tissue—unlike secondary cells, they have not been subcultured several times, Since primary cells retain many of the characteristics of the tissue from which they are derived, they ought to be the test system of choice. Unfortunately such cells are invariably phenotypically and genotypically unstable [66] and primary cell lines can only be sub-cultured a few times before they lose their characteristics or viability. This means that consistent results are not always obtained and long term or chronic effects cannot be investigated therefore they are of limited value in the investigation of food-related secondary natural products. One approach to overcoming some of the problems is to change the environment in which primary hepatocytes (and other primary cells are cultured). By using 3-D scaffolds rather than 2D cultures, the stability of the cell system can be maintained for longer [67]. Among the advantages noted in 3D rather than 2D cultures was the continuance of xenobiotic metabolising capability. This was maintained in 3D cultures for at least 3 weeks compared to 2D cultures where this capability was largely (90%) lost after 2 days. A range of 3D systems for primary cell culture have been developed and applied to toxicity testing following xenobiotic metabolism. Nonetheless the challenge of obtaining useful data on the consequences of long term exposure to secondary natural products at low levels for long periods (equivalent to lifetime exposure) and in complex mixtures remains problematic. The possibility of making use of biopsy samples directly has been examined. The comparison of expression of a range of genes including those responsible for production of xenobiotic metabolising enzymes, nuclear receptors and transcription factors in biopsy samples of parts of the intestinal mucosa and in commonly used cell lines (e.g. CaCo2, HT29, T84 and others) was carried out. Statistical analysis showed that two distinct clusters of expression profiles were obtained—one corresponding to all cultured cell lines and the other to all intestinal biopsies [68]. These results suggest that results from biopsy and cell line experiments are not directly comparable. One significant drawback to the use of biopsy material to assess biological effects is that it only remains viable for about 24 h so again medium and long term studies are not possible. In addition, the effects of circulating metabolites and the effects of reactions in one organ on the xenobiotic response in another is challenging to mimic. One approach is the use of multiple organ co-culture in which cells derived from one organ are physically separated from those of another but are connected by medium [69]—multiple organ co-culture. There have been significant advances in human in vitro co-culture models including microfluidic devices in which human cells can be co-cultured with bacterial cells in order to more accurately reflect the impact of infections or microbiome/host interactions [70] and 'organoid'-type cocultures [71].

Organoids are an important (and increasingly essential) tool to determine the biological effects of xenobiotics and secondary natural products in particular. Organoids are stem cell-derived, 3D culture systems [72]. They can be described as a significant part of the future of NAM approaches and are a viable (and better) alternative to animal systems since there is no species barrier to cross. The most significant aspect of human organoid systems is that they are similar to the in vivo properties of living organs both in terms of their architecture and their response to exogenous agents [73]. In addition to being as complex as living systems, organoids can be used to study inter-individual differences (since they can be generated from any individual) and the development of particular organs and tissues. One of the most important challenges in the use of organoids is the timeline over which they can be used. In part this may be met by manipulation of the balance between stem cell renewal and differentiation in organoids [74]. In some instances, long term culture of organoids has been demonstrated (up to 8 months) and extending this will be a future challenge [75]. Such techniques have also been shown to have potential in xenotransplantation [76] and this along with investigating developmental effects [77] is likely to be the more significant driver for further studies. While aspects of the use of organoids has been reviewed in terms of ethics and morality [78], the situation is not as simple as the development of nonanimal models. In addition, moving from 2 to 3D culture techniques has been suggested to require the recognition and acceptance of a paradigm shift [79] and a different philosophical basis. Ethical issues arise, in particular, in the use of brain organoids and animal/human chimeras [80]. In the case of the former, the importance of consciousness as a concept and its development as a whole brain activity, raises a number of questions regarding the gradual development of neural organelles with greater and greater autonomy and functionality [81]. The question arises, when does a brain organelle become sufficiently developed to be conscious? Other ethical issues arise when considering the use of organoids as a source of transplantation tissue [82]. Chimeric species have been imagined (for example the human/lion form of the sphinx) however this has not been a reality. In a molecular sense, the development of heterologous gene expression (see Chap. 5) could be considered to be the creation of a genomic molecular chimera however the more usual definition requires mixed phenotypes and not simply mixed genotypes.

While the use of advanced in vitro techniques has increased in popularity, it is important to remember that significant work is still being carried out that makes use of animal based models. In particular, animal models are still the preferred choice for assessing effects on whole organisms rather than cells or organs in isolation. They are still very important in developing a better understanding of some disease processes. In addition, until recently, the use of animal models has been enshrined in legislation. There have been significant advances in the use of 'humanised' animal models to try to bridge the gap between human physiology and responses and those of animals. Nonetheless, the trend for the replacement of animal models remains and is likely to lead to a gradual diminution in the numbers of solely animal-based conclusions being drawn. In addition, the use of animals to investigate nutrition and to carry out clinical studies with relevance to humans (particularly long-term investigations) has challenges that more appropriate (and cheaper) models are more likely to address.

The objective of many early nutritional studies was the development of protocols to investigate vitamin and essential nutrient deficiency diseases. Animals were, in most cases, an appropriate surrogate for humans since they responded in a similar way and exhibited dose/response characteristics. A greater challenge was the establishment of optimal levels (rather than deficiency-prevention levels) of particular nutrients and the investigation of the beneficial (or even, competing and hence deleterious) aspects of non-essential compounds. Many compounds that fall into this latter category are secondary natural products and their assessment and effect, particularly on long term health, has been challenging.

As noted previously, both primary and secondary cell systems have only limited value in the investigation of the long term effects of secondary natural products. Nutritionists are rather fond of saying that there is no 'one size fits all' aspect to nutrition and they use this as a justification for developing 'personalised nutrition' or 'stratified nutrition'. While at the same time, they accept results obtained from animal and other model systems. It is philosophically and intellectually dishonest to extol the uniqueness of the individual while accepting broad biochemical similarities, not only within species but also between species. Having accepted that some of the definitions and distinctions that are used in the study of the effects of ingestion of particular individual and classes of compounds may be of limited value, it has to be acknowledged that nutrition is one of the main modifying factors in the development of health and disease. Malnutition can be caused by inadequate intake and also by inadequate uptake (including digestion) and by a range of disease states. Dietary inadequacies occur when the intake (and/or absorption) of a particular nutrient is above the level at which clinical deficiency can be observed but below the recommended daily intake. One of the problems associated with dietary inadequacies is that they can be difficult to diagnose. While deficiency diseases are often related directly to primary metabolic functions and the consequences are clinically observable, dietary inadequacies of essential (or nonessential) nutrients often result in the onset of less observable conditions (for example, a sub-optimal immune response or gradual cognitive impairment).

In the development of methods to study inadequate intakes, there are two key factors—firstly, the measurement of intake by analysis of foods as consumed combined with estimating daily intake by monitoring or by using food diaries and secondly the measurement of uptake by analysis of clinical biomarkers in blood or urine. In the case of the former, while ingenious methods have been developed for the analysis of components from many food matrices and the effects of cooking or processing have been studied, the extrapolation to actual intake is somewhat less precise except under specific, controlled conditions. While estimates of uptake can be very effective by measuring levels clinically (generally in plasma or serum), a single measure may not give a measure of the nutrient in an optimally clinically available form. In addition some nutrients occur in a range of forms (for example, iron) and are subject to homeostatic regulation so the serum or plasma levels may not be representative of the whole body status. In addition, many secondary natural products are metabolised by Phase I or Phase II enzymes and these metabolites may not be captured in a general analysis of the specific compound of interest. Additionally, in many cases, the biological activity of the metabolites is substantially different to that of the ingested compound. In summary, while methods for estimating intake and uptake each have limitations, the combination of the two, together with an understanding of metabolism, can give an estimate of exposure to specific compounds and sets of compounds. The next challenge is to correlate these exposure levels with relevant physiological endpoints. In some cases, even with essential macro and micronutrients, chronic deficiency over a long period of time can have an effect that is different to that of acute deficiency—the consequences of acute deficiency are not explained by the additive cumulative effects of multiple chronic deficiencies. For example, in the case of vitamin A, acute deficiency causes night blindness and can lead to permanent blindness. A chronic (sub-clinical) deficiency is associated with higher infection susceptibility and reduced physical growth. In other cases, particular populations are at risk from the consequences of deficiencies in particular nutrients. In individuals, folate deficiency results in anaemia. In pregnancy, folate deficiency has been associated with preterm delivery, low birth weight, retardation of foetal growth, and neural tube defects. Sub-clinical effects are also evident as a consequence of ingestion of some other, non-essential natural products. For example, indole glucosinolates can be broken down by a co-occurring thioglucosidase enzyme (myrosinase), like other glucosinolates, to give an isothiocyanate. Unlike other glucosinolates, indolyl-3-methyl isothiocyanate is unstable and only transiently formed [83] and is rapidly hydrolysed to give the corresponding indole-3-carbinol and thiocyanate ion (Fig. 4.6).

Thiocyanate ion is goitrogenic (it competes directly with iodide in the thyroid) however the amount generated by a normal serving of *Brassica* vegetables is less than a quarter of what would need to be consumed to be clinically relevant to cause goitrogenicity in normal adults [84]. Thiocyanate ion has a half life of 7–9 days [85] so cumulative exposure to indole glucosinolates may enable the concentration to reach a critical threshold. It should be noted, however, that in a case of acute

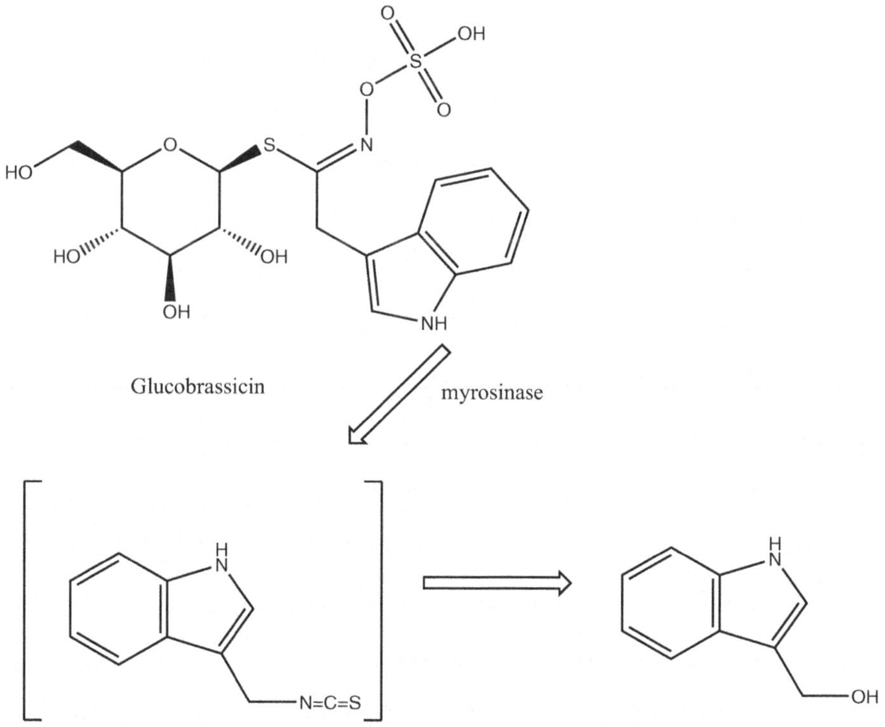

Fig. 4.6 Subsequent hypothesised breakdown of indolyl-3-methyl isothiocyanate

thiocyanate poisoning in an adult, a half life of 61 h was observed and it is eliminated by first order kinetics [86] so a simple extrapolation from acute to chronic ingestion may not be possible. The effect of the other part of the molecule involved in the production of thiocyanate ion, indole-3-carbinol has been studied as a potential cause of beneficial effects arising from consumption of vegetables. Indole-3-carbinol has no readily discernible clinical profile. It has, however, been implicated in a number of long term, chronic beneficial effects [87]. Some of these seem to be associated with synergistic effects with other plant components [88]. This will be discussed in more detail in Sect. 4.7. In addition, through its effect on pathways related to Nrf2, indole-3-carbinol has antioxidant and neuroprotective effects [89] through modulation of microglia inflammation [90]. Indole3-carbinol has an in vitro half life of 40 h in cell medium [91] and indole metabolites accumulate in the nucleus of cells. Indole-3-carbinol has a serum half life of greater than 48 h [92]. Cancer and diseases thought to be a consequence of inflammation are long term, chronic conditions and since it has activity on these types of disease pathways, the effect of indole-3-carbinol and its metabolites is likely to be long term and so chronic administration is likely a prerequisite for beneficial effects to be observed. In addition, there is some evidence of a cumulative effect and therefore, once a 'loading dose' has been administered, long term, chronic intakes may be responsible for the observational beneficial effects observed.

The use of NAMs has grown as economic and scientific reasons have interacted. Despite the enthusiasm being shown in many quarters, regulatory pressure for the continued use of animal models is likely to persist. It will be of little value to replace one inadequate model (animals) with another (cells or other in vitro methods) that is less well understood and which does not have the extensive methodological and research history supplemented by the clinical data generated as a result of product launches. It may be more fruitful to concentrate on areas where currently there are gaps. One of these is in defining the beneficial properties of long term exposure to secondary natural products. Long term animal studies are expensive and the measurement of relatively small effects with the likely presence of confounders will limit studies. However current models adopted by regulatory bodies in which the burden of evidence required to confirm health claims for foods by using a largely pharmacological model are fundamentally flawed. Pharmacology deals with organisms at an extreme state of actual or potential ill health and seeks ways to bring those systems back to healthy homeostasis. In fact, the concept of healthy homeostasis is incorrect. Healthy systems are in a state of (to varying degrees) unstable equilibrium and new tools are required to measure how to navigate between that succession of states without complete system collapse and decline into disarray. It is a bit like riding a bicycle. So long as the vehicle is moving then you can steer it towards where you want to go. When it stops (becomes homeostatic) then it is liable to fall over. Treating diet and health like a stationary bicycle with a specific and singular equilibrium position and supposing that is the best and healthiest option is an idea with a basic flaw at its core.

4.7 Assessing Complex Mixtures of Secondary Natural Products

Natural products are generally not designed with an effect on humans as their reason for existing. It has been supposed that the advantage that accrues from the production of secondary natural products is associated with a direct attack on competitors for resources (as seems to be the case for anti-bacterial species produced by fungi)—[93], the resistance to attack by microbial species as may be the case for production of quorum sensing inhibitors by some plants—[94] or as feeding inhibitors—[95]. Any physiological effects on humans (for example, anti-inflammatory activity) appears to be serendipitous. As noted previously, the reductionist chemistry approach to secondary natural products is focussed on steps to enable single molecules to be isolated and characterised. While this is important for studying the biosynthesis and molecular interrelationships of natural products, a main driver was the need to define precisely the efficacious (and sub-toxic) level of administration of a specific compound that would be beneficial. This was a major part of the reason for the development of isolation procedures for morphine (by Serturner, who was a pharmacist and whose business was, therefore, the dispensing of drugs) and quinine (by Pelletier and Caventou who were both pharmacists and had a laboratory at the rear of an apothecary).

The concept of complexity in the interactions of natural products has been explored. There are various types of interaction than can be envisioned. Some natural products exhibit multiple activities. For example, the polyphenolic, resveratrol shows a range of biological effects [96]. The extent to which agonist or antagonist activity may result can be a concentration-dependent effect. In other cases, multiple compounds with the same molecular target can be found in the same source species—or in different species. For example, as noted previously, *Brassica* vegetables can contain both goitrin and thiocyanate ion and these both interact with the thyroid [97]. Actively seeking synergy has also been studied including between natural products and pharmaceuticals [98], however, it is still a scientific pre-requisite that the nature of the synergistic effect, its mechanism and the dose/response relationship should be understood. It is scientifically and philosophically unsatisfactory to attempt to delineate additive or synergistic effects between highly purified and homogenous pharmaceutical agents and uncharacterised mixtures of natural products and the tendency will be to use activity-guided fractionation and isolation procedures to obtain a reproducible effect. Bioinformatic approaches to understanding both synergistic and antagonistic effects of natural products have been reviewed [99], however there remains a lack of consensus about the best way to proceed.

Different types of bioassay can be used in the investigation of secondary natural products. The development of NAMs will allow for more rapid progress than the use of expensive (and variably relevant) animal models. There will likely remain a role for humanised whole animal model systems however care should be taken with these and NAMs to ensure that replacement of one model system with another brings scientific understanding to a better place and that the drivers are not solely economic and ethical advantages. Given the regulatory challenges, it is possible that the use of NAMs will be adopted more rapidly for screening for beneficial effects than for defining levels of toxicity. Finally, the area of long-term benefit (or toxicity) is one that remains a challenge for current technology. Chronic toxicity models include both low level exposures and repeat exposure to bioactive agents. Population-based studies (epidemiological investigations) can sometimes provide some suggestion of an association between a particular set of exposures and both deleterious and beneficial endpoints however these generally lack mechanistic detail and so have limited value as predictive tools. In addition, the significance of confounders should not be underestimated. Scientists are perhaps even more prone than the general population to look for causative connections and Popperian testing for falsifiability should be incorporated when a conclusion regarding the nature of the link between two or more criteria is being proposed. Unfortunately, presuppositions abound in the area of the link between secondary natural products and health. This is particularly pronounced in the area of fruit and vegetable intake and long-term health [100]. It is dangerous to make dietary recommendations based purely on observational data [101] however the 'received wisdom' argument has tended to prevail and the quantification of the benefit and knowledge of which components are most important has tended to lag behind. This makes for poor and inadequate science and bad public policy recommendations.

References

1. Wu B, Eldeghaidy S, Ayed C, Fisk ID, Hewson L, Liu Y. Mechanisms of umami taste perception: From molecular level to brain imaging. Crit Rev Food Sci Nutr. 2022;62(25):7015–24. https://doi.org/10.1080/10408398.2021.1909532. Epub 2021 May 17 PMID: 33998842.
2. Korsmeyer C. Making sense of taste: Food and philosophy. Ithaca, NY: Cornell University Press; 1999.
3. O'Neill J, Brock C, Olesen AE, Andresen T, Nilsson M, Dickenson AH. Unravelling the mystery of capsaicin: a tool to understand and treat pain. Pharmacol Rev. 2012;64(4):939–71. https://doi.org/10.1124/pr.112.006163. PMID:23023032;PMCID:PMC3462993.
4. Abdel-Salam OME, Mózsik G. Capsaicin, The Vanilloid receptor TRPV1 agonist in neuroprotection: Mechanisms involved and significance. Neurochem Res. 2023;48(11):3296–315. https://doi.org/10.1007/s11064-023-03983-z. Epub 2023 Jul 26. PMID: 37493882; PMCID: PMC10514110.
5. Chu Y, Cohen BE, Chuang H. A single TRPV1 amino acid controls species sensitivity to capsaicin. Sci Rep. 2020;10:8038. https://doi.org/10.1038/s41598020-64584-2.
6. Nair KP. The chemistry of ginger. In: Turmeric (*Curcuma longa* L.) and ginger (*Zingiber officinale* Rosc.)—world's invaluable medicinal spices. Cham: Springer; 2019. https://doi.org/10.1007/978-3-030-29189-1_16
7. Ok S, Jeong WS. Optimization of extraction conditions for the 6-shogaol-rich extract from ginger (*Zingiber officinale* Roscoe). Prev Nutr Food Sci. 2012;17(2):166–71. https://doi.org/10.3746/pnf.2012.17.2.166. PMID:24471079;PMCID:PMC3866747.

8. Yoshitomi T, Oshima N, Goto Y, Nakamori S, Wakana D, Anjiki N, Sugimura K, Kawano N, Fuchino H, Iida O, Kagawa T, Jinno H, Kawahara N, Kobayashi Y, Maruyama T. Construction of prediction models for the transient receptor potential vanilloid subtype 1 (TRPV1)-stimulating activity of ginger and processed ginger based on LC-HRMS data and PLS regression analyses. J Agric Food Chem. 2017;65(17):3581–8. https://doi.org/10.1021/acs.jafc.7b00577. Epub 2017 Apr 19 PMID: 28398734.
9. Pertz HH, Lehmann J, Roth-Ehrang R, Elz S. Effects of ginger constituents on the gastrointestinal tract: role of cholinergic M3 and serotonergic 5-HT3 and 5-HT4 receptors. Planta Med. 2011;77(10):973–8. https://doi.org/10.1055/s-0030-1270747. Epub 2011 Feb 8 PMID: 21305447.
10. Kiyama R. Nutritional implications of ginger: chemistry, biological activities and signaling pathways. J Nutr Biochem. 2020;86: 108486. https://doi.org/10.1016/j.jnutbio.2020.108486. Epub 2020 Aug 19 PMID: 32827666.
11. Yin Y, Dong Y, Vu S, Yang F, Yarov-Yarovoy V, Tian Y, Zheng J. Structural mechanisms underlying activation of TRPV1 channels by pungent compounds in gingers. Br J Pharmacol. 2019;176(17):3364–77. https://doi.org/10.1111/bph.14766. Epub 2019 Jul 22. PMID: 31207668; PMCID: PMC6692589.
12. Rosenbaum T, Simon SA. TRPV1 receptors and signal transduction. In: Liedtke WB, Heller S, editors. TRP Ion channel function in sensory transduction and cellular signaling cascades. Boca Raton (FL): CRC Press/Taylor & Francis; 2007. Chapter 5. PMID: 21204507.
13. Andrei C, Zanfirescu A, Nițulescu GM, Olaru OT, Negreș S. Natural active ingredients and TRPV1 modulation: Focus on key chemical moieties involved in ligand-target interaction. Plants (Basel). 2023;12(2):339. https://doi.org/10.3390/plants12020339. PMID:36679051;PMCID:PMC9860573.
14. Pecoraro V, Moja L, Dall'Olmo L, Cappellini G, Garattini S. Most appropriate animal models to study the efficacy of statins: a systematic review. Eur J Clin Invest. 2014;44(9):848–71. https://doi.org/10.1111/eci.12304. PMID: 25066257.
15. Jung E, Kong SY, Ro YS, Ryu HH, Shin SD. Serum cholesterol levels and risk of cardiovascular death: A systematic review and a dose-response meta-analysis of prospective cohort studies. Int J Environ Res Public Health. 2022;19(14):8272. https://doi.org/10.3390/ijerph19148272. PMID:35886124;PMCID:PMC9316578.
16. Levine GN, Keaney JF Jr, Vita JA. Cholesterol reduction in cardiovascular disease. Clinical benefits and possible mechanisms. N Engl J Med. 1995;332(8):512–21. https://doi.org/10.1056/NEJM199502233320807. PMID: 7830734.
17. Diaz GJ, Murcia HW, Cepeda SM, Boermans HJ. The role of selected cytochrome P450 enzymes on the bioactivation of aflatoxin B1 by duck liver microsomes. Avian Pathol. 2010;39(4):279–85. https://doi.org/10.1080/03079457.2010.495109. PMID: 20706884.
18. Eaton DL, Williams DE, Coulombe RA. Species differences in the biotransformation of aflatoxin B1: Primary determinants of relative carcinogenic potency in different animal species. Toxins. 2025;17:30. https://doi.org/10.3390/toxins17010030.
19. Esber EC. An overview of the strengths and limitations of biological assays in quality control. J Pharm Biomed Anal. 1989;7(2):131–8. https://doi.org/10.1016/07317085(89)80076-7. PMID: 2488613.
20. Caruso G, Musso N, Grasso M, Costantino A, Lazzarino G, Tascedda F, Gulisano M, Lunte SM, Caraci F. Microfluidics as a novel tool for biological and toxicological assays in drug discovery processes: Focus on microchip electrophoresis. Micromachines (Basel). 2020;11(6):593. https://doi.org/10.3390/mi11060593. PMID:32549277;PMCID:PMC7344675.
21. Moragues T, Arguijo D, Beneyton T, et al. Droplet-based microfluidics. Nat Rev Methods Primers. 2023;3:32. https://doi.org/10.1038/s43586-023-00212-3.
22. Ismail BP, Ward RE. Glucose determination by enzyme analysis. In: Ismail BP, Nielsen SS, editors. Nielsen's food analysis laboratory manual. Food science text series. Cham: Springer; 2024. https://doi.org/10.1007/978-3-031-449703_25

23. Lorenz TC. Polymerase chain reaction: basic protocol plus troubleshooting and optimization strategies. J Vis Exp. 2012;63: e3998. https://doi.org/10.3791/3998. PMID:22664923;PMCID:PMC4846334.
24. Hoofnagle AN, Wener MH. The fundamental flaws of immunoassays and potential solutions using tandem mass spectrometry. J Immunol Methods. 2009;347(1–2):3–11. https://doi.org/10.1016/j.jim.2009.06.003. Epub 2009 Jun 16. PMID: 19538965; PMCID: PMC2720067.
25. Terzapulo X, Kassenova A, Bukasov R. Immunoassays: Analytical and clinical performance, challenges, and perspectives of SERS detection in comparison with fluorescent spectroscopic detection. Int J Mol Sci. 2024;25(4):2080. https://doi.org/10.3390/ijms25042080. PMID:38396756;PMCID:PMC10889711.
26. Tipper D, Strominger J. Mechanism of action of penicillins: A proposal based on their structural similarity to acyl-D-alanyl-D-alanine Proc. Natl Acad Sci USA. 1965;54:1133–41.
27. Zhang JM, Yang PL, Gray NS. Targeting cancer with small molecule kinase inhibitors. Nat Rev Cancer. 2009;9:28–39. https://doi.org/10.1038/nrc2559.
28. Alexander SPH, Fabbro D, Kelly E, Mathie AA, Peters JA, Veale EL, Armstrong JF, Faccenda E, Harding SD, Davies JA, Annett S, Boison D, Burns KE, Dessauer C, Gertsch J, Helsby NA, Izzo AA, Ostrom R, Papapetropoulos A, Pyne NJ, Pyne S, Robson T, Seifert R, Stasch JP, Szabo C, van der Stelt M, van der Vliet A, Watts V, Wong SS. The concise guide to pharmacology 2023/24: Enzymes. Br J Pharmacol. 2023;180(Suppl 2):S289–373. https://doi.org/10.1111/bph.16181. PMID: 38123154.
29. Brooks HB, Geeganage S, Kahl SD, et al. Basics of enzymatic assays for HTS. 2012 May 1 [Updated 2012 Oct 1]. In: Markossian S, Grossman A, Arkin M et al., editors. Assay guidance manual [Internet]. Bethesda (MD): Eli Lilly & Company and the National Center for Advancing Translational Sciences; 2004.
30. Ramsay RR, Tipton KF. Assessment of enzyme inhibition: A review with examples from the development of monoamine oxidase and cholinesterase inhibitory drugs. Molecules. 2017;22(7):1192. https://doi.org/10.3390/molecules22071192. PMID:28714881;PMCID:PMC6152246.
31. Stancu C, Sima A. Statins: mechanism of action and effects. J Cell Mol Med. 2001 Oct-Dec;5(4):378–87. https://doi.org/10.1111/j.1582-4934.2001.tb00172.x. PMID: 12067471; PMCID: PMC6740083.
32. Islam Z, Kumar P. Inhibitors of riboflavin biosynthetic pathway enzymes as potential antibacterial drugs. Front Mol Biosci. 2023;11(10):1228763. https://doi.org/10.3389/fmolb.2023.1228763. PMID:37496776;PMCID:PMC10366380.
33. Christopherson RI, Lyons SD, Wilson PK. Inhibitors of de novo nucleotide biosynthesis as drugs. Acc Chem Res. 2002;35(11):961–71. https://doi.org/10.1021/ar0000509. PMID: 12437321.
34. Hicks JL, Oldham KEA, McGarvie J, Walker EJ. Combatting antimicrobial resistance via the cysteine biosynthesis pathway in bacterial pathogens. Biosci Rep. 2022;42(10):BSR20220368. https://doi.org/10.1042/BSR20220368. PMID: 36148777; PMCID: PMC9547175.
35. Magden J, Kääriäinen L, Ahola T. Inhibitors of virus replication: recent developments and prospects. Appl Microbiol Biotechnol. 2005;66(6):61221. https://doi.org/10.1007/s00253-004-1783-3. Epub 2004 Dec 9. PMID: 15592828; PMCID: PMC7082807.
36. Chan JFW, Yuan S, Chu H., et al. COVID-19 drug discovery and treatment options. Nat Rev Microbiol 2024;22:391–407. https://doi.org/10.1038/s41579024-01036-y
37. Zhang J, Liu R, Niu L, Zhu S, Zhang Q, Zhao M, Liu W, Liu J. Determination of endocrine-disrupting potencies of agricultural soils in China via a battery of steroid receptor bioassays. Environ Pollut. 2018;234:846–54. https://doi.org/10.1016/j.envpol.2017.12.004. Epub 2017 Dec 21 PMID: 29248852.
38. Shiizaki K, Asai S, Ebata S, Kawanishi M, Yagi T. Establishment of yeast reporter assay systems to detect ligands of thyroid hormone receptors alpha and beta. Toxicol In Vitro. 2010;24(2):638–44. https://doi.org/10.1016/j.tiv.2009.10.001. Epub 2009 Oct 22 PMID: 19853653.

39. MacLachlan BJ, Greenshields-Watson A, Mason GH, Schauenburg AJ, Bianchi V, Rizkallah PJ, Sewell AK, Fuller A, Cole DK. Using X-ray crystallography, biophysics, and functional assays to determine the mechanisms governing Tcell receptor recognition of cancer antigens. J Vis Exp. 2017;120:54991. https://doi.org/10.3791/54991. PMID:28287509;PMCID:PMC5408581.
40. Chiba T, Tsuchiya T, Mori R, Shimokawa I. Protein reporter bioassay systems for the phenotypic screening of candidate drugs: A mouse platform for anti-aging drug screening. Sensors (Basel) 2012;12(2):1648–56. https://doi.org/10.3390/s120201648. Epub 2012 Feb 7. PMID: 22438730; PMCID: PMC3304132.
41. Baumann C, Chiang WC, Valsecchi R, Jurt S, Deluigi M, Schuster M, Rosengren KJ, Plückthun A, Zerbe O. Side-chain dynamics of the α1B -adrenergic receptor determined by NMR via methyl relaxation. Protein Sci. 2023;32(11): e4801. https://doi.org/10.1002/pro.4801. PMID:37805830;PMCID:PMC10593183.
42. Jin B, Thakur N, Wijesekara AV, Eddy MT. Illuminating GPCR signaling mechanisms by NMR spectroscopy with stable-isotope labeled receptors. Curr Opin Pharmacol. 2023;72:102364. https://doi.org/10.1016/j.coph.2023.102364. Epub 2023 Aug 21. PMID: 37612173; PMCID: PMC11221349.
43. Mulry E, Ray AP, Eddy MT. Production of a human histamine receptor for NMR spectroscopy in aqueous solutions. Biomolecules. 2021;11(5):632. https://doi.org/10.3390/biom11050632. PMID:33923140;PMCID:PMC8146376.
44. Geoffroy EF. Table des Differents Rapports Observés en Chimie entre Differentes Substances. Mem Acad R Sci. 1718:202–12.
45. Pelletier PJ, Caventou JB. “Recherches Chimiques sur les Quinquinas” Annales de Chimie et de Physique (in French). 1820:15.
46. Meth-Cohn O, Smith M. What did W. H. Perkin actually make when he oxidised aniline to obtain mauveine?. J Chem Soc Perkin Trans 1. 1994:5–7.
47. Travis AS, Hornix WJ, Bud R, Homburg E. The emergence of research laboratories in the dyestuffs industry, 1870–1900. Br J Hist Sci. 1992;25(1):91–111. https://doi.org/10.1017/S0007087400045349.
48. Ghiretti F. Bartolomeo Bizio and the rediscovery of Tyrian purple. Experientia. 1994;50:802–7. https://doi.org/10.1007/BF01956459.
49. Krishnamurti C, Rao SC. The isolation of morphine by Serturner. Indian J Anaesth. 2016;60(11):861–2. https://doi.org/10.4103/0019-5049.193696. PMID:27942064;PMCID:PMC5125194.
50. Hosztafi S. The discovery of alkaloids. Pharmazie. 1997;52(7):546–50 PMID: 9266593.
51. Hofmann AW. [Effect of heat on the ammonium bases. II. Co-niine.] Ber Chem 1881;14:705–13.
52. Giseke AL. “Ueber das wirksame Princip des Schierlings, Conium maculatum” (On the active component of hemlock, Conium maculatum). Archiv der Pharmazie. 1827;20(2):97–111.
53. Ladenburg A. [Approaches to the synthesis of coniine.] Ber Chem 1886;19:439–41.
54. Guenlickx I. Hanley AB. New approach methodologies. Sci Parliament. 2025;81:23–4.
55. Avey MT, Fenwick N, Griffin G. The use of systematic reviews and reporting guidelines to advance the implementation of the 3Rs. J Am Assoc Lab Anim Sci. 2015;54(2):153–62. PMID: 25836961; PMCID: PMC4382619.
56. Sun D, Gao W, Hu H, Zhou S. Why 90% of clinical drug development fails and how to improve it? Acta Pharm Sin B. 2022;12(7):3049–62. https://doi.org/10.1016/j.apsb.2022.02.002. Epub 2022 Feb 11. PMID: 35865092; PMCID: PMC9293739.
57. Wainwright M. Moulds in folk medicine. Folklore. 1989;100(2):162–66. http://www.jstor.org/stable/1260294
58. Anand U, Jacobo-Herrera N, Altemimi A, Lakhssassi N. A comprehensive review on medicinal plants as antimicrobial therapeutics: Potential avenues of biocompatible drug discovery. Metabolites. 2019;9(11):258. https://doi.org/10.3390/metabo9110258. PMID:31683833;PMCID:PMC6918160.

59. Pandey MM, Rastogi S, Rawat AK. Indian traditional ayurvedic system of medicine and nutritional supplementation. Evid Based Complement Alternat Med. 2013;2013:376327. https://doi.org/10.1155/2013/376327. Epub 2013 Jun 23. PMID: 23864888; PMCID: PMC3705899.
60. Vijay U, Gupta S, Mathur P, Suravajhala P, Bhatnagar P. Microbial mutagenicity assay: Ames test. Bio Protoc. 2018;8(6): e2763. https://doi.org/10.21769/BioProtoc.2763. PMID:34179285;PMCID:PMC8203972.
61. Swenson DH. Significance of electrophilic reactivity and especially DNA alkylation in carcinogenesis and mutagenesis. Dev Toxicol Environ Sci. 1983;11:247–54 PMID: 6677460.
62. Arzumanian VA, Kiseleva OI, Poverennaya EV. The curious case of the HepG2 cell line: 40 years of expertise. Int J Mol Sci. 2021;22(23):13135. https://doi.org/10.3390/ijms222313135. PMID:34884942;PMCID:PMC8658661.
63. Stanley LA, Wolf CR. Through a glass, darkly? HepaRG and HepG2 cells as models of human phase I drug metabolism. Drug Metab Rev. 2022;54(1):46–62. https://doi.org/10.1080/03602532.2022.2039688.
64. Landry JJ, Pyl PT, Rausch T, Zichner T, Tekkedil MM, Stütz AM, Jauch A, Aiyar RS, Pau G, Delhomme N, Gagneur J, Korbel JO, Huber W, Steinmetz LM. The genomic and transcriptomic landscape of a HeLa cell line. G3 (Bethesda). 2013;3(8):1213–24. https://doi.org/10.1534/g3.113.005777. PMID: 23550136; PMCID: PMC3737162.
65. Lucey BP, Nelson-Rees WA, Hutchins GM. Henrietta lacks, HeLa cells, and cell culture contamination. Arch Pathol Lab Med. 2009;133(9):1463–7. https://doi.org/10.5858/133.9.1463. PMID: 19722756.
66. Geraghty R, Capes-Davis A, Davis J, et al. Guidelines for the use of cell lines in biomedical research. Br J Cancer. 2014;111:1021–46. https://doi.org/10.1038/bjc.2014.166.
67. Vorrink SU, Ullah S, Schmidt S, Nandania J, Velagapudi V, Beck O, IngelmanSundberg M, Lauschke VM. Endogenous and xenobiotic metabolic stability of primary human hepatocytes in long-term 3D spheroid cultures revealed by a combination of targeted and untargeted metabolomics. FASEB J. 2017;31(6):2696–708. https://doi.org/10.1096/fj.201601375R. Epub 2017 Mar 6. PMID: 28264975; PMCID: PMC5434660.
68. Bourgine J et al. Gene expression profiling of systems involved in the metabolism and the disposition of xenobiotics: Comparison between human intestinal biopsy samples and colon cell lines. Drug Metabol Dispos. 2012;40(4):694–705.
69. Gayathri L, Dhanasekaran D, Akbarsha MA. Scientific concepts and applications of integrated discrete multiple organ co-culture technology. J Pharmacol Pharmacother. 2015;6(2):63–70. https://doi.org/10.4103/0976-500X.155481. PMID: 25969651; PMCID: PMC4419250.
70. Shah P, Fritz JV, Glaab E, Desai MS, Greenhalgh K, Frachet A, Niegowska M, Estes M, Jäger C, Seguin-Devaux C, Zenhausern F, Wilmes P. A microfluidics-based in vitro model of the gastrointestinal human-microbe interface. Nat Commun. 2016;11(7):11535. https://doi.org/10.1038/ncomms11535.PMID:27168102;PMCID:PMC4865890.
71. Liu J, Zhang B, Cui Y, Song H, Shang D. In vitro co-culture models for studying organoids-macrophages interaction: the golden technology of cancer immunotherapy. Am J Cancer Res. 2024;14(7):3222–40. https://doi.org/10.62347/BQFH7352. PMID:39113861;PMCID:PMC11301299.
72. Kim J, Koo BK, Knoblich JA. Human organoids: Model systems for human biology and medicine. Nat Rev Mol Cell Biol. 2020;21:571–84.https://doi.org/10.1038/s41580-020-0259-3
73. Lehmann R, Lee CM, Shugart EC, Benedetti M, Charo RA, Gartner Z, Hogan B, Knoblich J, Nelson CM, Wilson KM. Human organoids: a new dimension in cell biology. Mol Biol Cell. 2019;30(10):1129–37. https://doi.org/10.1091/mbc.E19-030135. PMID:31034354;PMCID:PMC6724519.
74. Yang L, Wang X, Zhou X, et al. A tunable human intestinal organoid system achieves controlled balance between self-renewal and differentiation. Nat Commun. 2025;16:315. https://doi.org/10.1038/s41467-024-55567-2.
75. Atamian A, Birtele M, Hosseini N, et al. Generation and long-term culture of human cerebellar organoids from pluripotent stem cells. Nat Protoc. 2024. https://doi.org/10.1038/s41596-024-01093-w.

76. Dekkers JF, van Vliet EJ, Sachs N, Rosenbluth JM, Kopper O, Rebel HG, Wehrens EJ, Piani C, Visvader JE, Verissimo CS, Boj SF, Brugge JS, Clevers H, Rios AC. Long-term culture, genetic manipulation and xenotransplantation of human normal and breast cancer organoids. Nat Protoc. 2021;16(4):1936–65. https://doi.org/10.1038/s41596-020-00474-1. Epub 2021 Mar 10. PMID: 33692550; PMCID: PMC8221035.
77. Corsini NS et al. Human organoids: New strategies and methods for analyzing human development and disease. Cell. 185(15):2756–69.
78. Barnhart AJ, Dierickx K. The many moral matters of organoid models: A systematic review of reasons. Med Health Care and Philos. 2022;25:545–60. https://doi.org/10.1007/s11019-022-10082-3.
79. Caianiello S, Bertolaso M, Militello G. Thinking in 3 dimensions: philosophies of the microenvironment in organoids and organs-on-chip. HPLS. 2023;45:14. https://doi.org/10.1007/s40656-023-00560-z.
80. de Jongh D, Massey EK; VANGUARD consortium; Bunnik EM. Organoids: A systematic review of ethical issues. Stem Cell Res Ther. 2022;13(1):337. https://doi.org/10.1186/s13287-022-02950-9. PMID: 35870991; PMCID: PMC9308907.
81. Hayashi Y, Sato R. The unity of consciousness and the practical ethics of neural organoid research. Neuroethics. 2025;18:3. https://doi.org/10.1007/s12152-024-09574-1.
82. Schneemann SA, Boers SN, van Delden JJM, Nieuwenhuis EES, Fuchs SA, Bredenoord AL. Ethical challenges for pediatric liver organoid transplantation. Sci Transl Med. 2020;12:8471. https://doi.org/10.1126/scitranslmed.aau8471.
83. Hanley AB, Parsley KR, Lewis JA, Fenwick GR. Chemistry of indole glucosinolates: intermediacy of indol-3-yl methyl isothiocyanates in the enzymic hydrolysis of indole glucosinolates. J Chem Soc Perkin Trans. 1990;I:2273.
84. Felker P, Bunch R, Leung AM. Concentrations of thiocyanate and goitrin in human plasma, their precursor concentrations in brassica vegetables, and associated potential risk for hypothyroidism. Nutr Rev. 2016;74(4):248–58. https://doi.org/10.1093/nutrit/nuv110.
85. Chandler JD, Day BJ. Thiocyanate: A potentially useful therapeutic agent with host defense and antioxidant properties. Biochem Pharmacol. 2012;84(11):1381–7. https://doi.org/10.1016/j.bcp.2012.07.029. Epub 2012 Aug 8. PMID: 22968041; PMCID: PMC3983948.
86. Watson CJ, Overbeek DL, Allegri-Machado G, Kellogg MD, Patterson A, McAlvin JB, Burns MM. Thiocyanate toxicity: A teaching case. Clin Toxicol (Phila). 2022;60(7):876–81. https://doi.org/10.1080/15563650.2022.2042012. Epub 2022 Mar 4 PMID: 35240906.
87. Katz E, Nisani S, Chamovitz DA. Indole-3-carbinol: A plant hormone combatting cancer. F1000Res. 2018;7:F1000 Faculty Rev-689. https://doi.org/10.12688/f1000research.14127.1. PMID: 29904587; PMCID: PMC5989150.
88. Bonnesen C, Eggleston IM, Hayes JD. Dietary indoles and isothiocyanates that are generated from cruciferous vegetables can both stimulate apoptosis and confer protection against DNA damage in human colon cell lines. Cancer Res. 2001;61(16):6120–30.
89. Singh AA, Yadav D, Khan F, Song M. Indole-3-carbinol and its derivatives as neuroprotective modulators. Brain Sci. 2024;14(7):674. https://doi.org/10.3390/brainsci14070674. PMID:39061415;PMCID:PMC11274471.
90. Peng L, Zhu X, Wang C, et al. Indole-3-carbinol (I3C) reduces apoptosis and improves neurological function after cerebral ischemia–reperfusion injury by modulating microglia inflammation. Sci Rep. 2024;14:3145. https://doi.org/10.1038/s41598-024-53636-6.
91. Staub RE, Feng C, Onisko B, Bailey GS, Firestone GL, Bjeldanes LF. Fate of indole-3-carbinol in cultured human breast tumor cells. Chem Res Toxicol. 2002;15(2):101–9. https://doi.org/10.1021/tx010056m. PMID: 11849035.
92. Arneson DW, Hurwitz A, McMahon LM, Robaugh D. Presence of 3.3'diindolylmethane in human plasma after oral administration of indole-3carbinol. Proc Annu Meet Am Assoc Cancer Res 1999;40:429.
93. Thomashow LS, Bonsall RF, Weller DM. Antibiotic production by soil and rhizosphere microbes in situ. In: Hurst CJ, Knudsen GR, MJ MI, Stetzenbach LD, Walter MV, editors. Manual of environmental microbiology. Washington: American Society of Microbiology; 1997.

94. Vattem DA, Mihalik K, Crixell SH, McLean RJ. Dietary phytochemicals as quorum sensing inhibitors. Fitoterapia. 2007;78(4):302–10. https://doi.org/10.1016/j.fitote.2007.03.009. Epub 2007 Apr 11 PMID: 17499938.
95. Koul O. Phytochemicals and insect control: An antifeedant approach. Crit Rev Plant Sci. 2008;27(1):1–24. https://doi.org/10.1080/07352680802053908.
96. Kim YJ, Chung SO, Kim JK, Park SU. Recent studies on resveratrol and its biological and pharmacological activity. EXCLI J. 2017;27(16):602–8. https://doi.org/10.17179/excli2017-253. PMID:28694761;PMCID:PMC5491918.
97. Felker P, Bunch R, Leung AM. Concentrations of thiocyanate and goitrin in human plasma, their precursor concentrations in brassica vegetables, and associated potential risk for hypothyroidism. Nutr Rev. 2016;74(4):248–58. https://doi.org/10.1093/nutrit/nuv110. Epub 2016 Mar 5. PMID: 26946249; PMCID: PMC4892312.
98. Pezzani R, Salehi B, Vitalini S, Iriti M, Zuñiga FA, Sharifi-Rad J, Martorell M, Martins N. Synergistic effects of plant derivatives and conventional chemotherapeutic agents: An update on the cancer perspective. Medicina (Kaunas). 2019;55(4):110. https://doi.org/10.3390/medicina55040110. PMID:30999703;PMCID:PMC6524059.
99. Caesar LK, Cech NB. Synergy and antagonism in natural product extracts: when 1 + 1 does not equal 2. Nat Prod Rep. 2019;36(6):869–88. https://doi.org/10.1039/c9np00011a. PMID:31187844;PMCID:PMC6820002.
100. Pem D, Jeewon R. Fruit and vegetable intake: Benefits and progress of nutrition education interventions—Narrative review article. Iran J Public Health. 2015;44(10):1309–21. PMID: 26576343; PMCID: PMC4644575.
101. Wallace TC, Bailey RL, Blumberg JB, Burton-Freeman B, Chen CO, CroweWhite KM, Drewnowski A, Hooshmand S, Johnson E, Lewis R, Murray R, Shapses SA, Wang DD. Fruits, vegetables, and health: A comprehensive narrative, umbrella review of the science and recommendations for enhanced public policy to improve intake. Crit Rev Food Sci Nutr. 2020;60(13):2174–211. https://doi.org/10.1080/10408398.2019.1632258. Epub 2019 Jul 3 PMID: 31267783.

Chapter 5
Biosynthesis—Approaches and Strategies

Contents

5.1 Recipe—Bolognese with Character (and Tomatoes) 136
5.2 Classes of Biosynthetic Pathways 140
5.3 Biosynthetic Pathways—From Molecules to Genes 144
5.4 Manipulation of Biosynthesis—Molecules and Mutation 148
5.5 Gene Editing—Scientific, Regulatory and Adoption Challenges 153
5.6 Extremophiles and Novel Biosynthetic Pathways 155
5.7 Marine Secondary Natural Products 156
References 158

Abstract An understanding of the link between biosynthetic pathways has offered the opportunity for the development of genetic manipulation however, in some cases, the driver has been the need for biologists to demonstrate their proficiency rather than a scientific, societal or economic advance. This has enabled those who have reservations about new technologies to criticise the development of sophisticated manipulation procedures. The biosynthesis of natural products has been an area of investigation for a considerable period and has enabled chemists to develop knowledge systems based on similarities in biogenesis and this has been both practically and philosophically important. It has helped to define the progress of the area. The investigation of natural products and how they might be assembled has aided the directed discovery of new enzyme systems (for example Diels Alderases) and of semi-synthesis of core natural product structures. The discovery of Biosynthetic Gene Clusters (BGCs) has enabled chemists to postulate the potential existence of novel natural products driven by biological plausibility rather than chemical isolation. Gene editing is one of the most significant tools that has been developed to biologically manipulate natural products in a planned and programmable way.

Keywords Biosynthesis · Diels Alderases · BGCs · Gene editing

B. Hanley, *Secondary Natural Products in Foods and Food Systems*,
https://doi.org/10.1007/978-3-032-00292-1_5

5.1 Recipe—Bolognese with Character (and Tomatoes)

The question of which compounds were present in natural product extracts was followed by two supplementary queries—where did they come from and how can they be used? Even before compounds were purified to a single entity, the discovery of the potential healing power of extracts was a significant motivation. As we have seen, the isolation of individual substances from classes of compounds was a major factor in ensuring that reproducible amounts of compounds could be used for treatments and other uses and the same effect observed. The need to understand the difference between effective dose and toxicity is illustrated by the example of opium. The use of opium as an analgesic was well known in the nineteenth century—as were its addictive properties [1]. Such were the problems associated with morphine that the diacetylated derivative was synthesised by Bayer in 1898 and marketed as a non-addictive cough suppressant [2]. This compound later became known as heroin. Toxicity and side effects were an important consideration when considering biological activity (for example, opioids and quinine) but it was also important in the reproducible isolation of dyestuffs and of oils and fats (to give just two examples). In the case of oils and fats, a major driver was the synthesis of soaps using a process called saponification. Mixing animal or vegetable fat with ashes from a fire to make soap was one of the earliest industrial processes [3] and is referred to in a medieval transcription of alchemical methods [4]. In addition, oils and fats became increasingly important for lubrication as the industrial revolution spread and machines replaced manual labour and horse power [5]. More recently, the development of surfactants rather than simple soaps has led parts of the oil and fat industry to look for higher value products [6] and at biosurfactants [7]. The use of specific oils and fats as a nutritional source has led to the development of new ways of measuring the biological effects of nutritional lipids such as long chain polyunsaturated fatty acids (LC-PUFAs) including in neurobiology [8], development [9] and in a range of human health outcomes [10]. Before the development of interest in the nutritional properties of lipids and their classification as 'secondary natural products', at the end of the nineteenth century and into the twentieth century, the reliance of industry on whale oil, in particular led to a significant loss of species and to the eventual banning of whale hunting [11]. More recently, the pressure on fish stocks (a major source of LC-PUFAs) [12] has increased interest in finding alternative ways of producing specific LC-PUFAs including the transfer of heterologous genes for the production of LC-PUFAs into oilseed crops [13]. In order to transfer LC-PUFA production into a suitable alternative producing species, knowledge of the biosynthetic pathway is a prerequisite. In the case of dyestuffs and their constituent colours, the relative scarcity of natural sources for some colours resulted in some hues commanding a premium price. This was perhaps most clearly seen in the price paid for Tyrian purple (only obtained from part of a type of sea snail—one gram of dye requires 10,000 sea snails and in Roman times one pound of dye cost 150,000 denarii or about three pounds of gold) and the limitations regarding its use [14]. The development of synthetic approaches to colours (beginning with Perkin's accidental synthesis of mauveine [15]) and continuing with other dyes and colours [16]. While synthetic dyes transformed the textile industry and new colours were rapidly adopted, the impact upon art and painting was no less profound. About half the colours in the palette of a typical impressionist painter was composed of synthetic materials [17]. Because of the success of synthetic chemistry (and the large profits that could be made) and the complexity of the biosynthetic pathways, biological (particularly gene manipulation) approaches to the synthesis of colours were slow to develop. Nonetheless, a focus on sustainability has encouraged scientists to move away from natural and synthetic sources and to investigate gene transfer technology. In the case of natural colours, the pigment is typically only a small part of the whole organism that the colour is extracted from and therefore there is considerable wastage of organic material which, in addition to being an inefficient process, must be disposed of later [18]. Synthesis of indigo—which replaced extraction of the natural product—requires the use of petrochemicals and harsh chemicals that must be subsequently disposed of [19]. This has encouraged researchers to investigate the possibility of using fermentation as a method for indigo production [20]. The use of biotechnology to produce valuable products on a scale that is economically viable requires a knowledge both of the biosynthetic pathway and of the market forces which will determine the production cost that must be met and will be responsible for the creation of a commercially significant demand. The best science may, in isolation, not be sufficiently compelling if the cost and scale of production is not competitive. This was illustrated by the development, initial success and eventual withdrawal of genetically modified tomato paste. Tomatoes are a staple ingredient of much cooking from Italy. In the case of Bolognese sauce (Ragù alla Bolognese), however, the more traditional approach has either a small amount of tomato paste or, indeed, no tomato at all. The term ragù probably derives from the same root as the French word 'ragout' which is a meat and vegetable stew cooked for a long time. The first written recipe that describes a Bolognese sauce was in Pellegrino Artusi's book, The Science of Cooking and the Art of Eating Well which was first published in 1891. The collaboration between cooking, science and art was, therefore, established in Italy (at least) in the nineteenth century. Probably the most important part of cooking a Bolognese sauce is to make sure you have enough time to cook it slowly [1, 2]. This is not a dish to be thrown together in 20 min.

Take two tablespoons of olive oil and heat in a pan (or casserole). Add a chopped onion. Stir and cook slowly until translucent. Add half a cup of pancetta cubetti (or just chopped bacon), a celery stick (chopped) and a chopped carrot. At this stage, if you want a fairly thick sauce, add a tablespoon of flour. I think it tends to make the flavour a bit muddy so I do not bother. Traditionally aromatic herbs are not added but a pinch of dried oregano can add a bit more flavour. The question of whether to add garlic is a matter of personal preference. If desired, crush and chop two cloves and add to the mixture. Add minced beef (500 g) and stir to break it up. Brown gently for 5 min. Add a tablespoon of tomato purée or paste and stir for 2 min. Add a glass of dry, white wine (red wine can be a bit overpowering and, due to the polyphenolics in it, can impart a bit of an astringent, tannic taste). Now the question is whether to add tomatoes or not. For a more 'authentic' taste leave them out but they give a more rounded flavour so, if desired, add a tin of chopped

tomatoes. Season with salt and pepper. Cook over a low heat for 90 min. Prepare your pasta of choice. Cook the pasta, drain and serve with sauce. If you wish, sprinkle with freshly grated Parmesan (I normally omit the cheese on the grounds that there is enough flavour going on already).

The introduction of genetically-modified tomato paste into the UK market was one of the first successes, then failures, of commercial biotechnology. It demonstrated both the potential and the adoption challenges faced by the application of science to a consumer-facing product.

The underlying principle and the desired outcome of genetic modification to improve tomatoes was conceptually quite simple. The ripening process for tomatoes is quite rapid and, if the fruit is harvested when ready, the shelf life is short and the fruit softens so there is considerable wastage. As a result, tomatoes were harvested when still unripe (green) and ripening stimulated by addition of ethylene [21]. This is, however, an inexact process [22] and, as green tomatoes cannot be processed, variability in the degree of ripeness is problematical. As tomatoes ripen, an enzyme (polygalacturonase) is produced. This enzyme digests pectin and is the major factor responsible for softening the fruit (Fig. 5.1) and for other changes in the properties of the fruit (see below). Although pectin is composed predominantly of galacturonic acid units, other sugars and other linkages can also occur.

Manipulation of the gene that codes for this enzyme was carried out in the 80s [23]. Changes in metabolism that occur specifically during ripening and involve polygalacturonase expression are responsible for the development of taste, colour and texture of tomatoes [24]. The genetically modified tomato was created using antisense technology to reduce the level of polygalacturonase produced [25]. While the use of antisense technology attracted much comment (much of it unencumbered by understanding or knowledge) an alternative approach using gene editing has been shown to also be effective [26]. It is unclear if this type of technology will be more acceptable to regulators and consumers (see below). While some initial success was noted (the genetically modified tomato product had a longer shelf life), it was still too soft for machine harvesting. Nonetheless by creating hybrids using conventional breeding, a commercial product with improved flavour and shelf life was created and commercialised in the USA. In the UK, a genetically modified tomato puree was produced and cost about 20% less than puree made from conventionally grown tomatoes. A number of adverse reports including considerable manipulation of public opinion, led to the withdrawal of all genetically modified tomato products in the UK. Despite this, researchers have continued to use tomatoes as a vector for the introduction of engineered characteristics. One of the most publicised of these has been a tomato with enhanced levels of anthocyanins—the genes were obtained from *Antirrhinum* (snapdragon). The resulting purple tomatoes have been marketed as being healthier than conventional unmodified tomatoes [27]. There are, however, two significant obstacles to be overcome. The first is the generic objection by consumer and other groups to genetic modification as being a safe technique. This is exacerbated by the fact that if new genes have been introduced from an exogenous source (heterologous genetic modification), as is the case with purple tomatoes, it is difficult to argue that the genetic modification has simply accelerated a natural plant breeding production

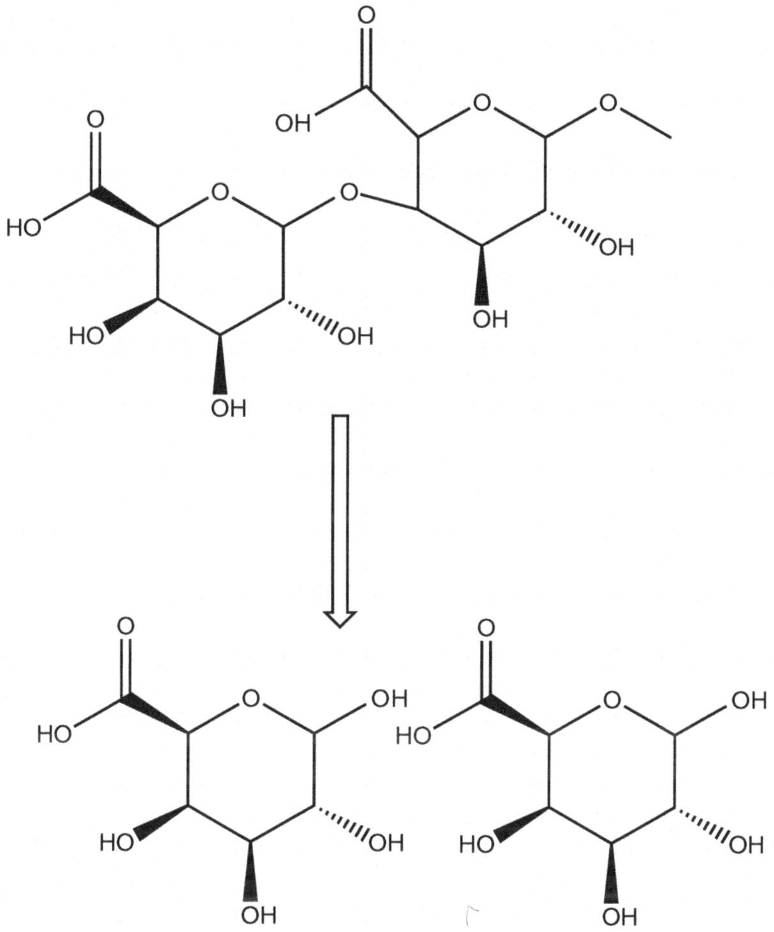

Fig. 5.1 The cleaved structure of pectin

process, as has been suggested in the case of gene editing [28]. The second obstacle is the rationale for the introduction of a product with enhanced levels of anthocyanins into the diet. There are a number of wholly natural sources of anthocyanins [29] and, while the health promoting qualities of anthocyanins have been studied, many of these investigations have been carried out on animal models (particularly rodents) and their relevance to long term exposure in humans remains unestablished and further studies are needed [30]. In addition, as noted previously (Chap. 4) the potential synergistic effects of different secondary natural products remains relatively unexplored and anthocyanins occur in a range of fruits and vegetables alongside other secondary natural products so the logic of genetic modification to introduce specific members of the class into tomatoes is not completely clear. There is always a danger that scientists carry out a set of experiments or a programme of work to demonstrate that they have the capability rather than to meet a societal, health or consumer need.

5.2 Classes of Biosynthetic Pathways

Empirical formulas which were developed in the eighteenth and nineteenth centuries as measurement techniques improved, only offered a degree of structural information—the relative proportions of the different elements that were present in a compound. While advanced isolative and purification techniques (including chromatography) were later developments, nonetheless, natural product chemists sought to find similarities between those isolated compounds that they were able to purify and to develop ways of finding out about similarities and differences (including, in particular, those related to their biological properties). This was a reasonable extension of the work on physicochemical similarities between elements that underpinned the Periodic Table. The success of methods for the production of neutral salts from mixtures of acids when combined with bases or with metals encouraged chemists to think in terms of similar attributes when dealing with organic natural products. This type of paradigm, based on similarities, (and, indeed underpinning many of the rules that have been developed) was rooted in the realisation that secondary natural products are biologically nonessential chemicals [31] So, at one extreme, there are physicochemical rules-based concepts that were developed by chemists to deal with how the (largely inorganic) elements of nature exist and combine and, on the other, there are guidelines that suggest that certain organic structures derived from living systems are likely to bear some relationship to one another based upon similarities in biological function. There is, then, measurable certainty co-existing with more organic fluidity. This rationalisation was not necessarily appreciated by non-chemists who may have had a more encompassing world view than chemists. For example, Kant sought to divide knowledge into sub domains—each with their own particular attributes and contributions (and, by implication with some overriding similarities). He accepted that chemistry was the science that explains the conversion of components into new substances and this fitted with his continualist view of the universe however he seemed to suggest this conversion required the potential for the division of matter an infinite number of times and this was contrary to atomist theories [32]. In addition, Kant's inclination towards a division of knowledge led him to develop the idea of a hierarchy of the scientific domains with physics occupying the premier position since it was, in his view, both a proper and a rational discipline. For Kant, this means that physics has laws which can be derived without their consequences having to be observed—they are intrinsic to nature since, for nature to exist at all, these laws must also exist. In the case of chemistry, the laws are not intrinsic (but they do exist) and are discoverable through observation and empirical testing. The fact that empirical testing is necessary means that chemistry cannot be reduced (or elevated) to physics. Empiricism is an essential aspect of chemistry and the laws of chemistry cannot be derived (unlike those of physics) by pure thought alone—they are not able to be considered purely in mathematical terms. One implication of this way of considering chemistry is that each empirical observation must be experienced by each practitioner. In the words of the motto of the Royal Society—*Nullius in Verba* [33]. This must be modified, however otherwise chemistry will be an ever-reinvented tabula rasa and progress will be severely limited. In order to avoid this, the empiricism that is the defining feature of chemistry must lead to empirically derived and scientifically accepted laws and these, then form the baseline from which further knowledge can be hypothesised and empirically derived. This codification means that continual reinvention and repopulation of the tabula rasa is not necessary. Knowledge is shared and accepted and used as a foundation for subsequent empirical observations and tests. There is, of course, a possibility that a law derived from empirical observation or extrapolation may be based on faulty data or on a misinterpretation. This means that laws in chemistry are not fixed points but rather are accommodations that are representative of the best data available at a point in time and have been subjected to an interpretation based on knowledge and experience. This rationale underpins the investigation of the biogenesis of natural products.

The biosynthesis of secondary natural products can be considered in terms of a series of interlinked schema including, in some cases, links to primary metabolism. The definition of the sub-classes of natural products can be considered from a number of different perspectives.

- Common structural characteristics implying pathway homogeneity and core molecular derivations.
- Biological derivation with features in common suggesting unified evolutionary pathways
- Biological activity either from the standpoint of the producing organism or the consuming organism.

Structural characteristics may be indicative of both physicochemical and biological properties. For example, organic acids can be separated according to their chemical property of differential solubility in basic solutions. In addition, acids are at the endpoint of oxidation and so are relatively chemically and biologically stable when isolated, purified and concentrated. Indeed, organic acids were among the first class of compounds that were (partially) characterised. In terms of biological activity, organic acids also have broadly similar taste characteristics to one another which suggests their interactions with taste receptors and activation of signal transduction pathways have a degree of equivalence. They also often demonstrate antimicrobial activity and, in more recent studies, have been found to be inhibitory towards biofilm formation [34] and this provides a rationale for their traditional use in the preservation of foods. The microbial growth inhibitory properties of organic acids—in particular those produced by other microorganisms—will give a growth advantage to the producing organism and inhibit pathogen growth. In addition, the production of organic acids by particular plants when under stress will help to prevent spoilage and loss of biological integrity as a result of invasion by microbial species [35]. In that way, the purposeful production of organic acids is a useful way of preventing microbial spoilage [36]. In addition, several pathogenic species are acid sensitive and their growth is inhibited in the presence of organic acids. The breadth of knowledge concerning organic acids was, however, largely driven by the ease with which they could be formed and isolated. Vegetable acids (and their salts) including citric, acetic, oxalic and tartaric acids were known in the eighteenth century however their biosynthesis remained unclear. More modern investigations have suggested that the key to understanding the mechanisms that underlie the biosynthesis and bioaccumulation of organic acids in plants lies in their link to primary metabolic pathways (pyruvate, citrate) and their role in the regulation of pH balance (malate) and in energy metabolism [37].

In the case of alkaloids, the driver for their isolation was the combination of chemical and biological properties. This was a consequence of the relative simplicity of their purification as a class (and sometimes, after crystallisation, as individual compounds) and their biological activity. This latter property was difficult to control when using crude extracts with varying amounts of active agent (and therefore with varying potencies) and this provided an important rationale for the isolation of the primary active principles from crude extracts.

The first approaches to understanding biosynthesis of secondary natural products (and how molecules might be related) was a consequence of their functionality (for example, the presence of acidic or basic groups) before there was even a clear idea of what those properties represented in molecular terms. This allowed separation techniques that were little different to those developed during centuries of alchemical investigations to be adopted. Indeed, many of the basic principles remained the same even if the nomenclature changed. For example, in 1700 chemists used the term vitriol which was later changed to sulphuric acid. While the chemical identity remained the same, the use to which the substance was put and how its properties were defined changed and this was reflected in the name. Vitriol refers to compounds derived from the quenching of sulphuric acid with metals or to sulphuric acid (oil of vitriol) itself. The type of vitriol is defined by its colour which, in turn, is dependent upon the metal used to quench the acid. For example blue vitriol is cupric sulphate and white vitriol is zinc sulphate. The use of the term sulphuric acid defines the substance as one member of a wider class of compounds which are linked by a specific molecular group—an acid functionality—rather than being defined as vitriols by their genesis from sulphuric acid. A similar change in how substances are defined applies to alkaloids. For example opioids were originally derived from the opium poppy. They were defined on the basis of their biological action and derivation from opium—in a similar way that various vitriols were defined in terms of their derivation from sulphuric acid. As an awareness of chemical functionality grew, opioids were defined as members of a wider class of substances whose similarities were derived from a chemical functionality responsible for a specific property—basicity. This property led to these substances being defined as 'alkali-like'—in other words, alkaloids.

The development of knowledge about specific biosynthetic pathways can be built up in different ways. The increasing awareness of chemical functionality-based classification enables core structural features to be identified—as was the case for organic acids and alkaloids. This approach develops a classification paradigm that begins with a distinct physicochemical structural feature (an acid or an ionisable N atom) which then defines a class. This is, however, of limited biosynthetic utility since the classification is very broad and different members of the class may not even be biosynthetically related. In order to reveal closer relationships, it is necessary to probe structures by classical analytical (degradative) techniques rather than physical functional properties. Isolation techniques—even into the twentieth century, relied on a relatively small number of nonchromatographic techniques such as precipitation, crystallisation, elemental analysis to give empirical formulas and colourimetry. Chemical degradation gave the core chemical structure and, in the case of isolable compounds such as alkaloids, the existence of a core structure enabled classification to be carried out. In this way the biogenesis of molecules could be speculated upon before methods were available to confirm proof of biosynthesis. Ultimately proof of structure, which was generally a necessary prerequisite to proof of biogenesis was accomplished by synthesis rather than by degradation although the isolation of biosynthetic or degradatively-derived intermediates may provide useful pathway indications which were not previously apparent even before a formal and full structural assignment can be completed. This can be encapsulated in a thought experiment and in a practical example. If a particular compound is synthesised then its subsequent reactions can be investigated. Before the compound is synthesised, the starting materials may combine in a host of ways to give a range of products which may then combine further. Once (under a specific set of conditions) a particular pathway is followed then a specific outcome (the synthesised compound is formed) and the options for further reactions are clarified (and reduced as well as being quantified). A similar rationale underpins the production of sulphur compounds by *Allium* species—which will be examined in more detail in Chap. 9. Starting with a small number of precursors, different combinations can create different groups and classes of compounds. In the case of terpene indole alkaloids, the pathway from tryptophan can be envisioned to take a number of routes. The biogenesis is, therefore, a matter for discussion and debate.

Once the intermediacy of secologanin was established (Fig. 5.2) then the options for the biosynthetic pathway were reduced. Interestingly secologanin itself is derived from geranyl pyrophosphate—a component of the mevalonate pathway. As the knowledge builds of structures and reaction that are not simply plausible but are actually found in nature then the hypotheses regarding biogenesis becomes more knowledge-based and less prone to individual preference. This has also led to the development of AI tools to optimise suggestions regarding likely biosynthetic pathways [38] through the analysis of data from known pathways and connections. Synthetic chemists have taken a different (although conceptually similar) approach—retrosynthesis to devise synthetic strategies (perhaps most well known is the 'Disconnection Approach' which involves breaking a molecule into synthetically feasible units and then combining this with functional group interconversions). By using AI coupled to databases of known biological pathways and interconversions, retro-biosynthesis can be carried out. The biological as well as the chemical plausibility of reactions can be probed and 'missing' intermediates postulated. Such methods are clearly limited by the constraints of the knowledge base.

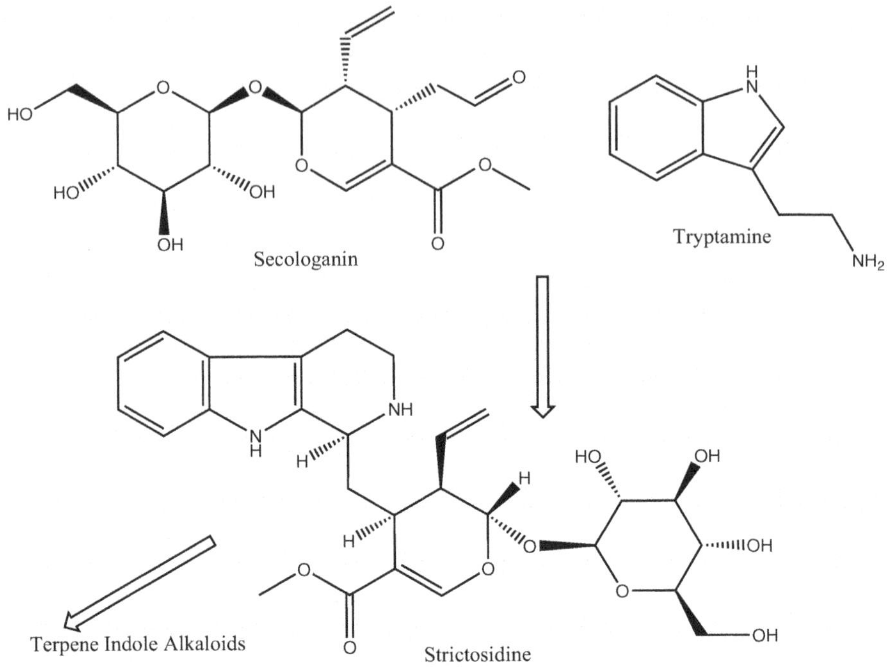

Fig. 5.2 The intermediacy of secologanin in biosynthesis

For example, until relatively recently, it was thought that enzymes that were able to calayse pericyclic Diels Alder reactions (Fig. 5.3) did not exist in nature.

The discovery of reactions—both conceptually and with specific examples—that support the existence of Diels-Alderases, has changed the theoretical landscape for retrobiosynthesis to investigate biogenesis [39]. Similarly a growing appreciation of the reactions that can be catalysed by enzymes operating in vivo rather than as isolated entities may change the view of natural product chemists regarding biosynthesis. For example, cytochrome P450 enzymes are well known to catalyse the hydroxylation of a range of secondary natural products [40] and it has been realised that they can operate using a range of substrates [41] and this may alter retrobiosynthetic strategies.

5.3 Biosynthetic Pathways—From Molecules to Genes

The realisation that secondary natural products shared some levels of structural similarity and may plausibly be connected through biogenesis, has led to significant amounts of speculation regarding potential biosynthetic pathways. Such speculation was possible (as mentioned earlier) even when complete structural assignment had not been carried out. Identification of core structures after degradation was one part

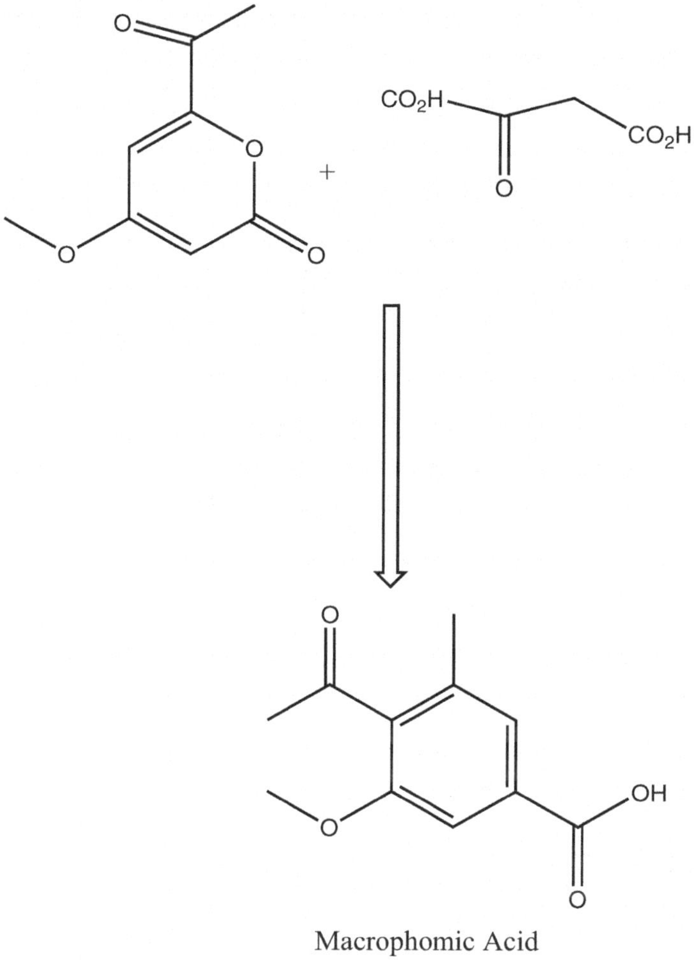

Fig. 5.3 The Diels Alderase reaction in nature

of this exercise while the identification of the attachment of certain functional groups also helped to suggest pathways. The use of reagents that reacted to specific groups with colour changes enabled chemists to identify particular molecular characteristics and this was a significant advance in the identification of functional groups. The development of methods for functional group analysis was a logical reductionist progression from the derivation of empirical formulas. Functional group analysis must be specific and sensitive. It is additionally useful if the environment of different functional groups can be determined. As was noted in Chap. 4, some spectroscopic techniques are particularly useful for the identification of certain molecular characteristics.

While the identification of specific functional groups remains an important aspect of chemistry and can provide a specific and rapid way of identifying classes of compounds in a mixture after separation by, for example, chromatography, the use of qualitative chemical analysis of functional groups has diminished in recent years as analytical instrumental methods have become more sophisticated. Nonetheless the use of reagents can signify the presence of, among others, phenols, flavonoids, reducing sugars, lipids and phospholipids, amines and indoles, aldehydes and ketones, alkaloids, amino acids, steroids, diols and glycosides [42]. These reagents are particularly useful for following the progress of reactions by thin layer chromatography—particularly if one functional moiety is being converted into another.

From a biosynthetic standpoint, as noted previously, being able to recognise classes of compounds is of particular value when building up a biogenesis hypothesis. Once an empirical formula has been established, successive removal of functional groups and an awareness of the structures of other molecules in the class, can enable biosynthetic links to be established. This then enables the natural product chemist to speculate about the type of reaction (and the enzyme) that is necessary for a particular transformation to be carried out. By comparison with other enzymes with similar characteristics, protocols for the isolation of an active enzyme can be developed [43]. In conceptual terms, the study of biosynthesis can be seen as a series of actions taken to reduce complexity by probing molecular structures, confirming the analytical supposition by synthesis, identification of intermediates or initial substrates in the pathway by inclusion of labelled compounds (radio or stable isotopically labelled) reduction of the biosynthetic pathway to a series of enzyme-catalysed steps followed by isolation of the enzymes and confirmation of the reaction carried out and its mechanism (although mechanistic studies are not always necessary). This then completes the cycle for the natural product chemist. By completing this sequence successfully, they will have determined the structure of a molecule, developed a knowledge of the starting materials (and some of the intermediates) in the biogenesis, created a plausible hypothesis regarding the biogenesis, confirmed the suggested biogenesis by carrying out biosynthetic studies and probed the biological system that is responsible. Two specific aspects of the process should be examined in greater detail and with a higher level of granularity—establishment of the pathway for the biogenesis of a specific secondary natural product and probing the biological mechanism which underlies the pathway including, in particular, enzyme isolation and characterisation.

The development of radio and stable isotope chemistry led to significant advances in the techniques available for the study of biosynthetic pathways. They helped to answer the question—what is the biological sequence of reactions that lead to the formation of a particular secondary natural product? What are the intermediates in that sequence? While the use of labelled compounds and their specific incorporation can provide significant information, methods to examine partial pathways can also be very useful. One way of examining partial pathways is through the use of blocked mutants [44]. A blocked mutant is an organism that does not produce the final secondary natural product but accumulates an intermediate. They can occur spontaneously in a population or can be produced by irradiation then the required clones selected [45]. An alternative method is to use chemically induced mutation [46]. Developments in molecular biology techniques have enabled mutations to specific genes to be carried out and this, combined with gene editing (CRISPR) has enabled biosynthetic pathways to be manipulated more precisely (see Sect. 5.4). One challenge to the use of mutants is the establishment of shunt pathways. These are pathways that result from the accumulation of a particular biosynthetic metabolite and convert the accumulated metabolite into another compound which is not on the actual biosynthetic pathway but is, rather, a metabolic dead end. This can be investigated by taking the accumulated metabolite and incubating it with a biosynthetically fully competent organism and looking for the disappearance of the accumulated metabolite and the increased synthesis of the secondary natural product. If neither occurs then it is a reasonable supposition that the accumulated compound is a shunt metabolite and is not an intermediate in the main biosynthetic pathway. As noted previously, examination of a complete biosynthetic pathway can be carried out by using labelled starting materials or intermediates. Labelled intermediates generally participate in enzyme-catalysed biosynthetic reactions to the same extent as the unlabelled analogues, although differences in the rates of enzyme-catalysed reactions due to isotope effects has been noted and is used as the basis for some types of authenticity confirmation (see Chap. 11). The most significant example of this occurs with the replacement of hydrogen in a substrate or intermediate with deuterium. In some reactions, the rate of the reaction is slower in the case of the deuterated molecule [47] and this may affect the flux through a pathway. One example of the use of a radiolabelled investigation protocol is the application of 14-C acetate in the investigation of fatty acid biosynthesis [48]. This method, called pulse chase labelling, involves incubating multiple samples of the biological system under investigation with 14-C acetate and then further incubating the samples for variable times and isolating lipids and related species at different time points. These are then separated by thin layer chromatography and the level of 14-C labelling in different metabolites is measured. A range of similar experiments can be carried out using different radioisotopes. In an analogous fashion, pulse chase experiments can be carried out using stable isotopically labelled (generally 13-C) starting materials that are ubiquitously incorporated into a number of products including secondary natural products and this incorporation followed by nuclear magnetic resonance (NMR) spectroscopy and mass spectrometry (MS). For example, both 13-C labelled glucose and carbon dioxide have been used to study biosynthetic pathways in plants using pulse chase techniques [49]. The relatively recent use of 14-C labelling in pulse chase experiments to examine lipid metabolism can be seen in the context of the use of the stable isotope, deuterium for the study of the metabolism of lipids [50]. This was the first published use of stable isotopes to investigate metabolic pathways. The use of deuterium as a specifically identifiable tracer is a reflection of the availability of this isotope of hydrogen and its specific incorporation into putative precursors and intermediates. The development of methods using 13-C would not be possible until a ready supply of 13-C became available using cryogenic distillation. However once this became possible, a whole new era of biosynthetic chemistry became feasible and the use of degradation as a way of examining biosynthetic pathways became redundant [51]. Frequent reviews of the use of stable isotopes—in conjunction with NMR and MS have illustrated the range of techniques and the increase in knowledge that has accrued [52]. While there have been significant advances in understanding, it is also the case that stable isotope labelling is a more significant confirmatory technique rather than a method for de novo investigations. More recently, the link between biosynthetic gene clusters (BGCs) and secondary natural products has been investigated using stable isotope labelling in conjunction with MS analysis [53]. In essence, the approach combines stable isotope labelling patterns of secondary natural products with BGC structure prediction in order to link the natural product with its BGC. This approach to integrating genomic and metabolomic data can confirm the biogenesis of known metabolites and, additionally, can facilitate the prediction of the occurrence of novel molecules from BGCs. In short, the biosynthesis of most classes of natural products are amenable to examination using stable isotopes in combination with NMR spectroscopy or MS. Although it lies out with the scope of this volume, the increasing use of stable isotopes to probe the interaction between secondary natural products and biological macromolecules has been noted [54]. One area of novel interest is the effect of deuterium labelled water on the activation of a sweet taste receptor [55]. An effect on binding to the histamine receptor has also been noted as a consequence of deuteration [56] and isotope effects in pharmacokinetics have been postulated [57]. Given that isotope labelling, as noted previously, can affect the rate of catalysed reactions then an effect of isotopic labelling on signal transduction pathways (including taste) or on molecular vibrations might be expected to operate although other examples will be needed to determine the mechanism of any effect. Stable isotope dilution analysis is a technique used in the determination of the effect of flavour compounds [58]. The idea that the presence of different isotopes in biological systems has been termed the 'isotopic labelling hypothesis' and it has been claimed to have been experimentally verified [59]. If this hypothesis can be verified and quantified and the effect can be shown to be a consequence of an impact on the kinetic properties of reactions in vivo then the use of stable isotope labels may not be the wholly non-invasive method that is claimed.

5.4 Manipulation of Biosynthesis—Molecules and Mutation

Making use of biological machinery to synthesise molecules has been a preoccupation for many years. There are three methods that are commonly adopted.

- Semi synthesis in which a key molecule is derived from a biological system and then modified to give an unnatural analogue which will usually have different properties to that of the biological analogue.
- Addition of novel substrates to a biological system. These are incorporated into the biosynthetic pathway and a novel product results.

- Modification of the biosynthetic pathway by addition, removal or altering (mutation or gene editing) specific genes to change the biosynthetic pathway.

While the concept of semi-synthesis of small molecules has been extensively adopted with varying degrees of success, the semisynthesis of larger molecules has proven to also have significant potential because of the challenges associated with pure synthesis, on the one hand, and pure biologically-driven approaches on the other. The example of protein synthesis will illustrate both the challenges and the opportunities. While the purely chemical synthesis of proteins, even if semi-automated, requires the use of protection group strategies and these can be complex, this may be the best option. If a purely biological approach is taken, there can be significant problems with the expression of heterologous proteins in host species due to protein misfolding and aggregation. In addition, the introduction of unnatural amino acids (or other moieties) can be carried out using elegant synthetic strategies rather than biological methods. The total synthesis of proteins has been the subject of frequent reviews [60] and methods such as automated and solid phase synthesis and peptide ligation strategies have been advocated. The semi-enzymic synthesis of proteins (including those containing non-genetically coded elements) is a particularly attractive area for those working at the interface of chemistry and biology [61]. The importance of both biological and chemical methods in the achievement of precision engineering and production of proteins and protein analogues has been acknowledged.

Semisynthesis of smaller secondary natural products is often through the application of older technologies. Some examples will be given to illustrate the principles involved with different classes of molecules. Unlike primary metabolism, the enzymes involved in secondary natural product biosynthesis can often make use of a range of substrates and this creates an opportunity for novel product formation. An alternative is the biosynthesis of a core molecule using a biological system and this is then isolated and chemically modified. A further iteration of semisynthetic methodology makes use of the chiral nature of many enzyme-catalysed reactions to prepare a particular enantiomer [62]. There are numerous examples of the use of semi-synthetic methodologies to prepare novel analogues of natural products. This includes anti-bacterial dialkyl resourcinols [63] and a range of Brefeldin A derivatives with antileukemic properties [64]. Advances in semi-synthetic methodology have contributed to a wider range of potential natural product derivatives being available [65]. One of the most significant areas of interest has been the semi-synthesis of penicillin and cephalosporin derivatives [66]. One of the most important intermediates in the semi-synthesis of penicillins is 6-amino penicillanic acid which is produced by removing the side chain of penicillin G or V with an enzyme, penicillin acylase (Fig. 5.4).

Similarly, semi-synthetic cephalosporin derivatives are often produced from 7-aminodeacetoxycephalosporanic acid (Fig. 5.5).

In addition to the production of specific molecules, the yields of bioactive precursors produced by microbial strains can be enhanced both by optimisation of the fermentation conditions and mutagenesis followed by screening of mutants. Strain and process improvements have also led to significant advances in yield [67]. Another

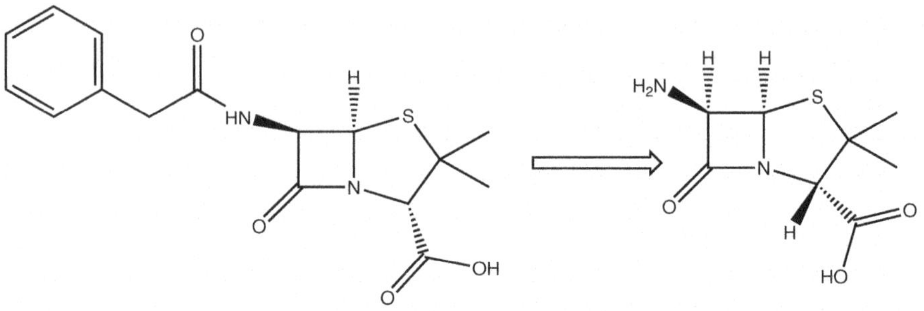

Fig. 5.4 Synthesis of 6-amino penicillanic acid

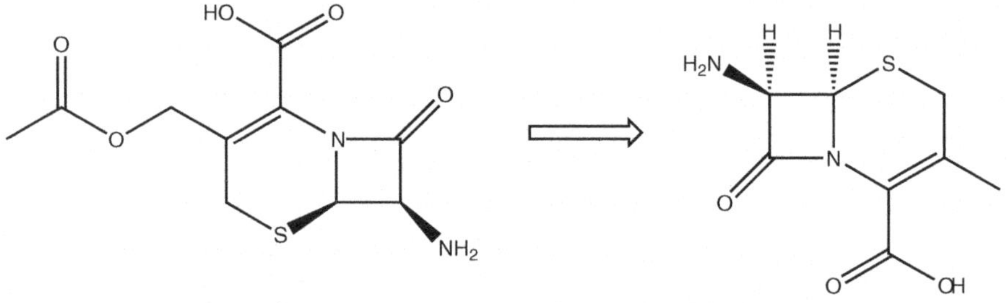

Fig. 5.5 Structure of 7-aminodeacetoxycephalosporanic acid

example of the use of semi-synthetic methods is the development of approaches to the anti-cancer drug, Paclitaxel (taxol). Taxol is isolated from the bark of the yew tree—unfortunately it is a slow growing source and taxol itself is present in gram quantities from tens of kg of bark. Taxol has a complex structure and synthesis de novo is not a viable option. It has been prepared in a semi-synthesis from a fairly late (but more biologically available) precursor, 10-deacetylbaccatin III [68] (Fig. 5.6).

Other approaches have been suggested based upon the isolation and modification of earlier biosynthetic precursors. Unlike penicillins and cephalosporins where a major driver for the semi-synthetic approach has been the synthesis of unnatural products, in the case of taxol, the semi-synthetic approach is designed primarily to produce the natural product itself. This has also encouraged investigations into modification of the biosynthesis and biotechnological solutions [69] including novel and potentially more amenable sources of the taxane core structure [70]. Semi-synthesis is relatively rare in the food sector since the value of many food components is insufficient to justify the use of advanced biotechnology. An exception is found in the synthesis of nature identical flavours and fragrances [71]. As noted previously, natural flavours and fragrances can command a premium price and synthesis using enzymes can mimic the biological system and give nature-identical compounds [72].

The use of semi-synthesis in enantiospecific synthetic strategies has already been mentioned and it has been adopted as a tool by bio-organic chemistry. While naturally occurring enzymes have been used, they can be limited in their use due to

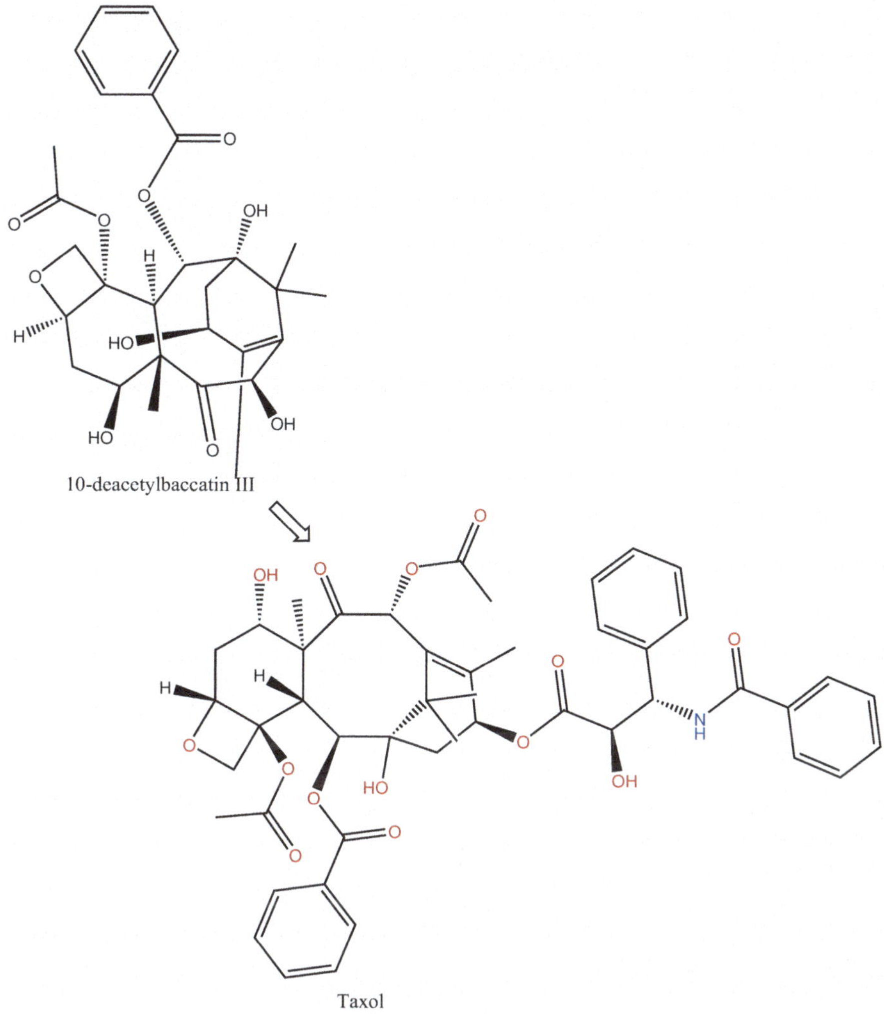

Fig. 5.6 Derivation of taxol from 10-deacetylbaccatin III

substrate specificity and therefore modification of naturally occurring enzymes has been carried out [73]. One of the major drivers for the use of biotransformations rather than synthetic chemistry is the preparation of enantiomerically pure bioactive agents for therapeutic use [74]. In the area of secondary natural products, a number of enantiospecifically pure compounds have been prepared including (S)-(+) linalool [75] and, as noted generically, previously, a range of flavour and fragrance compounds [76].

The introduction of novel (or even unnatural) substrates into biosynthetic pathways can result in the formation of unconventional molecules. This is in part a consequence of the wide substrate acceptance of the enzymes involved in secondary natural

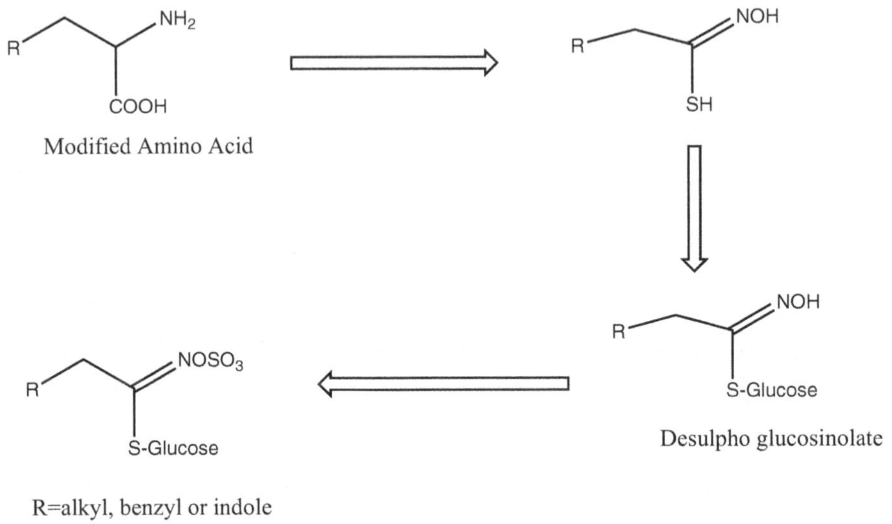

Fig. 5.7 Phases in the biosynthesis of glucosinolates

product synthesis and the driver to produce a range of products with different properties to overcome the resistance mechanisms of consuming or competing species. For example, the biosynthesis of glucosinolates involves primarily three phases—amino acid chain elongation (aliphatic (from alanine, valine, leucine, isoleucine, or methionine), benzenic (from phenylalanine or tyrosine) or indolic (from tryptophan)), synthesis of the core glucosinolate moiety and, finally, modification of the core structure (Fig. 5.7).

From this relatively simple sequence, the large (more than 130) and diverse structures of different glucosinolates are derived [77]. Part of the reason for this structural diversity is that glucosinolates (and their breakdown products) are not solely involved in the defence of a plant against predation, they are also part of the stress response of plants [78] and may be associated with the development of a protective leaf microbiome [79]. Another example of the capacity of biosynthetic enzymes to make use of unusual substrates occurs in *Penicillium* species. Synthesis of the δ-(l-α-aminoadipyl)-lcysteinyl-d-valine (ACV) tripeptide is the first step in the biosynthesis of penicillin and will accept a range of natural and unnatural substrates substrates [80]. Since it is the first reaction in the biosynthesis, the synthase enzyme is acting as a 'gatekeeper' for compounds that are able to enter the biosynthetic scheme. One of the main reasons for investigating unnatural substrates is to probe the mechanism of action of enzymes in the biosynthetic pathway [81]. A range of unnatural substrates have been found to be acceptable to isopenicillin-N-synthase [82] and this has been the basis of much investigation. The enzymes involved in the biosynthesis of penicillins and cephalosporins have been immobilised and have been used to synthesise a range of unnatural products [83].

The third opportunity for manipulation of a biosynthetic pathway to give novel products is the manipulation of the BGC (as can be carried out with, for example, modular pathways such as those of the polyketide pathway) or alteration of individual enzymes by mutation or gene editing. Polyketides are the most abundant secondary natural products. They are synthesised by modular enzyme clusters called polyketide synthases (PKS). The nature of a given polyketide is governed by the number and characteristics of the modules and the order in which they are assembled. The genes for polyketide biosynthesis are generally collected together in BGCs. The PKS enzymes that form the BGCs are modular and are responsible for the formation of the core carbon backbone of the polyketide. Removal or rearrangement of modules will affect the structure of the resulting polyketide. Despite the comparatively high level of understanding of the assembly of polyketides, engineering PKS enzymes to a specific end remains challenging [84] However, detailed reviews of the PKS structure and manipulation are available [85] and this detailed understanding has made the polyketide biosynthetic clusters useful targets for synthetic biology and engineering. An alternative approach is the combinatorial biosynthesis of polyketides. This approach was described as the genetic manipulation of two or more enzymes involved in polyketide biosynthesis [86]. Each multifunctional PKS contains one or more ketosynthases, acyltransferases and acyl carrier proteins in a defined order. In addition, cyclases, esterases and reductases may also be present. While engineering of whole domains has been the method of choice for many manipulative procedures, specific mutations within individual enzymes, affecting characteristics such as the binding pocket have also been noted [87]. Given the challenges of module manipulation and directed mutations for a specific purpose, the development of evolution-guided engineering with rapid screening may be a better (although intellectually less satisfying) approach [88] although analysis revealed the presence of a common amino acid motif for hybrid PKS formation.

5.5 Gene Editing—Scientific, Regulatory and Adoption Challenges

As noted previously, the route to commercial exploitation of genetic modification technologies has been somewhat laboured. While the science has been well founded, the societal impacts have been less easy to deal with. While the consequences of the use of genetic modification has been extensively tested in terms of safety, the spread of disinformation regarding ‘Frankenstein food’ has been extensive. In part this is a function of the lack of connectiveness to science experienced, in particular, in older sections of the population and those without a degree-level qualification. Public attitudes to different areas of science have been found to vary [89]. While acceptance of AI (particularly in areas such as health care) seems fairly high, there remains resistance to the use of genetic technologies to improve yield or the appearance of crops. The questions regarding the application of genetic modification (and even gene editing) remain and are likely to be recycled as each new wave of technology (or products) arise [90]. In such an environment, it is perhaps counter-productive for scientists to concentrate solely on solving the scientific problems associated with genetic modification rather that properly engaging with the ultimate consumers of that technology. One example of this is the development of genetically modified tomatoes with enhanced levels of anthocyanins [91]. One of the challenges is that anthocyanins are found in many edible fruits and vegetables [92] and the application of animal studies to humans and therefore the building of the evidence base remains open to question and interpretation [30]. There is not an overwhelmingly compelling narrative to suggest that the genetically modified product has beneficial properties and therefore the risk:benefit analysis, from the standpoint of the consumer, is unbalanced.

The attitudes of different populations to genetic modification have been reviewed [93] and the review suggests there are regional differences. In general, adoption of genetically modified food meets with less resistance in the US than in Europe and this, in turn, has encouraged certain types of modification compared to others. In terms of the development of gene editing technology compared to 'conventional' genetic modification, scientists had hoped that people would be more likely to align editing technologies [94] with conventional breeding (albeit in an accelerated form) than with genetic modification. Such an aspiration was expressed in the context of widening the available genetic pool by with positive results "it is now possible to catapult neglected, semi-domesticated and wild plants quickly into the focus of mainstream agriculture" [95]. However limited studies have suggested that, while gene editing is regarded with less suspicion that heterologous gene expression in crop plants, there remains a high level of mistrust "Although the lay public tended to have more favourable attitudes toward gene editing than toward genetic modification, such differences were much smaller than the differences between conventional breeding and genetic modification" [96]. In addition, consumers in the US (in contrast to Japan and Germany) did not differentiate between genetic modification and gene editing technologies [97]. Interestingly, the public perception and knowledge of gene editing, not surprisingly, lags behind that of genetic modification however those who are more technology averse have more knowledge of gene editing than those who are neutral or technology enthusiasts [98]. This might imply that, despite the attempts by scientists to equate gene editing with conventional breeding rather than genetic modification, the technologically averse sub-group of the population are more liable to be better informed. One final important aspect of gene editing is the detection of any alteration. If legislation and labelling is to be introduced then methods for the detection and quantification of gene edits will be necessary—as they were for genetic modification [99]. One approach is to use duplex PCR based methodology [100]. In addition to detection of gene edited final products, it is also necessary to follow the progress of the incorporation of gene edits at every stage in the modified production process and therefore analysis will be an important research tool [101].

Attempts to distinguish between gene editing and genetic modification (equating the latter with a form of 'natural' plant breeding) are unlikely to be successful. While people have a high degree of trust in science and scientists, this is not limitless and the alignment of gene editing with commercial interests is likely to erode that trust. In addition, the use of gene editing techniques for purposes that are considered by the public and regulators to be relatively trivial will simply serve to increase the impression that science is being done primarily for the benefit of scientists themselves and their commercial partners. The manipulation of secondary natural products in foods and food systems may fall into this category in many instances.

5.6 Extremophiles and Novel Biosynthetic Pathways

Organisms can be found in extreme environments. This can have advantages for the adapted organism. Competition for resources is often less and the organism can make use of novel substrates. In addition, as noted previously, secondary natural products can be synthesised as part of a stress response to environmental extremes. Extremophiles are classified according to the nature of the extreme conditions they encounter. Typical examples are extremes of heat, pressure, pH, radiation, salinity, lack of water and competition for resources. On a molecular level, organisms have evolved different ways of growing in extreme environments. Perhaps unsurprisingly, organisms that are only distantly related but exist in a common extreme environment, have been found to adopt similar strategies. This is reflected in patterns of expression of genes and even in specific gene sequences. Indeed the 'genomic signature' of some extremophiles have been found to share certain characteristics [102]. This combination of an environmental combined with a taxonomic characteristic in the genomes of extremophiles suggests that the adaptive response is present at a very basic level. There are, additionally, other characteristics of adaptive responses that occur on different levels and are associated with extremophiles.

One of the common responses to stress that occurs in both eukaryotes and prokaryotes is the production of heat shock proteins (HSPs). These are generally classified according to their molecular weight. They have a number of functions in a stressed cell [103] including acting as chaperones to ensure correct protein folding, remove misfolded proteins, involvement in cell apoptosis and modulation of the immune response [104]. In addition to being a response to stress, HSPs are also associated with particular disease states. One area of particular interest is the HSP response of plants placed under combinations of abiotic stresses [105] since this may have consequences for crops grown under sub-optimal agronomic conditions including drought, increased UV irradiation, temperature fluctuations and high salinity.

Living systems are, at a reductionist level, a series of reactions catalysed by enzymes. Many studies of enzymes have concentrated on measuring their properties (for example, reaction rates, the study of activators and inhibitors etc.) under standard conditions of pH, ionic strength and temperature. However in many cases (and certainly for extremophiles, these conditions are not physiologically relevant. Enzymes that are produced by organisms that thrive in extreme environments are particularly industrially useful [106]. One of the most commercially important extremozymes is Taq polymerase. This enzyme, isolated from the bacteria *Thermus aquaticus* is the key component in the polymerase chain reaction (PCR). It is a thermostable DNA polymerase and is optimally active at 70–75 °C. This means that it can duplicate specific single stranded DNA and produce billions of copies of a specific sequence. The use of PCR has hugely increased the sensitivity of DNA detection and has led to a large number of applications that were previously not possible. While the simplest way of obtaining thermostable enzymes is to isolate them from a thermostable organism, this is not always possible and can be difficult to reproduce on an industrial scale. Structural modification of nonthermostable enzymes to increase their structural stability while retaining heir catalytic activity is an optional route [107]. A range of attributes of enzymes, including salt bridges, disulphide linkages and hydrogen bonds, for example, can enhance stability. In addition, there are examples of extremophile and other enzymes maintaining structural integrity in low water systems where an ordered hydration shell around the protein is present [108]. Some other additives (e.g. poly glutamic acid) have been found to enhance the activity of some enzymes [109] which suggests that buffered aqueous solutions at neutral pH may not be the optimal conditions for all enzymes.

One of the consequences of the growth of extremophiles is their capacity to activate novel biosynthetic pathways and thereby to synthesise unusual products [110]. Such organisms have been the focus of considerable attention—particularly as a source of microbial therapeutics [111]. While much of the interest in extremophiles has focussed on microbes, other species, including particularly plants, are known to modulate enzymic [112] and secondary natural product production in response to extreme environments and/or biotic or abiotic stress [113] and some of these adaptations have a practical use [114].

5.7 Marine Secondary Natural Products

One of the most commonly studied areas of extremophiles is the link with marine natural products. Organisms that thrive in a marine environment are generally halophiles but, in addition, they may be able to function in low light levels and under pressure. There are a large number of reviews and databases of marine natural products, many of which are updated annually [115]. Some of these are open access and searchable [116]. While biological activity-guided isolation procedures remain popular, data mining for biosynthetic pathways has greatly increased the numbers of potential products [117]. In addition, marine natural products are opening new therapeutic areas that are not known to be approachable through terrestrial natural products. Nonetheless marine natural products remain an underexploited resource. One of the most researched and exploited area of marine natural products are omega-3 long chain polyunsaturated fatty acids (LC-PUFAs). These compounds (particularly docosahexaenoic acid (DHA) and eicosapentaenoic acid (EPA)) are found in the marine environment and are exploited commercially from algae [118]. The shorter chain homologue, alpha linolenic acid (Fig. 5.8) is found in some plants.

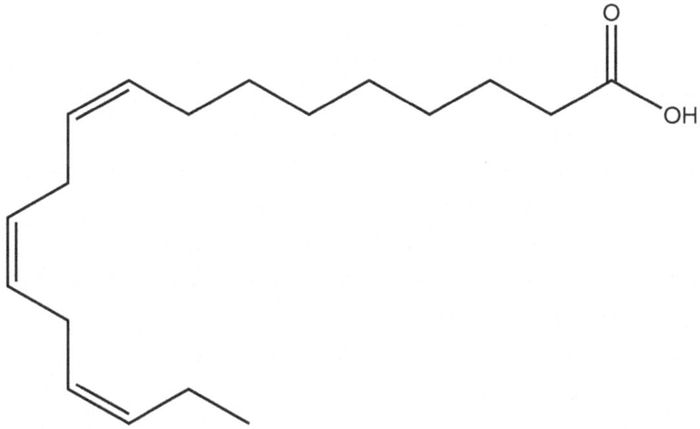

Fig. 5.8 Structure of alpha-linolenic acid

Neither DHA nor EPA are synthesised in mammals although they seem to be involved in a number of health-related biological effects. The precise benefit of DHA and EPA remains a matter of some conjecture however some beneficial aspects of consumption of EPA and DHA have been approved by the European Food Safety Authority (EFSA) Scientific Opinion on the substantiation of health claims related to docosahexaenoic acid (DHA), eicosapentaenoic acid (EPA) and brain, eye and nerve development (ID 501, 513, 540), maintenance of normal brain function (ID 497, 501, 510, 513, 519, 521, 534, 540, 688, 1323, 1360, 4294), maintenance of normal vision (ID 508, 510, 513, 519, 529, 540, 688, 2905, 4294), maintenance of normal cardiac function (ID 510, 688, 1360), "maternal health; pregnancy and nursing" (ID 514), "to fulfil increased omega-3 fatty acids need during pregnancy" (ID 539), "skin and digestive tract epithelial cells maintenance" (ID 525), enhancement of mood (ID 536), "membranes cell structure" (ID 4295), "anti-inflammatory action" (ID 4688) and maintenance of normal blood LDL cholesterol concentrations (ID 4719) pursuant to Article 13(1) of Regulation (EC) No 1924/2006 [119]. This type of approval in which pharmacological standards are claimed to be required, demonstrates a problem at the centre of claims relating specific food components to health outcomes. Except in the case of overt deficiency when a clinical endpoint can be clearly demonstrated, claims associated with dietary components are invalid since they refer to optimal health status when this is, in itself, undefined. In addition, the suggestion that a condition of 'healthy homeostasis' may exist is a fallacy. In practice, living systems progress through a series of non-equilibrium conditions rather than arriving at and maintaining a preferred status.

References

1. Cruz SL, Rafful C. A brief history of opioids and the evolution of concepts associated with substance use disorders. In: Cruz SL, editors. Opioids. Cham: Springer. https://doi.org/10.1007/978-3-031-09936-6_1.
2. Sneader W. The discovery of heroin. The Lancet. 352(9141):1697–1699.
3. Sakkaravarthi V. History of soap. CosmoDerma. 2022;2:133.
4. Smith CS, Hawthorne JG. Mappae Clavicula: a little key to the world of medieval techniques. Trans Am Philos Soc. 1974;64(4):1–128. https://doi.org/10.2307/1006317.
5. Anderson KJ. A history of lubricants. MRS Bull. 1991;16:69. https://doi.org/10.1557/S0883769400055895.
6. Myers D. An overview of surfactant science and technology. In: Myers D, editors Surfactant science and technology. https://doi.org/10.1002/047174607X.ch1.
7. Bjerk TR, Severino P, Jain S, Marques C, Silva AM, Pashirova T, Souto EB. Biosurfactants: properties and applications in drug delivery, biotechnology and ecotoxicology. Bioengineering (Basel). 2021;8(8):115. https://doi.org/10.3390/bioengineering8080115.PMID:34436118;PMCID:PMC8389267.
8. Sublette ME, Daray FM, Ganança L, et al. The role of polyunsaturated fatty acids in the neurobiology of major depressive disorder and suicide risk. Mol Psychiatry. 2024;29:269–86. https://doi.org/10.1038/s41380-023-02322-6.
9. Schuchardt JP, Huss M, Stauss-Grabo M, et al. Significance of long-chain polyunsaturated fatty acids (PUFAs) for the development and behaviour of children. Eur J Pediatr. 2010;169:149–64. https://doi.org/10.1007/s00431-0091035-8.
10. Zárate R, El Jaber-Vazdekis N, Tejera N, et al. Significance of long chain polyunsaturated fatty acids in human health. Clin Trans Med. 2017;6:25. https://doi.org/10.1186/s40169-017-0153-6.
11. Wilson DE, Ruff S, American Society of Mammalogists Smithsonian Institution. The Smithsonian book of north American mammals. Vancouver, BC: UBC Press; 1999.
12. Abedi E, Sahari MA. Long-chain polyunsaturated fatty acid sources and evaluation of their nutritional and functional properties. Food Sci Nutr. 2014;2(5):443–63. https://doi.org/10.1002/fsn3.121. Epub 2014 Jun 29. PMID: 25473503; PMCID: PMC4237475.
13. Sprague M, Betancor MB, Tocher DR. Microbial and genetically engineered oils as replacements for fish oil in aquaculture feeds. Biotechnol Lett. 2017;39(11):1599–1609. https://doi.org/10.1007/s10529-017-2402-6. Epub 2017 Jul 18. PMID: 28721583; PMCID: PMC5636849.
14. Cooksey C. Tyrian purple: the first four thousand years. Sci Prog. 2013;96(2):171–86. https://doi.org/10.3184/003685013X13680345111425.
15. Cova TFGG, Pais AACC, Seixas de Melo JS. Reconstructing the historical synthesis of Mauveine from Perkin and Caro: procedure and details. Sci Rep. 2017;7:6806. https://doi.org/10.1038/s41598-017-07239-z.
16. Mellor CM, Cardwell DSL. Dyes and dyeing 1775–1860. Br J Hist Sci. 1963;1(3):265–79. https://doi.org/10.1017/S0007087400001503.
17. Kalba LA. Color in the age of impressionism: commerce, technology, and art. University Park, USA: Penn State University Press; 2017. https://doi.org/10.1515/9780271079806.
18. Alegbe EO, Uthman TO. A review of history, properties, classification, applications and challenges of natural and synthetic dyes. Heliyon. 2024;10(13): e33646. https://doi.org/10.1016/j.heliyon.2024.e33646.PMID:39040261;PMCID:PMC11261106.
19. Kaplan G, Seferoğlu Z. The synthetic approaches for preparation of indigo and applications in denim industry. Curr Org Synth. 2023;20(4):361–4. https://doi.org/10.2174/1570179419666220830091956. PMID: 36043751.
20. Chandel N, Singh BB, Dureja C, et al. Indigo production goes green: a review on opportunities and challenges of fermentative production. World J Microbiol Biotechnol. 2024;40:62. https://doi.org/10.1007/s11274-023-03871-2.

21. Alexander L, Grierson D. Ethylene biosynthesis and action in tomato: a model for climacteric fruit ripening. J Exp Bot. 2002;53(377):2039–55. https://doi.org/10.1093/jxb/erf072.
22. Herrera-Ubaldo H. Big red: dissecting the role of ethylene in tomato fruit development and ripening. Plant Cell. 2022;34(9):3158–9. https://doi.org/10.1093/plcell/koac172. PMID: 35670756; PMCID: PMC9421465.
23. Bird CR, Smith CJ, Ray JA, Moureau P, Bevan MW, Bird AS, Hughes S, Morris PC, Grierson D, Schuch W. The tomato polygalacturonase gene and ripening specific expression in transgenic plants. Plant Mol Biol. 1988;11(5):651–62. https://doi.org/10.1007/BF00017465. PMID: 24272499.
24. Bennett AB, DellaPenna D, Fischer RL, Giovannoni J, Lincoln J. Regulation, maturation and function of tomato fruit polygalacturonase. In: Osborne DJ, Jackson MB, editors. Cell separation in plants. NATO ASI series, vol 35. Berlin, Heidelberg: Springer; 1989. https://doi.org/10.1007/978-3-642-74161-6_2.
25. Kramer MG, Redenbaugh K. Commercialization of a tomato with an antisense polygalacturonase gene: the FLAVR SAVR™ tomato story. Euphytica. 1994;79:293–7. https://doi.org/10.1007/BF00022530.
26. Nie H, Shi Y, Geng X, Xing G. CRISRP/Cas9-mediated targeted mutagenesis of tomato polygalacturonase gene (SlPG) delays fruit softening. Front Plant Sci. 2022;19(13): 729128. https://doi.org/10.3389/fpls.2022.729128. PMID: 35665160; PMCID: PMC9162796.
27. Gonzali S, Perata P. Anthocyanins from purple tomatoes as novel antioxidants to promote human health. Antioxidants (Basel). 2020;9(10):1017. https://doi.org/10.3390/antiox9101017. PMID: 33092051; PMCID: PMC7590037.
28. Saini H, Thakur R, Gill R, Tyagi K, Goswami M. CRISPR/Cas9-gene editing approaches in plant breeding. GM Crops & Food. 2023;14(1):1–17. https://doi.org/10.1080/21645698.2023.2256930.
29. Zhang J, Celli GB, Brooks MS, editors. Natural sources of anthocyanins in anthocyanins from natural sources: exploiting targeted delivery for improved health. The Royal Society of Chemistry; 2019, ch. 1. pp. 1–33.
30. Gonçalves AC, Nunes AR, Falcão A, Alves G, Silva LR. Dietary effects of anthocyanins in human health: a comprehensive review. Pharmaceuticals (Basel). 2021;14(7):690. https://doi.org/10.3390/ph14070690. PMID: 34358116; PMCID: PMC8308553.
31. Firn R. What are natural products? Nature's chemicals: the natural products that shaped our world. Oxford University Press; 2009.
32. McNulty MB, editor. Kant's metaphysical foundations of natural science. In: Kant's metaphysical foundations of natural science: a critical guide. Cambridge critical guides. Cambridge University Press; 2022. pp. i-i.
33. Sutton C. 'Nullius in Verba' and 'Nihil in Verbis': public understanding of the role of language in science. Br J Hist Sci. 1994;27(1):55–64.
34. Ji QY, Wang W, Yan H, Qu H, Liu Y, Qian Y, Gu R. The effect of different organic acids and their combination on the cell barrier and biofilm of Escherichia coli. Foods. 2023;12(16):3011. https://doi.org/10.3390/foods12163011. PMID: 37628010; PMCID: PMC10453431.
35. Panchal P, Miller AJ, Giri J. Organic acids: versatile stress response roles in plants. J Exp Botany. 72(11):4038–52. https://doi.org/10.1093/jxb/erab019
36. Pérez-Díaz IM, Altuntas EG, Juneja VK. Microbial fermentation in food preservation. In: Juneja V, Dwivedi H, Sofos J, editors. Microbial control and food preservation. Food microbiology and food safety. New York, NY: Springer. https://doi.org/10.1007/978-1-4939-7556-3_13.
37. Igamberdiev AU, Eprintsev AT. Organic acids: the pools of fixed carbon involved in redox regulation and energy balance in higher plants. Front Plant Sci. 2016;15(7):1042. https://doi.org/10.3389/fpls.2016.01042.PMID:27471516;PMCID:PMC4945632.
38. Zheng S, Zeng T, Li C, et al. Deep learning driven biosynthetic pathways navigation for natural products with BioNavi-NP. Nat Commun. 2022;13:3342. https://doi.org/10.1038/s41467-022-30970-9.

39. Watanabe K. Discovery and investigation of natural Diels-Alderases. J Nat Med. 2021;75(3):434–47. https://doi.org/10.1007/s11418-021-01502-4. Epub 2021 Mar 8. Erratum in: J Nat Med. 2021;75(3):448. https://doi.org/10.1007/s11418-021-01511-3. PMID: 33683566; PMCID: PMC8159843.
40. Podust LM, Sherman DH. Diversity of P450 enzymes in the biosynthesis of natural products. Nat Prod Rep. 2012;29(10):1251–66. https://doi.org/10.1039/c2np20020a. Epub 2012 Jul 23. PMID: 22820933; PMCID: PMC3454455.
41. Urlacher, Vlada B, et al. Cytochrome P450 monooxygenases in biotechnology and synthetic biology. Trends Biotechnol. 2019;37(8):882–97.
42. Waldi D. Spray reagents for thin-layer chromatography. In: Stahl E, editors. Thin-layer chromatography. Berlin, Heidelberg: Springer; 1965. https://doi.org/10.1007/978-3-662-01031-0_23.
43. Banerjee R. Isolation and purification of enzymes. In: Pandey A, Webb C, Soccol CR, Larroche C, editors. Enzyme technology. New York, NY: Springer; 2006. https://doi.org/10.1007/978-0-387-35141-4_26.
44. Rothstein DM, Love SF. Isolation of mutants blocked in calicheamicin biosynthesis. J Bacteriol. 1991;173(23):7716–8. https://doi.org/10.1128/jb.173.23.7716-7718.1991. PMID: 1938969; PMCID: PMC212544.
45. Hara O, Beppu T. Mutants blocked in streptomycin production in Streptomyces griseus—the role of A-factor. J Antibiot (Tokyo). 1982;35(3):349–58. https://doi.org/10.7164/antibiotics.35.349. PMID: 6804428.
46. Baltz RH, Seno ET. Properties of Streptomyces fradiae mutants blocked in biosynthesis of the macrolide antibiotic tylosin. Antimicrob Agents Chemother. 1981;20(2):214–25. https://doi.org/10.1128/AAC.20.2.214. PMID: 7283418; PMCID: PMC181666.
47. Chen CH. Biochemical effects of deuterium oxide and deuteration. In: Deuterium oxide and deuteration in biosciences. Cham: Springer; 2022. https://doi.org/10.1007/978-3-031-08605-2_7.
48. Allen DK, Bates PD, Tjellstrom H. Tracking the metabolic pulse of plant lipid production with isotopic labeling and flux analyses: Past, present and future. Prog Lipid Res. 2015;58:97–120.
49. Bacher A, Chen F, Eisenreich W. Decoding biosynthetic pathways in plants by pulse-chase strategies using 13-CO_2 as a universal tracer †. Metabolites. 2016;6(3):21. https://doi.org/10.3390/metabo6030021.PMID:27429012;PMCID:PMC5041120.
50. Schoenheimer R, Rittenberg D. Deuterium as an indicator in the study of intermediary metabolism. Science. 1935;82(2120):156–7. https://doi.org/10.1126/science.82.2120.156. PMID: 17811948.
51. Simpson TJ. Fungal polyketide biosynthesis—a personal perspective. Nat Prod Rep. 2014;31(10):1247–52. https://doi.org/10.1039/c4np00065j. PMID: 25053230.
52. Rinkel J, Dickschat JS. Recent highlights in biosynthesis research using stable isotopes. Beilstein J Org Chem. 2015;9(11):2493–508. https://doi.org/10.3762/bjoc.11.271.PMID:26734097;PMCID:PMC4685789.
53. McCaughey CS, van Santen JA, van der Hooft JJJ, Medema MH, Linington RG. An isotopic labeling approach linking natural products with biosynthetic gene clusters. Nat Chem Biol. 2022;18(3):295–304. https://doi.org/10.1038/s41589-02100949-6. Epub 2021 Dec 30. PMID: 34969972; PMCID: PMC8891042.
54. Jiang H, Favaro E, Goulbourne CN, Rakowska PD, Hughes GM, Ryadnov MG, Fong LG, Young SG, Ferguson DJ, Harris AL, Grovenor CR. Stable isotope imaging of biological samples with high resolution secondary ion mass spectrometry and complementary techniques. Methods. 2014;68(2):31724. https://doi.org/10.1016/j.ymeth.2014.02.012. Epub 2014 Feb 17. PMID: 24556558; PMCID: PMC4222523.
55. Ben Abu N, Ben Shoshan-Galeczki Y, Malach E, Niv MY. The T1R3 subunit of the sweet taste receptor is activated by D2O in transmembrane domaindependent manner. Chem Senses. 2023;48:bjad032. https://doi.org/10.1093/chemse/bjad032.
56. Liu M, Gao J. Deuterium isotope effects on acid-base equilibrium of organic compounds. Molecules. 2021;26(24):7687. https://doi.org/10.3390/molecules26247687. PMID: 34946769; PMCID: PMC8705040.

57. Hoehn R, Nichols D, Neven H, et al. Neuroreceptor activation by vibration-assisted tunneling. Sci Rep. 2015;5:9990. https://doi.org/10.1038/srep09990.
58. Wang J, Liu N, Yang S, Qiu G, Tian H, Sun B. Research progress in the synthesis of stable isotopes of food flavour compounds. Food Chem. 2024;435: 137635. https://doi.org/10.1016/j.foodchem.2023.137635. Epub 2023 Oct 4. PMID: 37813024.
59. Xie X, Zubarev R. Isotopic resonance hypothesis: experimental verification by Escherichia coli growth measurements. Sci Rep. 2015;5:9215. https://doi.org/10.1038/srep09215.
60. Brik A, Dawson P, Liu L, editors. The total chemical synthesis of proteins. Wiley; 2021.
61. Thompson RE, Muir TW. Chemoenzymatic semisynthesis of proteins. Chem Rev. 2020;120(6):3051–3126. https://doi.org/10.1021/acs.chemrev.9b00450. Epub 2019 Nov 27. PMID: 31774265; PMCID: PMC7101271.
62. Chen BS, Ribeiro de Souza FZ. Enzymatic synthesis of enantiopure alcohols: current state and perspectives. RSC Adv. 2019;9(4):2102–2115. https://doi.org/10.1039/c8ra09004a. PMID: 35516160; PMCID: PMC9059855.
63. Li J, Shi Y, Clark BR. Semi-synthesis of antibacterial dialkylresorcinol derivatives. J Antibiot. 2021;74:70–5. https://doi.org/10.1038/s41429-0200359-5.
64. Lu XX, Jiang YY, Wu YW, Chen GY, Shao CL, Gu YC, Liu M, Wei MY. Semi-synthesis, cytotoxic evaluation, and structure-activity relationships of brefeldin a derivatives with antileukemia activity. Mar Drugs. 2021;20(1):26. https://doi.org/10.3390/md20010026. PMID: 35049881; PMCID: PMC8777696.
65. Wang Z, Hui C. Contemporary advancements in the semi-synthesis of bioactive terpenoids and steroids. Org Biomol Chem. 2021;19(17):3791–812. https://doi.org/10.1039/d1ob00448d. PMID: 33949606.
66. Sekhri L, Kadri ML, Samira C, Senigra M. Synthesis of penicillin derivatives and study of their biological antibacterial activities. Biomed Pharmacol J. 2008;1(2).
67. Peñalva MA, Rowlands RT, Turner G. The optimization of penicillin biosynthesis in fungi. Trends Biotechnol. 1998;16(11):483–9. https://doi.org/10.1016/s01677799(98)01229-3. PMID: 9830157.
68. Wuts PG. Semisynthesis of taxol. Curr Opin Drug Discov Devel. 1998;1(3):329–37 PMID: 19649807.
69. McElroy C, Jennewein S. Taxol® biosynthesis and production: from forests to fermenters. In: Schwab W, Lange B, Wüst M, editors. Biotechnology of natural products. Cham: Springer; 2018. https://doi.org/10.1007/978-3-319-679037_7.
70. Mohammadi Ballakuti N, Ghanati F, Zare-Maivan H, et al. Taxoid profile in endophytic fungi isolated from Corylus avellana, introduces potential source for the production of Taxol in semi-synthetic approaches. Sci Rep. 2022;12:9390. https://doi.org/10.1038/s41598-022-13602-6.
71. Serra S et al. Biocatalytic preparation of natural flavours and fragrances. Trends Biotechnol. 23(4):193–8.
72. Vijayalakshmi S, Disalva X, Srivastava C, Arun A. Vanilla-natural vs artificial: a review. Res J Pharm Tech. 2019;12(6):3068–72. https://doi.org/10.5958/0974-360X.2019.00520.1.
73. Tian K, Sekar BS, Choo JPS, Li Z. Enzyme engineering for enantioselective biotransformations. Biomass, Biofuels, Biochemicals. 2020.
74. Yang H, Yu H, Stolarzewicz IA, Tang W. Enantioselective transformations in the synthesis of therapeutic agents. Chem Rev. 2023;123(15):9397–446. https://doi.org/10.1021/acs.chemrev.3c00010. Epub 2023 Jul 7. PMID: 37417731.
75. Lüddeke F, Harder J. Enantiospecific (S)-(+)-linalool formation from betamyrcene by linalool dehydratase-isomerase. Z Naturforsch C J Biosci. 2011;66(7–8):409–12. https://doi.org/10.1515/znc-2011-7-813. PMID: 21950166.
76. Braga A, Guerreiro C, Belo I. Generation of flavors and fragrances through biotransformation and De Novo synthesis. Food Bioprocess Technol. 2018;11:2217–28.
77. Mitreiter S, Gigolashvili T. Regulation of glucosinolate biosynthesis. J Exp Bot. 2021;72(1):70–91. https://doi.org/10.1093/jxb/eraa479.

78. Nguyen VPT, Stewart J, Lopez M, Ioannou I, Allais F. Glucosinolates: natural occurrence, biosynthesis, accessibility, isolation, structures, and biological activities. Molecules. 2020;25(19):4537. https://doi.org/10.3390/molecules25194537. PMID: 33022970; PMCID: PMC7582585.
79. Unger K, Raza SAK, Mayer T, et al. Glucosinolate structural diversity shapes recruitment of a metabolic network of leaf-associated bacteria. Nat Commun. 2024;15:8496. https://doi.org/10.1038/s41467-024-52679-7.
80. Tahlan K, Moore MA, Jensen SE. δ-(l-α-aminoadipyl)-l-cysteinyl-dvaline synthetase (ACVS): discovery and perspectives. J Industr Microbiol Biotechnol. 2017;44(4–5):517–24. https://doi.org/10.1007/s10295-016-1850-7.
81. Baldwin JE et al. Penicillin biosynthesis: dependence of substrate stereochemistry on the mode of second ring closure. Tetrahedron. 1990;46:6145–54.
82. Demain AL, Shen YQ, Jensen SE, Westlake DW, Wolfe S. Further studies on the cyclization of the unnatural tripeptide delta-(D-alpha-aminoadipyl)-L-cysteinylD-valine to penicillin N. J Antibiot (Tokyo). 1986;39(7):1007–10. https://doi.org/10.7164/antibiotics.39.1007. PMID: 3759642.
83. Wolfe S, et al. Enzymatic approach to syntheses of unnatural beta-lactams. Science. 1984;226(4681):1386–92.
84. Till M, Race PR. The assembly line enzymology of polyketide biosynthesis. In: Evans B, editors. Nonribosomal peptide and polyketide biosynthesis. Methods in molecular biology, vol 1401. New York, NY: Humana Press; 2016. https://doi.org/10.1007/978-1-4939-3375-4_2.
85. Cummings M, Breitling R, Takano E. Steps towards the synthetic biology of polyketide biosynthesis. FEMS Microbiol Lett. 2014;351(2):116–25. https://doi.org/10.1111/1574-6968.12365.
86. Wong FT, Khosla C. Combinatorial biosynthesis of polyketides—a perspective. Curr Opin Chem Biol. 2012;16(1–2):117–23. https://doi.org/10.1016/j.cbpa.2012.01.018. Epub 2012 Feb 16. PMID: 22342766; PMCID: PMC3328608.
87. Williams G J. Engineering polyketide synthases and nonribosomal peptide synthetases. Curr Opin Struct Biol. 2013;23(4):603–12. https://doi.org/10.1016/j.sbi.2013.06.012. Epub 2013 Jul 6. PMID: 23838175; PMCID: PMC4271453.
88. Mabesoone MFJ, Leopold-Messer S, Minas HA, Chepkirui C, Chawengrum P, Reiter S, Meoded RA, Wolf S, Genz F, Magnus N, Piechulla B, Walker AS, Piel J. Evolution-guided engineering of trans-acyltransferase polyketide synthases. Science. 2024;383(6689):1312–17. https://doi.org/10.1126/science.adj7621. Epub 2024 Mar 21. PMID: 38513027; PMCID: PMC11260071.
89. Weingart P, Joubert M, Connoway K. Public engagement with science-origins, motives and impact in academic literature and science policy. PLoS One. 2021;16(7):e0254201. https://doi.org/10.1371/journal.pone.0254201. Erratum in: PLoS One. 2024;19(6):e0306474. https://doi.org/10.1371/journal.pone.0306474. PMID: 34234382; PMCID: PMC8263305.
90. Naveen AK, Sontakke M. A review on regulatory aspects, challenges and public perception in acceptance of genetically modified foods. Food Sci Biotechnol. 2024;33(4):791–804. https://doi.org/10.1007/s10068-023-01481-0. PMID: 38371681; PMCID: PMC10866814.
91. Butelli E, Titta L, Giorgio M, Mock HP, Matros A, Peterek S, Schijlen EG, Hall RD, Bovy AG, Luo J, Martin C. Enrichment of tomato fruit with health-promoting anthocyanins by expression of select transcription factors. Nat Biotechnol. 2008;26(11):1301–8. https://doi.org/10.1038/nbt.1506. Epub 2008 Oct 26. PMID: 18953354.
92. Riaz M, Zia-Ul-Haq M, Saad B. Occurrence of anthocyanins in plants. In: Anthocyanins and human health: biomolecular and therapeutic aspects. SpringerBriefs in food, health, and nutrition. Cham: Springer; 2016. https://doi.org/10.1007/978-3-319-26456-1_3.
93. Waldhof G, Kopp RT. Our morality tells us what to eat—a systematic review on attitudes towards genetically modified foods. Outlook Agric. 2024;53(4):363–75. https://doi.org/10.1177/00307270241295398.

94. Xu Y, Li Z. CRISPR-Cas systems: overview, innovations and applications in human disease research and gene therapy. Comput Struct Biotechnol J. 2020;18:2401–15. https://doi.org/10.1016/j.csbj.2020.08.031.PMID:33005303;PMCID:PMC7508700.
95. Wolter F, Schindele P, Puchta H. Plant breeding at the speed of light: the power of CRISPR/Cas to generate directed genetic diversity at multiple sites. BMC Plant Biol. 2019;19:176. https://doi.org/10.1186/s12870-019-1775-1.
96. Kato-Nitta N, Maeda T, Inagaki Y, et al. Expert and public perceptions of geneedited crops: attitude changes in relation to scientific knowledge. Palgrave Commun. 2019;5:137. https://doi.org/10.1057/s41599-019-0328-4.
97. Kato-Nitta N, Tachikawa M, Inagaki Y, Maeda T. Public perceptions of risks and benefits of gene-edited food crops: an international comparative study between the US, Japan, and Germany. Sci Technol Human Values. 2023;48(6):1360–92. https://doi.org/10.1177/01622439221123830.
98. Paleologo M, Lanubile A, Camardo Leggieri M, Graffigna G, Gomarasca P, Barello S. Public perception of new plant breeding techniques and the psychosocial determinants of acceptance: a systematic review. Public Underst Sci. 2024;33(6):795–812. https://doi.org/10.1177/09636625241254981.
99. Hanley AB, Hird H, Johnson ML. Testing for genetically modified components in foods—current and future challenges. In: Clark S, editor. Rapid detection methods for food and water. Royal Society of Chemistry; 2001. pp. 225–30.
100. Peng C, Zheng M, Ding L, Chen X, Wang X, Feng X, Wang J, Xu J. Accurate detection and evaluation of the gene-editing frequency in plants using droplet digital PCR. Front Plant Sci. 2020;11: 610790. https://doi.org/10.3389/fpls.2020.610790. PMID: 33381141; PMCID: PMC7767858.
101. Shillito RD, Whitt S, Ross M, et al. Detection of genome edits in plants—from editing to seed. In Vitro Cell Dev Biol Plant. 2021;57:595–608. https://doi.org/10.1007/s11627-021-10214-z.
102. Arias PM, Butler J, Randhawa GS, et al. Environment and taxonomy shape the genomic signature of prokaryotic extremophiles. Sci Rep. 2023;13:16105. https://doi.org/10.1038/s41598-023-42518-y.
103. Hu C, Yang J, Qi Z, Wu H, Wang B, Zou F, Mei H, Liu J, Wang W, Liu Q. Heat shock proteins: biological functions, pathological roles, and therapeutic opportunities. MedComm 2022;3(3):e161. https://doi.org/10.1002/mco2.161. PMID: 35928554; PMCID: PMC9345296.
104. Tsan MF, Gao B. Heat shock proteins and immune system. J Leukoc Biol. 2009;85(6):905–10. https://doi.org/10.1189/jlb.0109005. Epub 2009 Mar 10. PMID: 19276179.
105. Kumar P, Paul D, Jhajhriya S, et al. Understanding heat-shock proteins' abundance and pivotal function under multiple abiotic stresses. J Plant Biochem Biotechnol. 2024;33:492–513. https://doi.org/10.1007/s13562-02400932-x.
106. Dumorné K, Córdova DC, Astorga-Eló M, Renganathan P. Extremozymes: a potential source for industrial applications. J Microbiol Biotechnol. 2017;27(4):649–59. https://doi.org/10.4014/jmb.1611.11006. PMID: 28104900.
107. Nezhad NG, Rahman RNZRA, Normi YM et al. Thermostability engineering of industrial enzymes through structure modification. Appl Microbiol Biotechnol. 2022;106:4845–66. https://doi.org/10.1007/s00253-022-12067-x.
108. Karan R, Capes MD, DasSarma S. Function and biotechnology of extremophilic enzymes in low water activity. Aquat Biosyst. 2012;8:4. https://doi.org/10.1186/2046-9063-8-4.
109. Lee EH, Tsujimoto T, Uyama H, et al. Enhancement of enzyme activity and stability by poly(γ-glutamic acid). Polym J. 2010;42:818–22. https://doi.org/10.1038/pj.2010.71.
110. Wilson ZE, Brimble MA. Molecules derived from the extremes of life. Nat Prod Rep. 2009;26:44–71. https://doi.org/10.1039/B800164M.
111. Uppar AP, Vanti GL, Poondla N, Kaulgud RS, Kurjogi MM. Applications of extremophiles in therapeutics. In: Shah MP, Dey S, editors. Trends in biotechnology of polyextremophiles. Cham: Springer; 2024. https://doi.org/10.1007/978-3-031-55032-4_10.

112. Menezes-Benavente L et al. Salt stress induces altered expression of genes encoding antioxidant enzymes in seedlings of a Brazilian indica rice (Oryza sativa L.). Plant Sci. 2004;166:323–31.
113. Yeshi K, Crayn D, Ritmejerytė E, Wangchuk P. Plant secondary metabolites produced in response to abiotic stresses has potential application in pharmaceutical product development. Molecules. 2022;27(1):313. https://doi.org/10.3390/molecules27010313. PMID: 35011546; PMCID: PMC8746929.
114. Shrivastava M, Kumar A, Saxena P. Bioactive compounds/metabolites from extremophiles for biocontrol and plant disease management. In: Ranjan A, Rajput VD, Chauhan A, Prazdnova EV, Minkina T, Zargar SM, editors. Extremophiles for sustainable agriculture and soil health improvement. Cham: Springer; 2024. https://doi.org/10.1007/978-3-031-70203-7_6.
115. Carroll AR, Copp BR, Grkovic T, Keyzers RA, Prinsep MR. Marine natural products. Nat Prod Rep. 2024;41(2):162–207. https://doi.org/10.1039/d3np00061c. PMID: 38285012.
116. Lyu C, et al. Nucleic Acids Res. 2021;49(D1):D509–15. https://doi.org/10.1093/nar/gkaa763.
117. Zhou Q, Hotta K, Deng Y, Yuan R, Quan S, Chen X. Advances in biosynthesis of natural products from marine microorganisms. Microorganisms. 2021;9(12):2551. https://doi.org/10.3390/microorganisms9122551. PMID: 34946152; PMCID: PMC8706298.
118. Patted PG, Masareddy RS, Patil AS, et al. Omega-3 fatty acids: a comprehensive scientific review of their sources, functions and health benefits. Futur J Pharm Sci. 2024;10:94. https://doi.org/10.1186/s43094-024-00667-5.
119. EFSA Journal 2011;9(4):2078 [30 pp.]. https://doi.org/10.2903/j.efsa.2011.2078.

Chapter 6
Epidemiology, Causation and Correlation

Contents

6.1 Recipe—Belly Pork 'Chinese Style' 166
6.2 Epidemiology 169
6.3 Causation and Correlation 172
6.4 Post-biosynthetic Modification of Natural Products 176
6.5 Formation of Isothiocyanates and Sulphides—Compare and Contrast 178
6.6 Modulation of Natural Product Effects 185
References 190

Abstract There is a reluctance to properly assess negative results in science in general. While this can be problematic in many areas and even lead to the survival of false theories, it is a particular concern in an area that is as full of complexity and confounders as secondary natural products in foods. While chemistry is very much empirical in nature with an emphasis on testing hypotheses, some of the core directions in secondary natural products have been established on the basis of observation and epidemiology rather than trying to distinguish between causality and correlation. This has led to a growth in nutritional epidemiology and the development of theories based upon dietary patterns rather than the identification of individual components and their interactions. In the case of secondary natural products that occur (and are consumed) in small amounts, there may not even be an identifiable correlation between components and endpoints—far less any causality. One approach is to use the tendency of living systems to drift towards higher entropic states with increasing age and to find a way of measuring the effects of components on 'entropy inhibition'. Another possibility is to use the principle of quantum entanglement to measure very weak interactions. The complexity of natural products is demonstrated by the S compounds in the genus *Allium* and *Brassica* species. While both give rise to a complex range of products, their biogenesis is based on very different principles. Quorum sensing and quorum quenching are biological endpoints that may amplify the effects of natural products and make them easier to detect.

Keywords Epidemiology · Correlation · Causality · Quorum sensing · Quorum quenching

B. Hanley, *Secondary Natural Products in Foods and Food Systems*,
https://doi.org/10.1007/978-3-032-00292-1_6

6.1 Recipe—Belly Pork ‘Chinese Style’

While the philosophy of reductionism lies at the core of Western chemistry, the Chinese approach to science comes from a very different tradition. Western science is largely a process of developing a set of hypotheses based upon examples and knowledge and then testing these using empirical methods. Sometimes the tools required are not available or have not been developed and science then has to either be stalled or it sets off in a different direction. There is a tendency to try to do something rather than nothing. Indeed one of the features of science is that its practitioners are obsessed with generating some form of empirical evidence—even if they are posing the wrong questions, have little patience and take insufficient time to think and can confuse activity with progress. There is a tradition that even a badly designed or executed experiment will still provide some facsimile of progress. In addition to advocating any form of activity rather than none, science frequently does not embrace negative results as potential contributors to an improved understanding. Demonstrating the falsification and testability of a hypothesis is an essential aspect of a Popperian philosophy of science [1]. Unfortunately, this approach is rarely adopted and scientists more often design experiments with limited scope in order to confirm rather than to disprove hypotheses. The role of falsification in science is neglected in the pursuit of verifiable truths, the accumulation of supporting data and the making of measurements with a high degree of accuracy. The development of more and more sensitive and accurate instrument means that analytical chemistry, in the perspective of Western chemists, has become a goal in itself rather than a tool to help gain an understanding of the world. Natural products are measured with greater and greater precision and at levels where their significance is lost in a flurry of decimal places. The more holistic view that underpins the approach of Chinese chemists to natural product chemistry [2] is based around the traditions of Chinese medicine and bioactivity [3] and is less reductionist in outlook. This may mean that the disconnect experienced by Western chemists when their extreme reductionism collides with the complex reality of living systems may be less prevalent in the Chinese tradition. Because data and results are easier to assemble than thought and conjecture, the application of Western scientific analytical logic has become the dominant characteristic of global scientific efforts. This even applies in an area of science where observation rather than empirical experimentation ought to be more prevalent—epidemiology. This will be considered in more detail in Sect. 6.2.

It ought to be axiomatic that a combination of reductionism and empiricism should lead to a higher proportion of verifiable results than a more holistic approach since the former will seek to simplify and concentrate on key, important facts while the latter will embrace the many contributory factors that comprise the complexity of a whole system. The role of secondary natural products is a consequence of the collision between reductionist/empirical approaches and complex biological systems. This interface is an important aspect of biomedical science. In fact, as has been noted, many research findings in biomedical science are arguably false [4], perhaps in part as a consequence of the incompatibility of the simple/complex views outlined above.

One of the reasons for this falsity is the considerable weight given to perceived statistical significance (the rule of weight of data as a deciding factor in the proof of a hypothesis). The prevalence of bias (intentional as well as unintentional) and the drive to prove a particular hypothesis in which a researcher is invested mean that studies based on results from an individual investigator should be treated with care. The author who doubted the veracity of biomedical research also suggested a number of corollaries to his initial supposition. These are particularly worrying for anyone who is well versed in Western scientific thought and traditions and include the following.

"The smaller the studies conducted in a scientific field, the less likely the research findings are to be true." This seems logical. The larger the sample size, the more likely the results are to be meaningful and widely applicable. This should encourage different approaches and not be used as an excuse for repeating similar experiments multiple times to build a weight of evidence.

"The smaller the effect sizes in a scientific field, the less likely the research findings are to be true." This is particularly problematic for natural product chemists seeking to demonstrate that a given biological effect is caused (primarily or exclusively) by a minor dietary component. It also means that the area is likely to be subjected to significant numbers of false positive claims as researchers attempt to build an evidence base for their own favoured substance and, in addition, the cumulative effects of confounders will be ignored.

"The greater the number and the lesser the selection of tested relationships in a scientific field, the less likely the research findings are to be true." In effect, this statement implies that the better the hypothesis and the more precisely it is formulated, the more likely it is that the experiment will provide confirmation. Science also requires that some relationships are shown to be false. If a large number of relationships are tested but not subjected to selection then the research findings are more likely to be untrue.

"The greater the flexibility in designs, definitions, outcomes, and analytical modes in a scientific field, the less likely the research findings are to be true." In other words, if there is ambiguity in the choice of endpoints or in the construction of experiments then any research findings are more likely to also be ambiguous. A clinical trial that fails to meet an agreed primary objective but is considered a success based on secondary endpoints should be greeted with derision and laughter.

"The greater the financial and other interests and prejudices in a scientific field, the less likely the research findings are to be true." This does not only apply to commercial interests for which a particular result is the desired economic outcome—it can also apply to publicly funded research where the opportunity to receive future funding is dependent upon 'success' criteria.

"The hotter a scientific field (with more scientific teams involved), the less likely the research findings are to be true." This may seem counter-intuitive however it should be remembered that results do not necessarily oscillate around the 'true' outcome. It also can mean that a particular group may emphasise its most 'positive' results and discard anything that detracts from their main hypothesis—or as Swift put it, a lie is "like a physician, who has found out an infallible medicine, after the patient is dead." It is only when the alternative theories have been killed off that the remainder (which may or may not be correct) is left standing. The more alternative theories you start with, the more valid your theory appears to be as each alternative is, in turn, vanquished.

Belly Pork, Chinese Style

In the spirit of holistic complexity, this recipe starts out with something that is deceptively simple and then the level of complexity increases as more ingredients are added. Take a piece of belly pork (about 700 g) and salt and pepper it. Leave to sit at room temperature for 30 min. Place the pork in a stove top casserole dish or a saucepan and cover with water. Bring to the boil and simmer slowly for 30 min. Remove from the pot and place on a chopping board to rest and cool. Cut the pork into pieces of about 2–3 cm square. Heat a wok and add 2 tablespoons of rapeseed oil. Brown the pork for 2 min—it is likely to spit a bit and you can cover with a lid if this bothers you. Safety glasses are not necessary but do not peer into the wok. Prepare peeled root ginger cut into matchsticks and add these and three whole star anise to the wok. Add a glass of dry white wine, five or six Sichuan peppercorns. Sichuan pepper produces a numbing effect, attributed to the presence of hydroxy-alpha-sanshool ((2E,6Z,8E,10E)-N-(2-Hydroxy-2methylpropyl)dodeca-2,6,8,10-tetraenamide) **(Fig. 6.1)**. ***Sichuan pepper is a member of the Zanthoxylum family and is not botanically related to pepper. Add 3 tablespoons of dark brown sugar and 200 ml of water or chicken stock. Stir to dissolve the sugar. Add a teaspoon of dark soya sauce and two dried birds eye chillies. Cover with a tight fitting lid. The lid can be a frying pan lid, for example and it does not matter if it is not big enough to fit on the rim of the wok, so long as it covers the sauce and pork. Cook over a low/medium heat for about 45 min. By this time the liquid should have reduced and thickened and should be red/brown in colour. Remove the chilli peppers, the peppercorns and the star anise. Serve with white rice and a steamed green vegetable.***

This recipe starts with a long cook of a piece of pork. This is necessary to tenderise the pork and to remove some of the fat. One of the problems with cooking is the heterogeneous nature of food [5]. Pre-cooking the pork means that all the parts of the

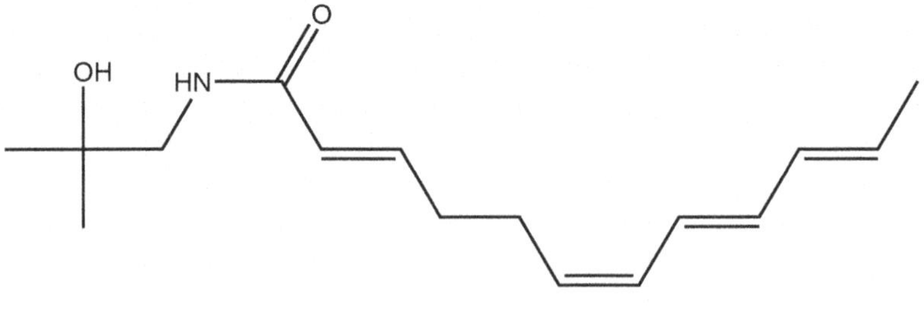

Fig. 6.1 Structure of hydroxy alpha sanshool

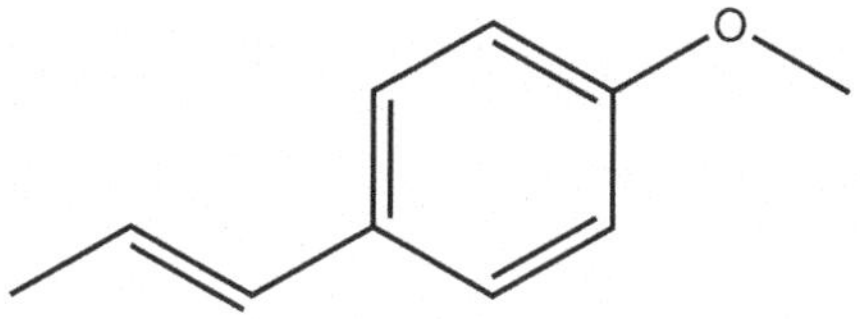

Fig. 6.2 Structure of trans-anethol

meat are equally tender and the second part of the cooking process becomes more about imparting flavours than actual cooking.

The addition of sugar and wine changes the properties of the cooking liquor and increases the extraction of trans anethol (Fig. 6.2), a key flavour component of star anise which is virtually insoluble in water but is miscible with ethanol.

6.2 Epidemiology

In principle, epidemiology ought to be the ultimate observational scientific technique. Epidemiology is the study of disease occurrence and the factors that influence it. It is implicit in the study of epidemiology that the importance of a given factor (and whether it is causal or correlative) should be tested by an intervention and/or the development of a mechanistic hypothesis. By understanding about how and why diseases occur, the practical corollary of epidemiology is the improved prevention and treatment of diseases. Perhaps the simplest way of considering epidemiology is to take mortality (death) and morbidity (the state of having a disease or the numbers of people who have a particular disease) as endpoints and to then look for factors that are either associated with or are directly responsible for either and their interrelationship. A given disease will have a mortality rate associated with it and knowing the spread of the disease will enable public health bodies to assess impact. The differentiation between causality and correlation will be dealt with in more detail in Sect. 6.3. While the occurrence and spread of infectious diseases is, in principle, simpler than investigating factors that influence non-communicable diseases (NCDs), differences in susceptibility and outcomes as well as the influence of confounders can create complications.

The link between diet and health outcomes is often referred to as nutritional epidemiology [6]. As noted previously (Sect. 6.1), all clinical or biomedical studies are subject to a number of confounders that may negate both the results and the conclusions that are drawn from them. In some cases, historical evidence is available that illustrates erroneous conclusions. The link between the air and various infectious diseases (for example 'malaria') was assumed until it was realised that some of the causes of mortality and morbidity were a consequence of water-borne vectors. The example of Dr John Snow and the linking of sewage to cholera [7] is perhaps the first recorded scientific demonstration of a link between an identifiable disease outcome and a specific environmental event [8]. In nutritional epidemiology, challenges to the development of causal links between diet and specific diseases have included statistical validation and measurement [9], measurement errors [10] and the complexity inherent in dietary studies because of the multifactorial nature of nutrition-related NCDs and interrelationships between dietary components [11]. The inherent scientific unacceptability of many nutritional epidemiological studies has been noted [12], however too many careers are created by building an evidence base to support spurious associations for the poor practices to be completely abandoned. There is a prevailing wisdom that the weak nature of associations, the prevalence of confounders and the high cost of carrying out studies has encouraged a 'groupthink' approach which, while acknowledging the difficulties and lack of rigour, contests that 'the best that is practically achievable' is better than nothing. One approach which has gained some level of credibility is the anti-reductionist approach of investigating dietary patterns rather than individual components of the diet [13]. Unfortunately, combining the complexity of whole diets with the complexity of living systems is often responsible for confusing and scientifically incoherent results. That is not to say that all approaches are equally likely to end in failure, however sometimes the recommendations that arise from such investigations (even when they claim to be 'hypothesis-driven') are so generic as to be the precise opposite of illuminating. In the case of secondary natural products, the situation is, in some ways even less promising. Many secondary natural products occur in very small amounts and can often exhibit a range of effects. There are, however, some methods that are a departure from the use of analytical chemistry as the only tool. Typically nutritional epidemiology starts with the analysis of foods for individual components (or, in some cases, groups of components such as estrogens). Dietary intakes are then either measured or self-reported. This gives a figure for how much of a given dietary constituent is taken in. At this level of reporting, the information generated is of little use. Not only is the dietary intake not linked to a health endpoint, there is no consideration of either the actual intake or, in the case of self-reported studies, the accuracy of recall. In order to establish a coherent baseline, duplicate diet samples have to be analysed, a clinical endpoint has to be hypothesised and some idea of uptake and metabolism/excretion must be obtained. In essence, simply measuring the presence of a particular constituent is of very limited value. It is more useful to measure changes in function however, many of the effects of dietary constituents are relatively small and, in order to ensure statistical significance, a large cohort and/or a lengthy study period is required. This is a resource intensive route and there are also issues with compliance. For these reasons, researchers often consider predictive biomarkers. A predictive biomarker is a clinical measurement snapshot which may be correlated with the risk of a disease or health outcome. One example of this is the presence of high levels of low density lipoprotein (LDL)/cholesterol which is indicative (although not invariably) of an increased risk of subsequently developing cardiovascular disease. The same problems occur with predictive biomarkers as with clinical markers (small effects and the presence of confounders) however, in addition, there are two other factors. The first of these is the strength of the link between the biomarker and the clinical condition. The second is the predictive power of the biomarker. A measurement at a particular time point is insufficient to show a trend and may not be sufficiently pronounced compared to a control to demonstrate the likelihood of a future event. Observation without an intervention to demonstrate the relevance of the measurement is of limited value. An intervention can be useful however it may simply demonstrate a link between a given intervention and a biomarker and not necessarily between the biomarker and an eventual clinical endpoint.

While some researchers have looked at the way living systems are perturbed or resist deleterious effects, this in again an approach with a considerable risk due to the existence of confounders. In addition, it relies on a particular paradigm (that of healthy homeostasis). Living systems do not exist in a state of healthy homeostasis but rather transition between a succession of states—some of which may be more likely to degrade into disease with an epigenetic [14] increase in the disorder (entropy) of the system while others will move to another metastable entropic state with broadly the same characteristics as before. Indeed, if resisting the age-related increase in entropy is a measure of 'health' then methods to measure the effect of interventions on changes in the entropy of a system (using epigenetic age-related DNA methylation as a proxy [15]) may be a more effective measurement of later outcomes. The entropy of a living system cannot increase indefinitely and there is an energetic cost in resisting this. This is captured in the development of systems theories that replace the notion of systems in equilibrium with entropic systems theory [16]. Such an approach shows a major weakness in the use of pharmacological models to investigate nutritional health claims associated with dietary components. Pharmaceutical investigations do not generally deal with systems that are in a metastable equilibrium state. They deal instead with systems with a dysfunctional tendency to move rapidly towards an increase in entropy. The role of pharmaceuticals is to discontinuously move the system to a state of lowered entropy (hopefully permanently). The role of dietary components is to lower or reverse the gradual drift of a living system towards higher entropic states. One of the problems is that the clinical markers that are used are inappropriate to measure this. They are either linked to a disease state directly (and diet does not cure disease) or are so generic as to be virtually useless. Measurement of antioxidant potential falls into this latter category. Another example of the use of epidemiology will illustrate this. The definition of molecular epidemiology is that it is the study of factors that influence the pathogenesis of disease. These can include genetic factors, specific molecules and pathways [17]. To an extent it deals with causation but it is concerned, at its core, with disease pathway and, in the case of infectious diseases (where it seems to have the greatest potential) with transmission and impact upon populations [18]. In conclusion, epidemiology as applied to secondary natural products is very challenging. The challenges include the following.

- The difficulty in defining a clinical endpoint that is responsive to a change in a minor dietary component.
- The ubiquity of confounders and the interactions between dietary components.
- The lack of systematic mechanistic knowledge of chronic disease states that enables hypotheses to be proposed.
- The small size of the effect in most cases and the concomitant need for large cohort sizes.

- The illusion that a pharmacological model or an investigation based upon the maintenance of a 'healthy equilibrium' is valid.
- The lack of valid predictive biomarkers that are able to be used to extrapolate from a snapshot (or a limited exposure) to an eventual outcome.
- The funding mechanisms that reward 'positive' results and the importance attached to trying to change eating habits at a societal level.

Dietary (nutritional) epidemiology is, therefore, a challenging area and, while associations are relatively easy to propose, particularly where there appears to be a weight of evidence, the line between association (which may be simply coincidental and may not actually be valid) and causality can become blurred [19]. Finally, there is a core problem with epidemiological studies which lies within the central philosophical principle of the application of epidemiological approaches. The testing of theories using Popperian logic and hypothetical/deductive reasoning has greater weight than the inductive reasoning that underpins epidemiology [20]. If it is accepted that an important aspect of Popperian approaches relies on distinguishing between science and nonscience (or pseudoscience), by the possibility of falsification then epidemiology must pass this test. If a theory that is built upon epidemiological association or causality can be tested and shown to be false, then it qualifies as science. This implies that science involves a degree of risk since it must be open to the possibility that it could be proven to be false. This is, however, as noted below, a concept that does not appear to be central to epidemiological thought. While falsifiability can enable a theory to achieve greater apparent credibility, no theory is able to attain absolute certainty. Epidemiology seeks to verify rather than falsify associations and causality and, in its use of statistical models, creates a paradigm that is the inverse of Popperian deductive inference. It remains a matter for debate whether a Popperian approach to epidemiology and inferential logic can complement each other. The apparent lack of an agreed evidential and inferential framework for epidemiology will, eventually, limit its growth and relevance.

6.3 Causation and Correlation

Two of the most important concepts when considering the interaction of secondary natural products with living systems (either on an individual or a collective level) are causation and correlation. These ideas provide the underpinning for associative and hypothesis-driven epidemiology.

While correlation may appear to be a relatively simple concept—the association between two things—there are divisions within this uncomplicated idea. Correlation can be positive or negative and can be linear or non-linear. In addition, it is possible to measure the strength and type of correlation and these are expressed through the use of 'correlation coefficients'. Correlation is a natural extension of measuring the extent to which things change. If a single component changes with time then that is referred to as a univariate distribution. When two components vary and the variation in one is reflected in the variation of the other then that is referred to as a bivariate distribution and the two components are referred to as being correlated. Measuring correlation is a statistical technique for defining the extent of the relationship between the variables. It does not probe the nature of the relationship—simply its existence. While in many cases, a correlation is obvious, it can be complex and may be a hidden by confounders and other variables. Quantum entanglement is a form of correlation, but the two terms are distinct and have conceptual differences. While a detailed discussion of the difference between quantum entanglement and classical correlation is complex, one particular feature may serve as an illustration. If two quantum particles are entangled then they cannot be described separately but only as a complete whole. Classical correlation allows for the measurement of just one variable and the observation of the effect upon another variable. The corollary of this is that a classical variable can exist outside of its correlated 'partner'. Indeed, hidden correlations are an important aspect of natural variables and contribute to our lack of knowledge. In the case of quantum entanglement, the non-entangled particle has no capability to exist so entanglement is a necessary pre-requisite for the quantum particle to exist. This could lead to the conclusion that, if a quantum particle or event can be measured as a distinct entity, then it cannot be entangled. A more detailed description of quantum entanglement is available [21]. It is known that, in addition to correlation, biological systems also exhibit some aspects of quantum behaviour in, for example, photosynthesis [22] and quantum tunnelling in enzymology [23]. The concept of quantum entanglement has been used to explain the complexity of bird migration [24]. The apparent observation of non-targeted effects in biological systems which may be explained by quantum biology [25] suggests that quantum entanglement may present an opportunity for measuring the effect of very weak interactions (such as are postulated to occur as a consequence of the interaction of secondary natural products upon biological systems) in a controlled environment. Indeed, quantum biology has been used to explain the effects of radical oxygen species on bioenergetics in the cell [26]. A similar application may enable the investigation, using quantum biology, of the effects of antioxidants, for example, which have been widely promoted as being one of the principal effectors of the protective effects of secondary natural products [27].

As noted previously, classical correlation is a way of measuring the statistical relationship between two variables. Correlation can be of the 'cause and effect' variety where a change in one thing leads to an observable change in another. It can also be the result of a third factor having an influence on two outcomes—the third factor is then a confounder. The final possibility is that an apparent correlation is actually a pure chance event. For example, the growing of vegetables appears to correlate with health but this may not simply be a consequence of increased consumption. People who grow vegetables are more likely to spend more time outdoors and their endogenous vitamin D synthesis will be increased compared to those who stay indoors [28]. In addition, there is a tendency for positive results that fit with a particular hypothesis to be reported more than negative ones [29] so a review of any area will tend to be biased towards positive results. This applies particularly to diet/health correlations

[30]. The use of correlation should be carefully examined for bias and the reporting of results that suggest a correlation should be treated with caution. Correlations may, however, be an early indication of a trend that could reward further study but they should not be used as a sole evidentiary knowledge base. For example, populations of storks in Europe appears to correlate with the birth rate in parts of Europe [31]. While the link between storks and the delivery of babies is a commonly held belief, the statistical correlation between birth rate and the number of storks is an example of a spurious causality. The link between macronutrients or micronutrients and health is more assured because clinical causality can often be demonstrated, at least in the case of deficiencies (see below). The evidence for a positive correlation between secondary natural product intake and a positive health outcome is much more difficult to confirm. A Popperian, falsification route should be part of any such attempt rather than a total preoccupation with a verification strategy. It is also important to realise that, unlike pharmaceuticals, most secondary natural products are not designed with positive health benefits for consuming organisms as their primary biological purpose (indeed, if anything, the reverse may apply). This means that, while some of the benefits (for example, antimicrobial activity) are common to the producer and the consumer through each having a common 'enemy', most are designed for a different purpose and any positive dietary effect on the health of the consumer is a coincidence—indeed it is more likely to be a pure chance event.

In the case of causality, the basis for a link between two factors is underpinned by a mechanistic or logical understanding. The difference between correlation and causation is conceptually fairly simple. Correlation means that there is a statistically proven relationship between variables. As mentioned previously, this connection can be positive, negative or random. In order to demonstrate causation, the linked change in variables is not simply correlative, it is interdependent. The change in one variable causes the change in another as a result of a defined mechanistic interdependence. One important aspect of correlation and causation is implicit in their definition. If two (or more) things are correlated, it simply means that a change in one is reflected in a change in another. It does not imply a cause-and-effect relationship. A correlative relationship need not be causal. As a consequence of the interaction between them, a causal relationship will result in a change in the linked variable and so causal relationships are correlative but the reverse is not necessarily true. Causal relationships are more difficult to prove than correlative relationships. To prove a causal relationship, potential confounders have to be discounted and, generally, the nature of the relationship has to be understood. Popperian guidelines are useful in confirming causal relationships since understanding the nature of the relationship between two variables allows experiments that can predictably disrupt that relationship to be designed.

Unlike secondary natural product/health links, the link between macro and micronutrient intake and health can more easily be demonstrated to be causal. For example, riboflavin (vitamin B2) has defined physiological roles. It is a biological precursor of flavin adenine dinucleotide (FAD) and flavin mononucleotide (FMN). Any physiologically relevant enzymic processes that rely on these cofactors may be affected by riboflavin deficiency [32]. Supplementation with riboflavin can be demonstrated to impact on plasma biomarkers of vitamin B2 [33]. While deficiency of riboflavin is causally linked to anaemia, it is also involved more generally in energy metabolism and the difference between avoidance of deficiency and optimal intake should be noted [34]. The clinical endpoint of the presence of clinical symptoms of anaemia may be preceded by other symptoms of a sub-optimal intake [35] and clinical symptoms may only occur after a lengthy period of deficiency. In addition, anaemia is a multifactorial process so, as noted previously, causality may not be a simple matter of a single cause and effect process. The implication of this is that, even in the case of essential dietary components where a deficiency has a clinical consequence, the causality of a relationship may not be absolute. In some cases of secondary natural products, the proof of the level of causality may be a consequence of deleterious physiological effects. For example, there are very specific effects resulting from the ingestion of some microbial toxic entities (e.g. some marine toxins—for example Amnesic Shellfish Poisoning (ASP), Paralytic Shellfish Poisoning (PSP), Diarrheic Shellfish Poisoning (DSP) and Azaspiracid Shellfish Poisoning (AZP)—are named according to their physiological function [36]. The toxin, goitrin which occurs in *Brassica* species has a distinct physiological effect—it blocks tyrosine iodination and inhibits T4 formation [37], however it is not simply a monofunctional toxin since it also appears to be protective against septic shock [38].

In summary, the nature of secondary natural products is complex—even when chemical reductionism is applied and single compounds are examined in their interaction with living systems. The clear distinction between correlation and causality can be blurred although a complete absence of mechanistic information means that there may be an absence of testable hypotheses. Toxins may be more likely to be vectors of causality since their removal may have a more defined effect on clinical outcomes however the role of confounders remains significant. It can be argued that, like mathematical laws, the laws of causality do not apply, in any precise sense in nature [39]. When causality is applied to natural phenomena, it is often unprovable due to the complexity of the systems. There is, nonetheless, a degree of intellectual value in seeking causation—even if the perfect situation is elusive and complete predictivity is unavailable, there are still opportunities to develop mathematics in biology and to thereby obtain at least a partial vision and some level of quantification [40]. The ability to measure the intricacies of a living system can, on occasion, be mistaken for the actual truth of the universe. In many cases, the observation and measurement of specific characteristics is an incomplete surrogate for reality. The enthusiasm for unbiased observation has enabled scientists (and non-scientists) to develop a set of characteristics that are said to define the 'scientific method'. The promulgation of this method suggests that all science shares certain characteristics in common. In essence, because science is united by the use of this single, underlying method, it must represent unambiguous and objective truth. This obsession with the methodological aspects of measurement has, at times, obscured the importance of what is being measured rather than how it is evaluated and led to an obsession with higher and higher levels of accuracy and greater and greater sensitivity.

6.4 Post-biosynthetic Modification of Natural Products

The post-biosynthetic modification of natural products is responsible for some of the structural diversity that is observed among molecules in nature [41]. For example, in the case of polyketides, glycosylation and other modifications are known to occur once the core structure has been synthesised. Numerous examples of post-biosynthetically modified polyketide-derived natural products including griseorhodins (Fig. 6.3) and others [42].

Another group of natural products comprise complex peptides. Two main routes have been identified for the biosynthesis of complex peptides—either ribosomally synthesised (and often post-translationally modified) or non-ribosomally synthesised [43]. In the former group are included anti-microbial peptides that are amphiphilic and cationic while the latter group display a wide range of activities and are synthesised through the action of non-ribosomal peptide synthases (NRPS). In addition, by combining enzymes from different pathways, novel natural products can be prepared. Ribosomally synthesized and post-translationally modified peptides are also referred to as RiPPs products. Data from genome sequencing has led to a rapid increase in the number and variety of RiPPs that have been identified. It is often the case that, while a significant number of post-translational modifications can be carried out on core RiPPs, the number of different enzymes involved may be limited to a fairly small collection [44]. Dehydration and heterocyclisation are two common reactions whose products (derived from serine and threonine in the case of dehydration or the production of azol(in)es in the case of heterocyclisation, as can be seen in the structure of Bottromycin A2 (Fig. 6.4)), are found as structural motifs in many RiPPs.

A range of other particular structural features are commonly found including cyclic peptides in bacteria [45] and in plants [46], Lantibiotics [47] which are found

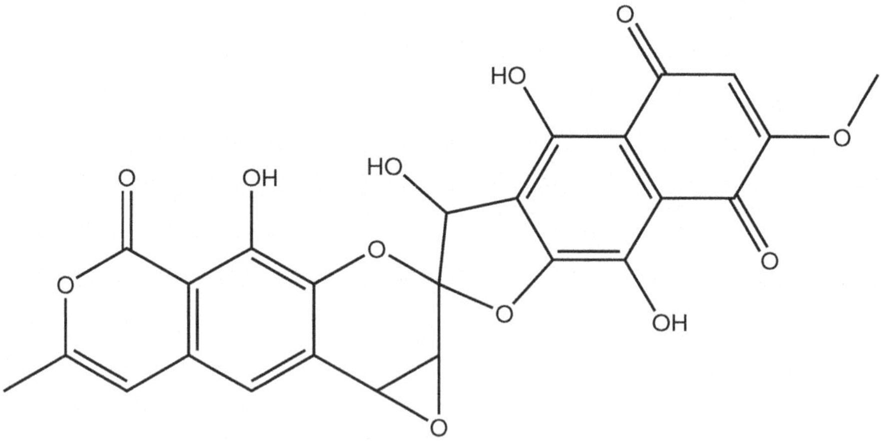

Fig. 6.3 Structure of Griseorhodin A

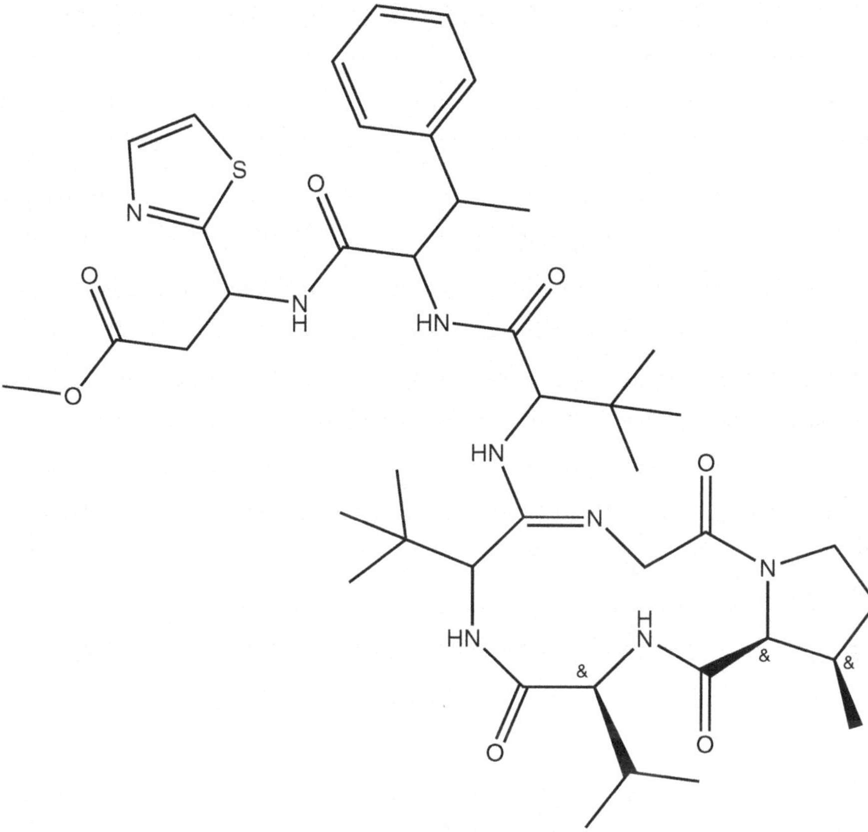

Fig. 6.4 Structure of Bottromycin A2—a representative RiPP

in food systems as preservatives [48] and other peptides are subject to modifications using cytochrome P450 type enzymes [49]. Post-translational modification, some examples of which were noted above for polyketides and peptides, generally refers to the addition of exogenous moieties or the formation of new structures (e.g. cyclisations). The formation of isothiocyanates and sulphides are not considered to be post-translational modifications although they result from post-biosynthetic, enzyme-catalysed reactions and they result from the conversion of precursors into novel molecules. These will be considered in more detail in Sect. 6.5—particularly in the context of food and food systems.

6.5 Formation of Isothiocyanates and Sulphides—Compare and Contrast

Isothiocyanates and sulphides are produced by *Brassica* and *Allium* plants respectively. In both cases, the toxic entities are stored by the plant as a relatively inactive precursor which, upon the plant experiencing cellular damage, is brought into contact with an enzyme which cleaves the precursor to give rise to a range of products. In the case of *Brassica* plants, only a limited number of structures are formed and most of the diversity in the formation of products arises from the existence of a wide range of precursors. In the case of plants of the genus *Allium* there are only a limited number of precursors and the diversity in product formation arises from reactions between the post-enzymic products. These two different approaches to protecting the plant from damage use the same general mechanism but with different routes to essentially the same result—the formation of a range of bioactive secondary natural products.

Glucosinolates were first characterised (by isolation of their enzymically produced breakdown products) as mustard oils and later as crystalline compounds that were known to contain a sugar component and so were called mustard oil glycosides. The glucosinolates in plants are stored separately from the enzyme (myrosinase) that breaks them down [50] to give initially an aglucone which then undergoes a rearrangement to give the corresponding isothiocyanate and other products (Fig. 6.5).

Isothiocyanates are toxic electrophilic compounds which can bind to thiol groups (for example in glutathione) and have both anti-microbial activity [51] and mechanistically proposed, anti-cancer activity—partially as a consequence of their enzyme inhibitory activity [52]. The propensity of isothiocyanates to bind to thiol groups in proteins and other biological molecules may be a factor in their storage as an inactive glucosinolate which is only converted to the biologically active isothiocyanate under stress (damage to plant structures) conditions [53]. Several reports have suggested that the tricarboxylic acid cycle enzyme, succinate dehydrogenase has a cysteine-derived thiol residue in its active site although other isoforms of the enzyme may also exist [54]. The capacity of isothiocyanates to bind covalently to cysteine residues [55] suggests that free isothiocyanates may be deleterious to energy metabolism in living systems [56, 57]. In addition, part of the action of isothiocyanates involves membrane disruption [58] and while this has been observed in bacteria, if it were also to occur in plants then initial damage and release of isothiocyanates would further disrupt the cell wall separating the enzyme myrosinase from the inactive glucosinolate leading to an accelerated production of isothiocyanates.

The post cellular damage formation of isothiocyanates has been referred to as the 'mustard oil bomb' [59]. The two components of the bomb—the biologically inactive glucosinolates and the thioglucosidase enzyme, myrosinase, are both variables that can affect predation of plants by insects. For example, plants with low myrosinase activity suffer greater damage from the crucifer flea beetle (*Phyllotreta cruciferae*) than those with high myrosinase activity [60] and the glucosinolate amount and profile can affect the types of insects that consume Brassicaceous plants [61]. In the development of resistance to the adverse effects of isothiocyanates, insects have

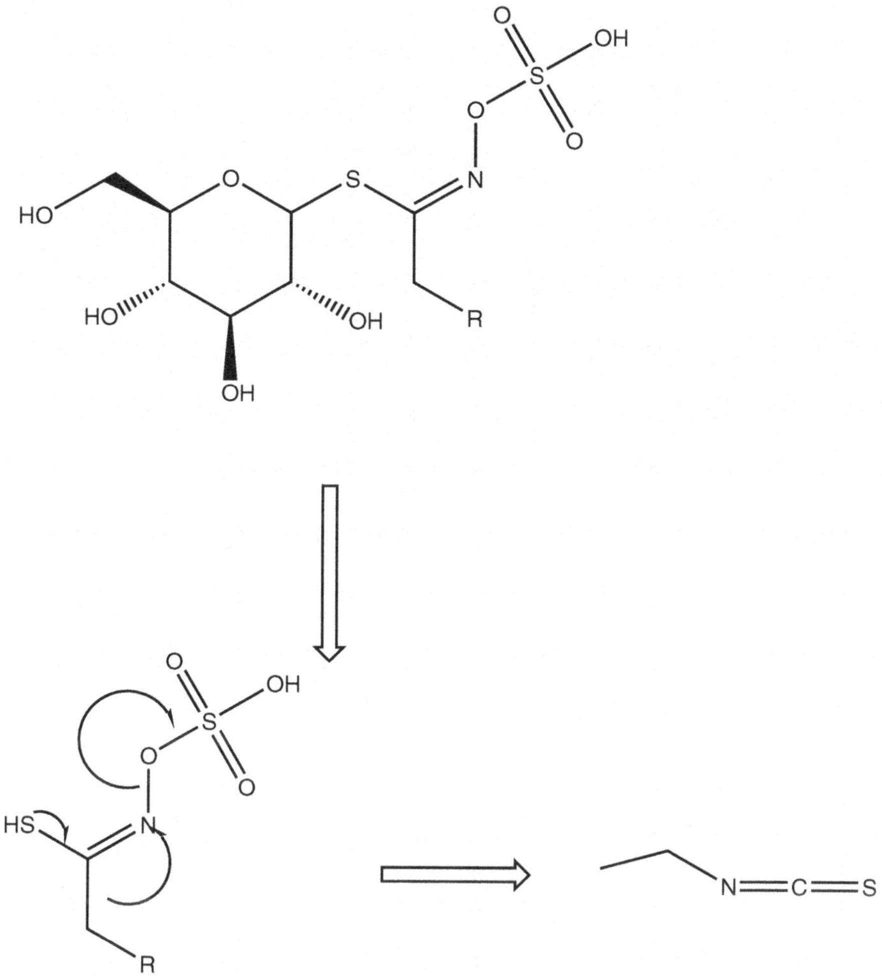

Fig. 6.5 Rearrangement of glucosinolate aglucone

developed a number of strategies including removal of the sulphate group using insect sulphatases [62] and sequestering glucosinolates or diverting their breakdown towards less toxic nitriles rather than isothiocyanates [63]. In the case of production of desulphated glucosinolates, the products are not simply unable to be hydrolysed by myrosinase, they act as inhibitors of the enzyme [64].

Glucosinolates are derived from amino acids and the core structure is built up following various modifications including chain extension, oxidation and alkylation. As noted previously, these modifications account for the large number of different glucosinolates that exist in nature. The breakdown of glucosinolates, which is responsible for the manifestation of much of the biological profile of plants that contain glucosinolates, appears to follow one of three distinct, post enzymic pathways. In the case of most of the glucosinolates, the isothiocyanates that are formed are the final metabolic product. These compounds are primarily employed by the host plant to deter predation [65]. Their biological activity is in part dependent upon the electrophilicity of the C atom of the isothiocyanate group which in turn is influenced by the nature of the amino acid derived side chain. Indole glucosinolates follow a different reaction pathway. It is postulated that they form an unstable isothiocyanate which is hydrolysed to give, as a primary product, indole-3-carbinol. This compound is relatively stable but can selfcondense to give a range of products (Fig. 6.6) including, in particular, 3,3′-diindolylmethane and indolo[3,2-b]carbazole.

While the intermediate isothiocyanate has not been isolated and fully characterised, its transient occurrence has been noted [66] and the formation of thiocyanate ion [67] and ascorbigen (a condensation product of indole-3-carbinol with ascorbic acid) [68] is further, indirect evidence of the pathway. It has been suggested that indole glucosinolates are precursors for the formation of phytoalexins such as Brassinin (Fig. 6.7) or may share a common precursor [69].

The third group of glucosinolates are those whose enzyme products spontaneously cyclise to form oxazolidinethiones (Fig. 6.8).

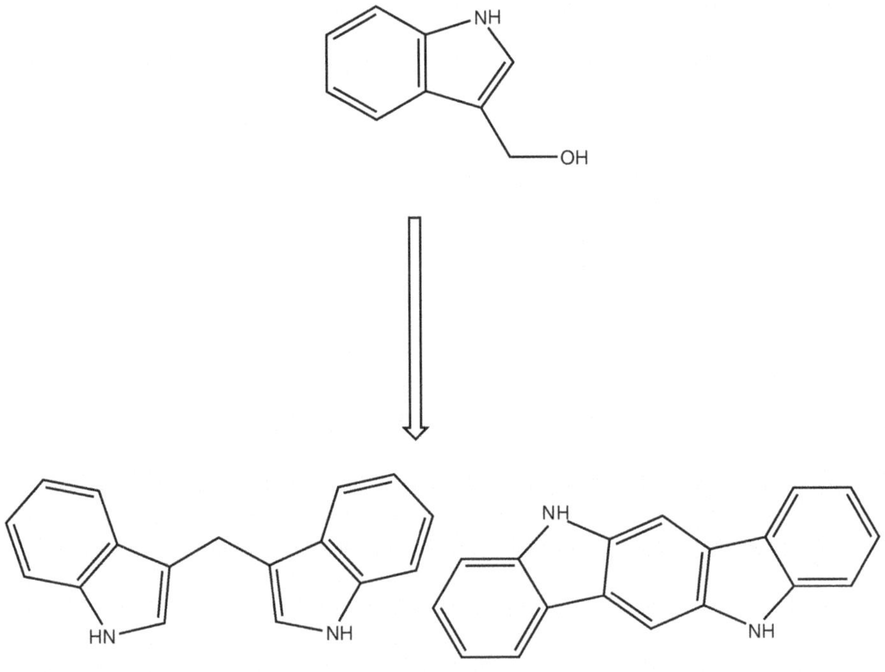

Fig. 6.6 Products from Indole-3-carbinol

Fig. 6.7 Structure of brassinin

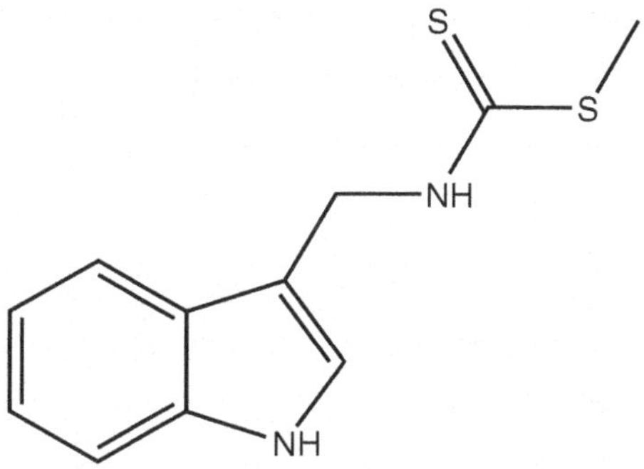

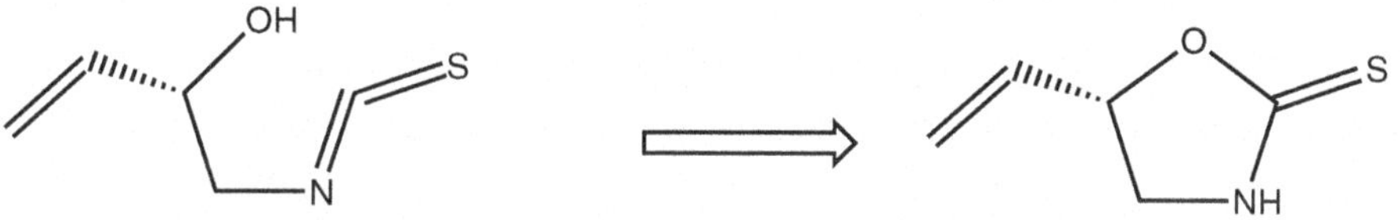

Fig. 6.8 Structure of goitrin and its precursor

These compounds, the most common of which is the vinyl analogue called goitrin. This compound is responsible for a proportion of the bitter taste of *Brassica* vegetables [70] and therefore may also be an anti-feedant principle. The different biological functions of different glucosinolates (or, more particularly, their breakdown products) is reflected in their occurrence in different parts of plants. In addition, myrosinases with different specificities will also affect the profile of different glucosinolate breakdown products formed at different plant locations and at different times.

The formation of sulphides by plants of the genus *Allium* follows a different pathway and rationale to the formation of glucosinolates, isothiocyanates and the 'mustard oil bomb'. While a number of species of the genus *Allium* exist, the most significant, in terms of use both medicinally and in cooking are onion (*Allium cepa*) and garlic (*Allium sativum*). The key starting materials for the wide range of S-containing compounds are a small group of S-alk(en)yl-L-cysteine sulphoxides (Fig. 6.9).

Fig. 6.9 Structure of S-Alk(en)yl-L-cysteine Sulphoxides

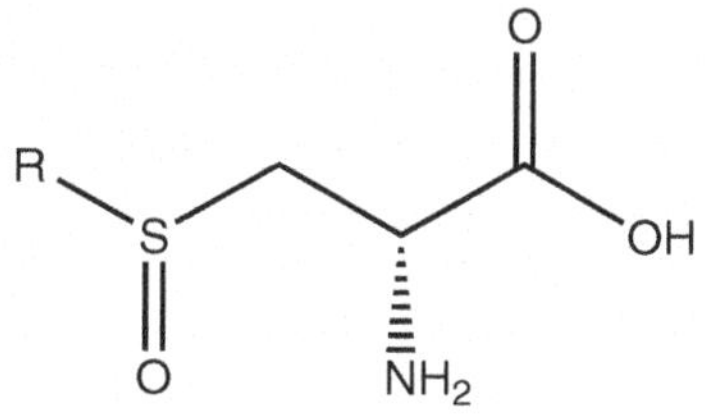

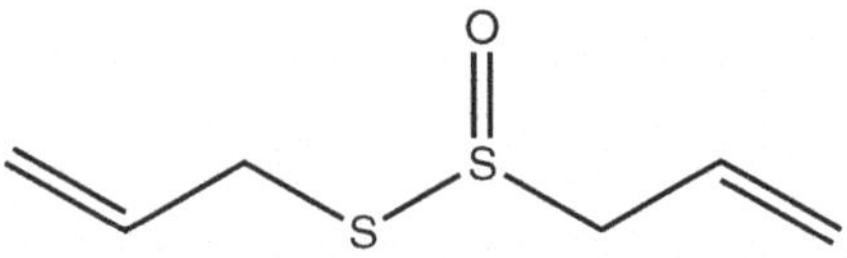

Fig. 6.10 Structure of allicin

These precursors are cleaved by an enzyme, alliinase to give sulphenic acids which then undergo a range of reactions to give S-containing metabolites including (from allyl-Lcysteine sulphoxide) allicin (Fig. 6.10).

By analogy with glucosinolates and myrosinase, the enzyme and substrate are separated in the intact plant and it is only after damage to the cells that contact is made and the reaction occurs. The similarity between plants of the genus *Allium* and those that are identified as being of the genus *Brassica* extends to the unstable nature of the initially formed enzyme product—sulphenic acids or aglycones. The first of the precursors isolated was S-allyl cysteine sulphoxide which was isolated in 1948 [71] and given the trivial name 'alliin'. It is interesting to note that the first glucosinolate (sinigrin) was isolated considerably earlier in 1840 [72] and a structure (later found to be incorrect) proposed in 1897 [73, 74]. The reason that glucosinolates were isolated over 100 years before the first alk(en)yl cysteine sulphoxide was obtained in purified form is likely because sinigrin occurs as a single isolable compound in black mustard seed while alliin is present in plant tissue along with other members of the class. This relatively late isolation had one particular advantage. The structure of alliin was confirmed by synthesis shortly after its isolation [75] unlike sinigrin whose correct structure was not proposed until 1956 [76] and it was not synthesised until 1965 [77].

Three additional members of the class were later isolated from onion (S-methylcysteine sulfoxide (methiin), S-1-propenylcysteine sulfoxide (isoalliin—also isolated as the lachrymatory precursor, see below), and S-n-propylcysteine sulfoxide (propiin)) and these have been found to be the main precursors present in nature. While butyl and ethyl analogues have been found, these are minor components from *Allium* species.

Two of the most significant products from the further reactions of sulphenic acids are ajoene ((1E)-3-(Prop-2-ene-1-sulfinyl)-1-[(prop-2-en-1-yl) disulfanyl]prop-1-ene) and the lachrymatory factor of onion (propanethial-S-oxide). In addition, while thiosulphinates are relatively unstable, as noted previously, and readily form thiosulphonates and disulphides, they can, like isothiocyanates, react with -SH groups to form a range of adducts. This may make a significant contribution to their observed biological activity [78]. The formation of a 'lachrymatory factor' in cut onion is one of the defining characteristics of the vegetable. The formation of the lachrymatory factor and its consequences have been extensively reviewed [79]. The starting material is (+)-S-(E)-1propenyl-L-cysteine S-oxide which is cleaved by alliinase and the intermediate sulphenic acid acted on by lachrymatory factor synthase to give the final product (Fig. 6.11).

The crystal structure of lachrymatory factor synthase has been reported [80]. Originally it was assumed that the formation of the lachrymatory factor was a spontaneous

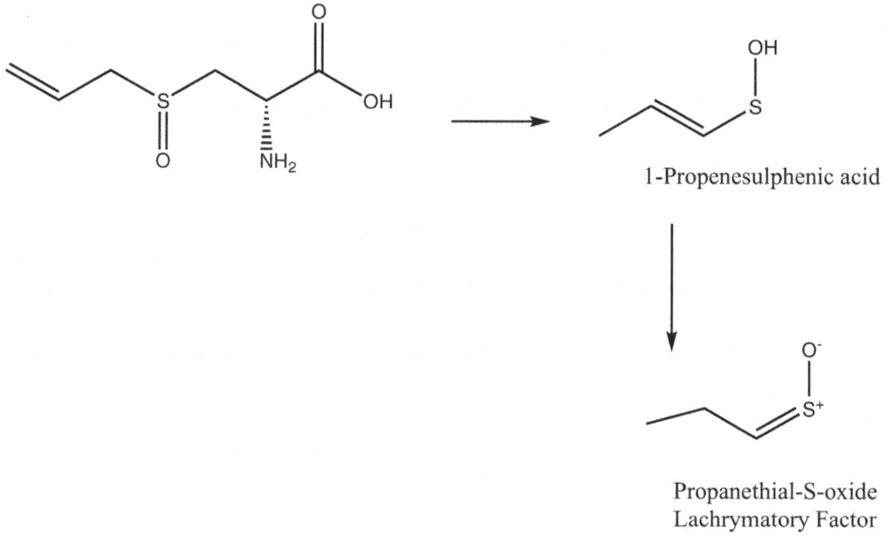

Fig. 6.11 Formation of onion lachrymatory factor

process arising from the rearrangement of the precursor (E)-1-propenesulphenic acid [81], however it was later realised that the reaction required enzymic catalysis [82]. The original isolation of the chemical entity responsible for formation of the lachrymatory effect was a fractionation/bioactivity study that relied upon the use of the human eye to test fractions for lachrymatory potential [83]. The investigators realised that the lachrymatory factor was unstable and postulated that it might be formed by the enzymic degradation of an inactive precursor (it is no coincidence that the Principal Investigator in the study was Professor A I Virtanen who carried out the isolation of a number of glucosinolates—compounds that also acted as precursors for the formation of volatile plant compounds). After having prepared a freeze dried, crude alliinase preparation from onion, they tested fractions using the following protocol "4–6 drops of the aqueous solution to be tested was placed in a small weighing vial and the pH was adjusted to 6–7. A small amount of the enzyme powder—usually about 5 mg—was added to the solution, the suspension was stirred for 10 s and the vial held tightly, but without pressure, to the eye. Depending on the concentration, the lachrymatory effect was perceived after 15–45 s, usually becoming unendurable after an additional 5 s." The lachrymatory factor precursor was isolated and proposed on the basis of further tests to be S-(prop-1-enyl)-cysteine sulphoxide. The authors also describe a series of elegant chemical and enzymic transformations to confirm the structures of the lachrymatory precursor and the lachrymatory factor itself. Completion of the structural confirmation by synthesis (by analogy with the synthesis of alliin) was not able to be carried out.

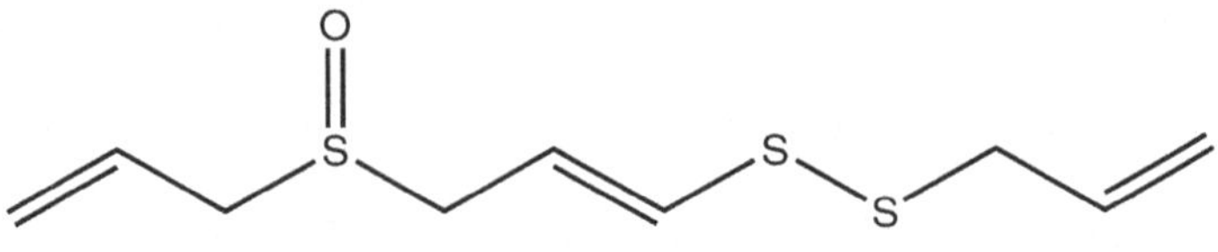

Fig. 6.12 Structure of ajoene

A second important compound formed from garlic is ajoene (Fig. 6.12). This compound is formed from allicin and was first isolated in 1983 [84] on the basis of its anti-platelet activity.

It was determined to be (E,Z)-4,5,9-trithiadodeca-1,6,11-triene-9-oxide [85]. While Zajoene is the predominant form in fresh macerated garlic extract, it is gradually converted to the E form on storage. Both forms of ajoene are more stable than allicin and therefore production in large amounts became an important objective in order to confirm its potential as a pharmacological agent. However the natural levels of ajoene are fairly low (600 μg/g of garlic) and the biomimetic synthesis of it from allicin gives poor yield and a range of side products. A short total synthesis has been carried out [86] and the product biologically verified by demonstration of its bacterial anti-quorum sensing properties [87]. In addition to anti-microbial, anti-fungal and anti-viral properties ajoene is also effective against platelet aggregation and as an anti-cancer agent [88].

The examples of compounds from two major groups of plant species (*Alliums* and *Brassicas*) demonstrates the importance of sulphur-containing compounds as precursors of bioactive plant protective secondary natural products. In both cases the biological system is activated following cellular damage which facilitates the mixing of a precursor with a very specific enzyme. The initial product is, in both the examples considered, relatively unstable and engages in further reactions. However the approaches are also divergent. In the case of glucosinolates, the functional structural diversity is limited by the different side chains. These can have an effect on the reactivity (and hence the biological profile) of the isothiocyanates and nitriles that are the major products of the decomposition of the initially formed aglycone. In the case of the cysteine sulphoxides, the genesis of the various diverse products is driven from a narrow base (principally just 4 precursors) and is a consequence of the chemical reactivity of the initial products and the various sulphides that are then formed. It is interesting to note that both thiosulphinates and isothiocyanates can react with thiol groups and may give rise to a range of other products. For example, the reaction of isothiocyanates with thiols (such as are formed in *Allium* species) will give rise to dithiocarbamates which may also have an interesting biological profile [89] including neuroprotective activity against Alzheimer's disease [90].

6.6 Modulation of Natural Product Effects

While many secondary natural products are designed to deter predation/consumption, there are some species that directly compete with each other for resources and 'chemical warfare' seems to be designed to confer a growth and survival advantage. In other cases the growing realisation that taste receptors may have a number of biological functions and therefore compounds that bind to them may have a range of biological activities [91] has suggested alternative reasons for organisms making an investment in secondary natural product synthesis. In addition, some secondary natural products may be designed to protect organisms from invasion by microbial species, while the microbes also develop strategies to overcome anti-microbial defences Some of the uses of secondary natural products will be considered in the following categories.

- Anti-bacterial compounds and resistance to them
- Quorum sensing inhibition as a strategy to prevent pathogen growth
- Effects on the immune system
- Anti-feedant and inhibitors of digestion.

Many of the effects ascribed to secondary natural products are those which have been found to be most useful to humans (particularly medicinal effects) and are not the uses for which the compounds were synthesised by the hosts.

The 'Golden Age' of natural product chemistry is considered by many to be synonymous with the isolation and development of anti-microbials—particularly those that incorporate a beta lactam or a macrolide structural feature. Some of the major classes of antibiotics, their origin and their primary mode of action are given in Table 6.1.

Table 6.1 The major classes of antibiotics, their origin and their primary mode of action

Antibiotic type and example	Origin	Mode of action
Beta lactams (penicillin, cephalosporin)	*Penicillium, Cephalosporium* fungi	Inhibit bacterial cell wall synthesis
Amino glycosides (streptomycin, kanamycin)	*Streptomyces* bacteria	Inhibit bacterial protein synthesis
Macrolides (erythromycin)	*Saccharopolyspora erythraea,* bacteria	Inhibit bacterial protein synthesis
Tetracyclines (doxycycline)	*Streptomyces* bacteria	Inhibit bacterial protein synthesis
Glycopeptides (vancomycin)	*Amycolatopsis orientalis* bacteria	Inhibit bacterial cell wall biosynthesis
Quinolones (ciprofloxacin)	Synthetic	Interferes with bacterial DNA replication and transcription
Lipopeptides (daptomycin)	*Streptomyces* filamentosus bacteria	Disrupts bacterial cell membrane function

One of the major problems associated with the widespread use of antibiotics is the development of resistance. Antimicrobial resistance (AMR) is recognised as a global problem by the World Health Organisation. It does not just apply to antibacterials but also to anti-fungal, anti-viral and anti-parasitic drugs. In addition to finding new antimicrobials with novel and specific properties (in order to minimise side effects), another approach is to use combinations to target a specific bacterial function in a synergistic fashion. For example, it has been suggested that amoxycillin-resistant *Streptococcus pneumoniae* can be resensitised to the antibiotic by using a mevalonate pathway inhibitor [92]. The mechanism of this action is the downregulation of the mevalonate pathway leading to cell elongation. By using an inhibitor of undecaprenyl phosphate (clomiphene), resistant *Streptococcus pneumoniae* are resensitised to amoxycillin. Similar combination therapies may prove valuable in the cases of other resistance mechanisms. It is worth noting that the mevalonate target was identified using a genome-wide, single-cell based, gene silencing screen using CRISPR interference.

The discovery of quorum sensing has opened a further area for the investigation of control of the growth of microorganisms. The modulation of the activity of antimicrobial agents by their target pathogens can take many forms. Some of these are outlined below. Quorum sensing is a cell to cell communication system using small molecules. If a microbial community is of a sufficient size, then the production of quorum sensing molecules can stimulate a number of effects including virulence, swarming, biofilm formation and AMR. As noted previously, AMR can take a number of forms. These include the production of enzymes to break down the antimicrobial agent (e.g. beta-lactamases), efflux pumps to remove antimicrobial compounds from the cell and the formation of biofilms to prevent microbial cells being exposed to anti-microbials. Quorum sensing molecules are involved in each of these mechanisms so the discovery or development of quorum sensing inhibitors has the potential to regulate pathogenic microbial growth [93]. In addition, because some quorum sensing mechanisms and molecules are common to several species, quorum sensing inhibition strategies may also control polymicrobial infections [94]. Quorum sensing inhibition using specific secondary natural products (also called quorum quenching) may be a means for plant species to develop resistance to microbial pathogens. Indeed, many of the antimicrobial properties of plants may be a consequence of the production of secondary natural products with quorum disrupting properties. A range of sources of putative plant quorum quenching activities have been investigated. These include phenolics, quinones, flavonoids, alkaloids, terpenoids, and polyacetylenes as well as specific compounds (Fig. 6.13) and ajoene from garlic [95].

Further work on the modulation of AMR through the use of quorum quenching is likely to be a growing area of interest.

One of the most significant areas of interest in terms of the modulation of the effects of natural products is that of changes to the immune system. Immunomodulation by secondary natural products is a first line of defence against foreign compounds and organisms. While many of the mechanisms and the structure/activity relationships are often unclear, the evidence base is building [96]. In simple terms, the immune system can be considered to consist of two parts—innate immunity and adaptive

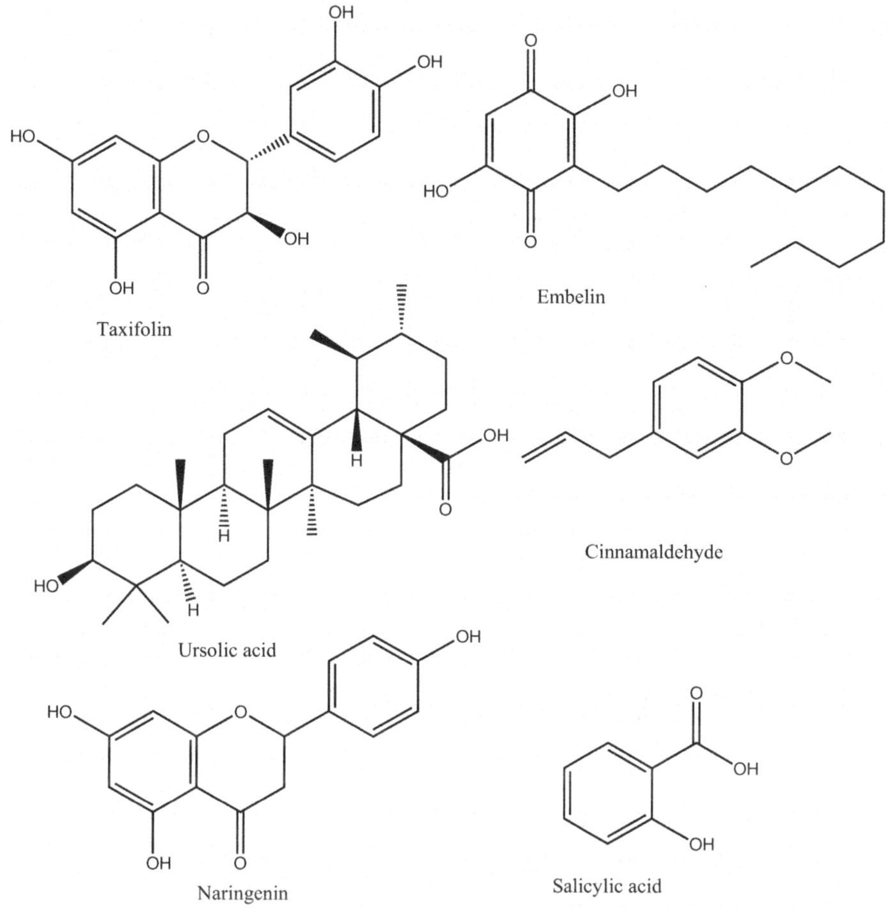

Fig. 6.13 Some phytochemicals with microbial quorum sensing inhibitor activity

immunity. The first of these is a non-specific response and leads to effects such as phagocytosis and mitogen-activated protein kinase (MAPK) pathways. The adaptive immune response takes longer but this system retains a memory of the exposure and is activated when reexposed and leads to the production of antigen-specific antibodies by B cells. There are a number of other aspects to the immune response which are detailed in several publications. The way that food components can provoke an immunostimulatory response is varied and can include the stimulation of cytokine production, induction of immune cell production and modulation of T cell response. Among the specific entities that appear to have immunostimulatory activity are beta glucans, poly-γ-glutamate, polysaccharides in *Chlorella* and exopolysaccharides of some *Lactobacillus* species. In addition to an immunostimulatory response, there has been considerable interest in the immunomodulatory effects of secondary natural products [97] and their capability to control inflammatory responses. There has been considerable interest in China in traditional biopharmaceuticals however this is often through the study of complex mixtures and extracts rather than the reductionist, isolative approaches taken by pharmacology [98].

One further class of secondary natural products that have provoked some interest are those which inhibit digestion or are resistant to providing nutrients. Some of these that are related to processed macronutrients (e.g. starch and resistant starch and prebiotics) will be discussed in Chapter Eight however some anti-nutrients will be considered here. The broad definition of anti-nutrients are those dietary factors which inhibit the absorption of nutrients. They are generally considered to include lectins, goitrogens, phytates and oxalates. In addition, some dietary constituents such as saponins and tannins are sometimes classified as anti-nutrients. The conclusion that some dietary components are harmful as a result of their capacity to reduce nutrient uptake is debatable for two reasons. First of all, some of these components have been found to have beneficial properties in addition to their anti-nutritive profile. Secondly, humans have adapted over many years to the consumption of certain foods. The processing (and/or cooking) of food which may be designed to remove anti-nutritive aspects is a relatively recent phenomenon and therefore the consumption of nutrientrich food may have unintended consequences.

Lectins are glycoproteins that contain carbohydrate binding sites. By binding to specific glycoproteins (monosaccharide and polysaccharide sites), they can agglutinate red blood cells. In addition lectins can interfere with the absorption of nutrients. The structure of lectins depends on their source and they are found throughout living species including microbes, plants and animals. Some isolated lectins appear to have anti-proliferative activity and also are anti-microbial and anti-viral. Immunomodulation has also been ascribed to some lectins although mechanistic information for most effects has not been assembled.

The goitrogenic activity of thiocyanate ion and vinyl oxazolidine thione (goitrin) which are both derived from glucosinolates has been described elsewhere. The mechanism may involve inhibition of TH4 (goitrin) or competition with uptake of iodide (thiocyanate ion). As with lectins, heating can remove the biological potency of glucosinolates although in this case such treatment also inactivates the thioglucosidase enzyme, myrosinase thereby preventing formation of the goitrogenic principles. Many of the studies on glucosinolates and their metabolites have been carried out using rodent models or in vitro cell culture. The undifferentiated nature of secondary cell lines means that any organotypic effects should be treated with caution. In addition, it has been found that many of the cell lines commonly used are contaminated with rapidly growing HeLa cells (a cancer derived cell line) and therefore produce results that are intrinsically unsafe [99].

Phytates (myo-inositol hexaphosphate or IP6) are found in a number of plant species—particularly beans and pulses. The activity of phytates is a consequence of their capability to form soluble complexes at acid pH in the stomach with divalent cations (Zn, Fe, Ca). When they pass into the neutral pH of the intestine, these complexes precipitate and the cations cannot be absorbed. Phytates can also, in some circumstances, form complexes with proteins. As with other anti-nutrients mentioned above, heating (and even just irrigation with water) can significantly

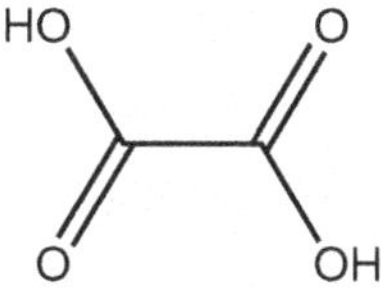

Fig. 6.14 Structure of oxalates

reduce the levels of phytate ingested. Phytates have been postulated to have a positive metabolic effect. Indeed, as inhibitors of 3-hydroxy-3-methylglutaryl-coenzyme A (HMGCoA) reductase, they may function a in a similar way to statins. If one effect of phytates is to inhibit the mevalonate pathway then, as noted previously, they may also re-sensitise amoxycillinresistant bacterial strains or have co-antibacterial activity, as may also be the case for statins [100].

Oxalates are another common type of anti-nutrient (Fig. 6.14).

Oxalates can bind to a variety of cations however the link between oxalate consumption and specific clinical endpoints (for example, the risk of developing kidney stones) remains questionable. It has been hypothesised that a combination of oxalate together with Ca leads to the formation of calcium oxalate stones however there is little to indicate a link between dietary constituents that are high in oxalates (e.g. spinach) and stone formation.

Two other postulated anti-nutrients are saponins and tannins. Saponins are a widely occurring series of compounds comprising a hydrophobic portion (triterpene or sterol) and a hydrophilic (carbohydrate) residue [101]. The hydrophobic part of the molecule derives, biosynthetically from the mevalonate pathway. While saponins can affect nutrient uptake and often have a bitter or astringent taste, they are also useful medicinal agents and have been associated with cardioprotective and antimicrobial effects. They are also used as natural surface active foaming agents/emulsifiers and in other food uses [102]. While a large number of saponin structures with diverse biological activities have been found, as with many secondary natural products they generally occur in small amounts and this limits their commercial application and increases the opportunity for synthesis of useful natural and unnatural members of the class [103]. Tannins are polyphenols present in many foods. In plants, one of their natural functions is as antimicrobials to prevent pathogen growth. As with other complex natural products, the association of tannins with anti-nutritional effects is uncertain [104]. Compounds such as catechins, proanthocyanidins and epicatechins (Fig. 6.15) have been associated with a number of health benefits including increased blood flow to the brain [105] and cardioprotective effects [106].

While observational studies can, as noted previously, help to suggest directions for future research, epidemiology (and in particular nutritional epidemiology) are difficult to interpret and to develop meaningful interventions to demonstrate correlation and causality due to the size of the effects observed and the ubiquitous presence of confounders. While some adaptive responses have helped producers of natural products to benefit from their development, target species have also adapted and this can lead to the rise in AMR, for example. In addition, some target species may benefit from the development of a change in use of secondary natural products to

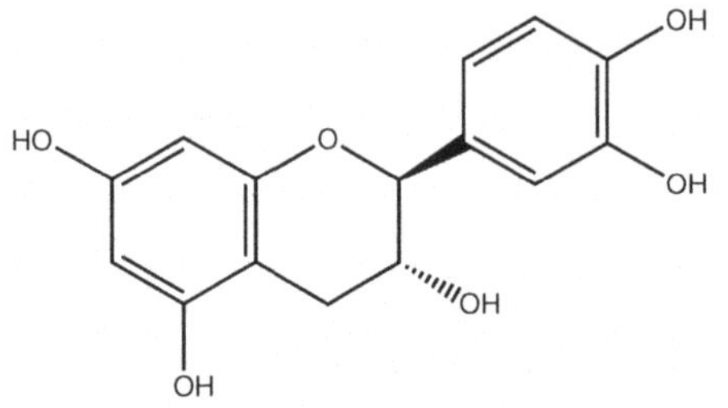

Catechin

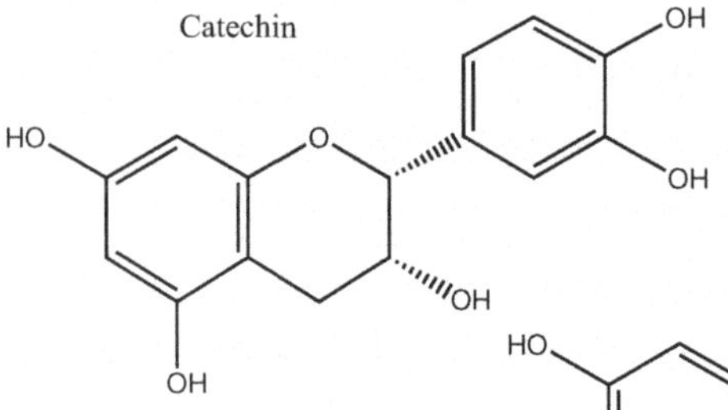

Epicatechin

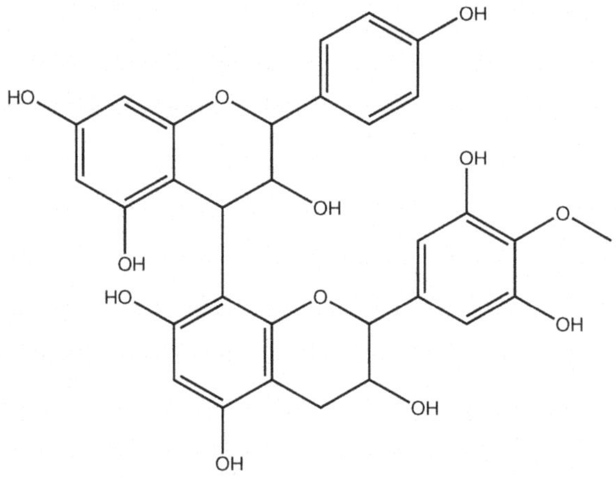

Proanthocyanidin

Fig. 6.15 Structures of catechins, proanthocyanidins and epicatechins

confer a marginal advantage for the consuming species. In this way, secondary natural products may be a factor in the evolutionary development of species.

References

1. Parusniková Z. Karl Popper: his philosophy and science. In: Parusniková Z, Merritt D, editors. Karl Popper's science and philosophy. Cham: Springer; 2021. https://doi.org/10.1007/978-3-030-67036-8_1.
2. Adolph WH. The history of chemistry in China. Scient Monthly. 1922;14(5):441–6.
3. Ball P. Synthetic organic chemistry in China: building on an ancient tradition—an interview with Qi-Lin Zhou and Xiaoming Feng. Natl Sci Rev. 2017;4(3):437–40. https://doi.org/10.1093/nsr/nwx035.
4. Ioannidis JP. Why most published research findings are false. PLoS Med. 2005;2(8):e124. https://doi.org/10.1371/journal.pmed.0020124. Epub 2005 Aug 30. Erratum in: PLoS

Med. 2022 Aug 25;19(8):e1004085. https://doi.org/10.1371/journal.pmed.1004085. PMID: 16060722; PMCID: PMC1182327.
5. Di Lorenzo E, Romano F, Ciriaco L, et al. Periodic cooking of eggs. Commun Eng. 2025;4:5. https://doi.org/10.1038/s44172-024-00334-w.
6. Shim JS, Oh K, Kim HC. Dietary assessment methods in epidemiologic studies. Epidemiol Health. 2014;22(36): e2014009. https://doi.org/10.4178/epih/e2014009. PMID: 25078382; PMCID: PMC4154347.
7. Tulchinsky TH. John snow, cholera, the broad street pump; waterborne diseases then and now. Case Stud Public Health. 2018:77–99. https://doi.org/10.1016/B978-0-12-804571-8.00017-2. Epub 2018 Mar 30. PMCID: PMC7150208.
8. National Research Council (US) Committee on Indicators for Waterborne Pathogens. Indicators for waterborne pathogens. Washington (DC): National Academies Press (US); 2004. 1, Introduction and historical background. Available from: https://www.ncbi.nlm.nih.gov/books/NBK215658/.
9. Bennett DA, Landry D, Little J, et al. Systematic review of statistical approaches to quantify, or correct for, measurement error in a continuous exposure in nutritional epidemiology. BMC Med Res Methodol. 2017;17:146. https://doi.org/10.1186/s12874-017-0421-6.
10. Hu Y et al. The impacts of measurement errors on a dietary pattern analyses: a simulation study based on dietary data from the China Multi-Ethnic Cohort (CMEC) study. Am J Clin Nutr. 116(2):523–30.
11. Willett W. Nutritional epidemiology: issues and challenges. Int J Epidemiol. 1987;16(2):312–7. https://doi.org/10.1093/ije/16.2.312. PMID: 3610460.
12. Brown AW, Aslibekyan S, Bier D, Ferreira da Silva R, Hoover A, Klurfeld DM, Loken E, Mayo-Wilson E, Menachemi N, Pavela G, Quinn PD, Schoeller D, Tekwe C, Valdez D, Vorland CJ, Whigham LD, Allison DB. Toward more rigorous and informative nutritional epidemiology: the rational space between dismissal and defense of the status quo. Crit Rev Food Sci Nutr. 2023;63(18):3150–67. https://doi.org/10.1080/10408398.2021.1985427. Epub 2021 Oct 22. PMID: 34678079; PMCID: PMC9023609.
13. Schulz CA, Oluwagbemigun K, Nöthlings U. Advances in dietary pattern analysis in nutritional epidemiology. Eur J Nutr. 2021;60(8):4115–30. https://doi.org/10.1007/s00394-021-02545-9. Epub 2021 Apr 25. PMID: 33899149; PMCID: PMC8572214.
14. Ying K, Liu H, Tarkhov AE, Sadler MC, Lu AT, Moqri M, Horvath S, Kutalik Z, Shen X, Gladyshev VN. Causality-enriched epigenetic age uncouples damage and adaptation. Nat Aging. 2024;4(2):231–46. https://doi.org/10.1038/s43587-023-005570. Epub 2024 Jan 19. PMID: 38243142; PMCID: PMC11070280.
15. Tarkhov AE, Lindstrom-Vautrin T, Zhang S, et al. Nature of epigenetic aging from a single-cell perspective. Nat Aging. 2024;4:854–70. https://doi.org/10.1038/s43587-024-00616-0.
16. Gunaratne SA. Understanding systems theory: transition from equilibrium to entropy. Asian J Commun. 2008;18(3):175–92. https://doi.org/10.1080/01292980802207033.
17. Honardoost M, Rajabpour A, Vakili L. Molecular epidemiology; new but impressive. Med J Islam Repub Iran. 2018;28(32):53. https://doi.org/10.14196/mjiri.32.53. PMID: 30175079; PMCID: PMC6113584.
18. Eybpoosh S, Haghdoost AA, Mostafavi E, Bahrampour A, Azadmanesh K, Zolala F. Molecular epidemiology of infectious diseases. Electron Physician. 2017;9(8):5149–58. https://doi.org/10.19082/5149. PMID: 28979755; PMCID: PMC5614305.
19. Confounding by familial factors influences the associations between maternal health and autism. Nat Med. 2025. https://doi.org/10.1038/s41591-025-03537-6.
20. Banegas JR, Rodríguez Artalejo F, del Rey Calero J. Popper y el problema de la inducción en epidemiología [Popper and the problem of induction in epidemiology]. Rev Esp Salud Publica. 2000;74(4):327–39. Spanish. PMID: 11031841.
21. Terhal BM, Wolf MM, Doherty AC. Quantum entanglement: a modern perspective. Phys Today. 2025;78(1):40–46. https://doi.org/10.1063/pt.rzbf.mjwy.

22. Li Q, Orcutt K, Cook RL, Sabines-Chesterking J, Tong AL, Schlau-Cohen GS, Zhang X, Fleming GR, Whaley KB. Single-photon absorption and emission from a natural photosynthetic complex. Nature. 2023;619(7969):300–4. https://doi.org/10.1038/s41586-023-06121-5. Epub 2023 Jun 14. PMID: 37316658; PMCID: PMC10338339.
23. Allemann RK, Scrutton NS, editors. Quantum tunnelling in enzyme-catalysed reactions. Royal Society of Chemistry; 2009.
24. Hore P, Mouritsen H. The quantum nature of bird migration: migratory birds travel vast distances between their breeding and wintering grounds. New research hints at the biophysical underpinnings of their internal navigation system. Sci Am. 2022;326(4):26. https://doi.org/10.1038/scientificamerican0422-26. PMID: 39016862.
25. Matarèse BFE, Rusin A, Seymour C, Mothersill C. Quantum biology and the potential role of entanglement and tunneling in non-targeted effects of ionizing radiation: a review and proposed model. Int J Mol Sci. 2023;24(22):16464. https://doi.org/10.3390/ijms242216464. PMID: 38003655; PMCID: PMC10671017.
26. Usselman R, Chavarriaga C, Castello PR, Procopio M, Ritz T, Dratz EA, Singel DJ, Martino CF. The quantum biology of reactive oxygen species partitioning impacts cellular bioenergetics. Sci Rep. 2016;6:38543. https://doi.org/10.1038/srep38543.
27. Xu DP, Li Y, Meng X, Zhou T, Zhou Y, Zheng J, Zhang JJ, Li HB. Natural antioxidants in foods and medicinal plants: extraction, assessment and resources. Int J Mol Sci. 2017;18(1):96. https://doi.org/10.3390/ijms18010096. PMID: 28067795; PMCID: PMC5297730.
28. Howarth M, Brettle A, Hardman M, Maden M. What is the evidence for the impact of gardens and gardening on health and well-being: a scoping review and evidence-based logic model to guide healthcare strategy decision making on the use of gardening approaches as a social prescription. BMJ Open. 2020;10(7): e036923. https://doi.org/10.1136/bmjopen-2020-036923. PMID: 32690529; PMCID: PMC7371129.
29. Mlinarić A, Horvat M, Šupak SV. Dealing with the positive publication bias: why you should really publish your negative results. Biochem Med (Zagreb). 2017;27(3): 030201. https://doi.org/10.11613/BM.2017.030201. PMID: 29180912; PMCID: PMC5696751.
30. Yamamoto N, Ejima K, Zoh RS, Brown AW. Bias in nutrition-health associations is not eliminated by excluding extreme reporters in empirical or simulation studies. Elife. 2023;12: e83616.
31. Matthews R. Storks deliver babies (p= 0.008). Teach Stat. 2000;22:36–8. https://doi.org/10.1111/1467-9639.00013.
32. McNulty H, Pentieva K, Ward M. Causes and clinical sequelae of riboflavin deficiency. Annu Rev Nutr. 2023;21(43):101–22. https://doi.org/10.1146/annurev-nutr061121-084407. PMID: 37603429.
33. Hustad S, McKinley MC, McNulty H, Schneede J, Strain JJ, Scott JM, Ueland PM. Riboflavin, flavin mononucleotide, and flavin adenine dinucleotide in human plasma and erythrocytes at baseline and after low-dose riboflavin supplementation. Clin Chem. 2002;48(9):1571–7 PMID: 12194936.
34. Aljaadi AM, Devlin AM, Green TJ. Riboflavin intake and status and relationship to anemia. Nutr Rev. 2022;81(1):114–32. https://doi.org/10.1093/nutrit/nuac043. PMID: 36018769.
35. Mosegaard S, Dipace G, Bross P, Carlsen J, Gregersen N, Olsen RKJ. Riboflavin deficiency-implications for general human health and inborn errors of metabolism. Int J Mol Sci. 2020;21(11):3847. https://doi.org/10.3390/ijms21113847. PMID: 32481712; PMCID: PMC7312377.
36. Gerssen A, Pol-Hofstad IE, Poelman M, Mulder PP, van den Top HJ, de Boer J. Marine toxins: chemistry, toxicity, occurrence and detection, with special reference to the Dutch situation. Toxins (Basel). 2010;2(4):878–904. https://doi.org/10.3390/toxins2040878. Epub 2010 Apr 23. PMID: 22069615; PMCID: PMC3153220.
37. Babić Leko M, Gunjača I, Pleić N, Zemunik T. Environmental factors affecting thyroid-stimulating hormone and thyroid hormone levels. Int J Mol Sci. 2021;22(12):6521. https://doi.org/10.3390/ijms22126521. PMID: 34204586; PMCID: PMC8234807.

38. Ruan D, Yang J, Luo Q, Shi Y, Ding L, Wang Z, Wang R, Yang L. The protective effects of goitrin on LPS-induced septic shock in C57BL/6J mice via caspase-11 non-canonical inflammasome inhibition. Molecules. 2023;28(7):2883. https://doi.org/10.3390/molecules28072883. PMID: 37049646; PMCID: PMC10096381.
39. Garte S, Marshall P, Kauffman S. The reasonable ineffectiveness of mathematics in the biological sciences. Entropy (Basel). 2025;27(3):280. https://doi.org/10.3390/e27030280. PMID: 40149204; PMCID: PMC11941032.
40. Borovik A. A mathematician's view of the unreasonable ineffectiveness of mathematics in biology. Biosystems. 2021;205: 104410. https://doi.org/10.1016/j.biosystems.2021.104410. Epub 2021 Mar 23 PMID: 33766624.
41. Chen M, Liu J, Duan P, Li M, Liu W. Biosynthesis and molecular engineering of templated natural products. Natl Sci Rev. 2017;4(4):553–75. https://doi.org/10.1093/nsr/nww045.
42. Walsh CT, Gehring AM, Weinreb PH, Quadri LE, Flugel RS. Post-translational modification of polyketide and nonribosomal peptide synthases. Curr Opin Chem Biol. 1997;1(3):309–15. https://doi.org/10.1016/s1367-5931(97)80067-1. PMID: 9667867.
43. Tajbakhsh M, Karimi A, Fallah F, Akhavan MM. Overview of ribosomal and nonribosomal antimicrobial peptides produced by Gram positive bacteria. Cell Mol Biol (Noisy-le-grand). 2017;63(10):20–32. https://doi.org/10.14715/cmb/2017.63.10.4. PMID: 29096754.
44. Bhushan A, Egli PJ, Peters EE, Freeman MF, Piel J. Genome mining- and synthetic biology-enabled production of hypermodified peptides. Nat Chem. 2019;11:931–9.
45. Nguyen TQN, Tooh YW, Sugiyama R, et al. Post-translational formation of strained cyclophanes in bacteria. Nat Chem. 2020;12:1042–53. https://doi.org/10.1038/s41557-020-0519-z.
46. Chooi YH. RiPP-ing through the plant kingdom. Nat Chem Biol. 2022;18:2–3. https://doi.org/10.1038/s41589-021-00901-8.
47. McAuliffe O, Ross RP, Hill C. Lantibiotics: structure, biosynthesis and mode of action. FEMS Microbiol Rev. 2001;25(3):285–308. https://doi.org/10.1111/j.15746976.2001.tb00579.x. PMID: 11348686.
48. Xin B, Zheng J, Xu Z, Li C, Ruan L, Peng D, Sun M. Three novel lantibiotics, Ticins A1, A3, and A4, have extremely stable properties and are promising food biopreservatives. Appl Environ Microbiol. 2015;81(20):6964–72. https://doi.org/10.1128/AEM.01851-15. Epub 2015 Jul 31. PMID: 26231642; PMCID: PMC4579438.
49. Li Z, Zhang W, Li S. Cytochrome P450 enzymes and microbial drug development—a review. Wei Sheng Wu Xue Bao. 2016;56(3):496–515 Chinese PMID: 27382792.
50. Kissen R, Rossiter JT, Bones AM. The 'mustard oil bomb': not so easy to assemble? Localization, expression and distribution of the components of the myrosinase enzyme system. Phytochem Rev. 2009;8:69–86. https://doi.org/10.1007/s11101-008-9109-1.
51. Romeo L, Iori R, Rollin P, Bramanti P, Mazzon E. Isothiocyanates: an overview of their antimicrobial activity against human infections. Molecules. 2018;23(3):624. https://doi.org/10.3390/molecules23030624. PMID: 29522501; PMCID: PMC6017699.
52. Kumar G, Tuli HS, Mittal S, et al. Isothiocyanates: a class of bioactive metabolites with chemopreventive potential. Tumor Biol. 2015;36:4005–16. https://doi.org/10.1007/s13277-015-3391-5.
53. Nakamura T, Abe-Kanoh N, Nakamura Y. Physiological relevance of covalent protein modification by dietary isothiocyanates. J Clin Biochem Nutr. 2018;62(1):11–19. https://doi.org/10.3164/jcbn.17-91. Epub 2017 Dec 12. PMID: 29371751; PMCID: PMC5773839.
54. Rutter J, Winge DR, Schiffman JD. Succinate dehydrogenase—assembly, regulation and role in human disease. Mitochondrion. 2010;10(4):393–401. https://doi.org/10.1016/j.mito.2010.03.001. Epub 2010 Mar 10. PMID: 20226277; PMCID: PMC2874626.
55. Olayanju JB, Bozic D, Naidoo U, Sadik OA. A comparative review of key isothiocyanates and their health benefits. Nutrients. 2024;16(6):757. https://doi.org/10.3390/nu16060757. PMID: 38542669; PMCID: PMC10974736.
56. Na G, He C, Zhang S, Tian S, Bao Y, Shan Y. Dietary isothiocyanates: novel insights into the potential for cancer prevention and therapy. Int J Mol Sci. 2023;24(3):1962. https://doi.org/10.3390/ijms24031962. PMID: 36768284; PMCID: PMC9916827.

57. Calmes B, N'Guyen G, Dumur J, Brisach CA, Campion C, Iacomi B, Pigné S, Dias E, Macherel D, Guillemette T, Simoneau P. Glucosinolate-derived isothiocyanates impact mitochondrial function in fungal cells and elicit an oxidative stress response necessary for growth recovery. Front Plant Sci. 2015;3(6):414. https://doi.org/10.3389/fpls.2015.00414. PMID: 26089832; PMCID: PMC4452805.
58. Nowicki D, Maciąg-Dorszyńska M, Bogucka K, Szalewska-Pałasz A, HermanAntosiewicz A. Various modes of action of dietary phytochemicals, sulforaphane and phenethyl isothiocyanate, on pathogenic bacteria. Sci Rep. 2019;9(1):13677. https://doi.org/10.1038/s41598-019-50216-x. PMID: 31548597; PMCID: PMC6757042.
59. Luthy B, Matile P. The mustard oil bomb: rectified analysis of the subcellular organization of the myrosinase system. Biochem Physiol Pflanzen. 1984;179:5–12.
60. Siemens DH, Mitchell-Olds T. Glucosinolates and herbivory by specialists (Coleoptera: Chrysomelidae, Lepidoptera: Plutellidae): consequences of concentration and induced resistance. Environ Entomol. 1996;25:1344–53. https://doi.org/10.1093/ee/25.6.1344.
61. Poelman EH, Van Dam NM, Van Loon JJ, Vet LE, Dicke M. Chemical diversity in Brassica oleracea affects biodiversity of insect herbivores. Ecology. 2009;90(7):1863–77. https://doi.org/10.1890/08-0977.1. PMID: 19694135.
62. Malka O, Shekhov A, Reichelt M, et al. Glucosinolate desulfation by the phloem-feeding insect Bemisia tabaci. J Chem Ecol. 2016;42:230–5.
63. Jeschke V, Zalucki JM, Raguschke B, Gershenzon J, Heckel DG, Zalucki MP, Vassão DG. So much for glucosinolates: a generalist does survive and develop on brassicas, but at what cost? Plants (Basel). 2021;10(5):962. https://doi.org/10.3390/plants10050962. PMID: 34066079; PMCID: PMC8150600.
64. Hanley AB, Kwiatkowski CA, Fenwick GR. Enzymic hydrolysis of glucosinolates in a low water system. J Sci Food Agric. 1990;51:417.
65. Halkier BA, Gershenzon J. Biology and biochemistry of glucosinolates. Annu Rev Plant Biol. 2006;57:303–33. https://doi.org/10.1146/annurev.arplant.57.032905.105228.
66. Hanley AB, Parsley KR, Lewis JA, Fenwick GR. Chemistry of indole glucosinolates: intermediacy of indol-3-yl methyl isothiocyanates in the enzymic hydrolysis of indole glucosinolates. J Chem Soc Perkin Trans. 1990;I:2273.
67. Gmelin R, Virtanen AI. Glucobrassicin, the precursor of SCN, 3indolylacetonitrile and ascorbigen in Brassica oleracea species. Annales Academiae Scientiarum Fennicae. 1961;107:3–23.
68. Kutacek M, Prochazka Z, Veres K. Biogenesis of glucobrassicin, the in vitro precursor of ascorbigen. Nature 1962;194:393–394. https://doi.org/10.1038/194393b0.
69. Ramirez D, Abellán-Victorio A, Beretta V, Camargo A, Moreno DA. Functional ingredients from brassicaceae species: overview and perspectives. Int J Mol Sci. 2020;21(6):1998. https://doi.org/10.3390/ijms21061998. PMID: 32183429; PMCID: PMC7139885.
70. Wooding S, Gunn H, Ramos P, Thalmann S, Xing C, Meyerhof W. Genetics and bitter taste responses to Goitrin, a plant toxin found in vegetables. Chem Senses. 2010;35(8):685–92. https://doi.org/10.1093/chemse/bjq061.
71. Stoll A, Seebeck E. Allium compounds. I. Alliin, the true mother compound of garlic oil. Helv Chim Acta. 1948;31:189–210.
72. Bussy A. J. Note sur la formation de I'huile essentielle de moutarde. Pharmac. Chim., 1840;26:39.
73. Gadamer J. Über die Bestandteile des schwarzen und des weissen Senfsamens. Arch Pharm. 1897;235:44.
74. Gadamer J. Uber das Sinapine Ber. deutsch. chem. Ges. 1897;30:2322.
75. Stoll A, Seeback E. Chemical investigations on alliin, the specific principle of garlic. Adv Enzymol Relat Subj Biochem. 1951;11:377–400.
76. Ettlinger MG, Lundeen AJ. The structures of Sinigrin and Sinalbin; an enzymatic rearrangement. J Am Chem Soc. 1956;78:4172–3.
77. Benn MH, Ettlinger MG. The synthesis of sinigrin. Chem Commun. (London) (1965):445–7.
78. Fenwick GR, Hanley AB. The genus Allium, Part 2. CRC Crit Rev Food Sci Nutr. 1985;22:273.

79. Mu X, Yin S, Su F, Jiang R, Liang S, Liu Y, Li N, et al. Understanding the onion lachrymatory factor: structure, formation, and flavor implications. Food Rev Intl. 2024;40(10):3938–63. https://doi.org/10.1080/87559129.2024.2377639.
80. Silvaroli JA, Pleshinger MJ, Banerjee S, Kiser PD, Golczak M. Enzyme that makes you cry-crystal structure of lachrymatory factor synthase from allium cepa. ACS Chem Biol. 2017;12(9):2296–2304. https://doi.org/10.1021/acschembio.7b00336. Epub 2017 Jul 26. PMID: 28708375; PMCID: PMC5693239.
81. Block E, Bayer T, Naganathan S, Zhao SH. Allium chemistry: synthesis and sigmatropic rearrangements of alk(en)yl 1-propenyl disulfide s-oxides from cut onion and garlic. J Am Chem Soc. 1996;118:2799–810.
82. Imai S, Tsuge N, Tomotake M, Nagatome Y, Sawada H, Nagata T, Kumagai H. Plant biochemistry: an onion enzyme that makes the eyes water. Nature. 2002;419:685. https://doi.org/10.1038/419685a.
83. Spåre C-G, et al. On the lachrymatory factor in onion (Allium cepa) vapours and its precursor. Acta Chem Scand. 1963;17:641–50.
84. Apitz-Castro R, Cabrera S, Cruz MR, Ledezma E, Jain MK. Effect of garlic extract and of three pure components isolated from it on human platelet aggregation, arachidonate metabolism, release reaction and platelet ultrastructure. Thromb Res. 1983;32:155–69.
85. Block E, Ahmad S, Jain MK, Crecely WW, Apitz-Castro R, Cruz MR. (E, Z)-Ajoene: a potent antithrombotic agent from garlic. J Am Chem Soc. 1984;106:8295–6.
86. Silva F, Khokhar SS, Williams DM, Saunders R, Evans GJS, Graz M, Wirth T. Short total synthesis of Ajoene. Angew Chem Int Ed Engl. 2018;57(38):1229012293. https://doi.org/10.1002/anie.201808605. Epub 2018 Aug 19. PMID: 30079981; PMCID: PMC6221122.
87. Jakobsen TH, van Gennip M, Phipps RK, Shanmugham MS, Christensen LD, Alhede M, Skindersoe ME, Rasmussen TB, Friedrich K, Uthe F, Jensen PØ, Moser C, Nielsen KF, Eberl L, Larsen TO, Tanner D, Høiby N, Bjarnsholt T, Givskov M. Ajoene, a sulfur-rich molecule from garlic, inhibits genes controlled by quorum sensing. Antimicrob Agents Chemother. 2012;56(5):2314–25. https://doi.org/10.1128/AAC.05919-11. Epub 2012 Feb 6. PMID: 22314537; PMCID: PMC3346669.
88. Ahmed N, Laverick L, Sammons J, Zhang H, Maslin DJ, Hassan HT. Ajoene, a garlic-derived natural compound, enhances chemotherapy-induced apoptosis in human myeloid leukaemia CD34-positive resistant cells. Anticancer Res. 2001;21(5):3519–23. PMID: 11848518.
89. Shinde SD et al. An insight into medicinal attributes of dithiocarbamates: bird's eye view. Bioorg Chem. 2020;105(2020):104346.
90. He Q, Liu J, Lan JS, Ding J, Sun Y, Fang Y, Jiang N, Yang Z, Sun L, Jin Y, Xie SS. Coumarin-dithiocarbamate hybrids as novel multitarget AChE and MAO-B inhibitors against Alzheimer's disease: design, synthesis and biological evaluation. Bioorg Chem. 2018;81:512–28. https://doi.org/10.1016/j.bioorg.2018.09.010. Epub 2018 Sep 11 PMID: 30245233.
91. Behrens M, Lang T. Extra-oral taste receptors-function, disease, and perspectives. Front Nutr. 2022;4(9): 881177. https://doi.org/10.3389/fnut.2022.881177. PMID: 35445064; PMCID: PMC9014832.
92. Dewachter L, Dénéréaz J, Liu X, de Bakker V, Costa C, Baldry M, Sirard JC, Veening JW. Amoxicillin-resistant Streptococcus pneumoniae can be resensitized by targeting the mevalonate pathway as indicated by sCRilecs-seq. Elife. 2022;24(11): e75607. https://doi.org/10.7554/eLife.75607. PMID: 35748540; PMCID: PMC9363119.
93. Zhao X, Yu Z, Ding T. Quorum-sensing regulation of antimicrobial resistance in bacteria. Microorganisms. 2020;8(3):425. https://doi.org/10.3390/microorganisms8030425. PMID: 32192182; PMCID: PMC7143945.
94. Cui S, Kim E. Quorum sensing and antibiotic resistance in polymicrobial infections. Commun Integr Biol. 2024;17(1):2415598. https://doi.org/10.1080/19420889.2024.2415598. PMID: 39430726; PMCID: PMC11487952.
95. Jakobsen TH, Warming AN, Vejborg RM, et al. A broad range quorum sensing inhibitor working through sRNA inhibition. Sci Rep. 2017;7:9857. https://doi.org/10.1038/s41598-017-09886-8.

96. Kim JH, Kim DH, Jo S, et al. Immunomodulatory functional foods and their molecular mechanisms. Exp Mol Med. 2022;54:1–11. https://doi.org/10.1038/s12276-022-00724-0.
97. Catanzaro M, Corsini E, Rosini M, Racchi M, Lanni C. Immunomodulators inspired by nature: a review on curcumin and echinacea. Molecules. 2018;23(11):2778. https://doi.org/10.3390/molecules23112778. PMID: 30373170; PMCID: PMC6278270.
98. Kang Q, He L, Zhang Y, Zhong Z, Tan W. Immune-inflammatory modulation by natural products derived from edible and medicinal herbs used in Chinese classical prescriptions. Phytomedicine. 2024;130:155684. https://doi.org/10.1016/j.phymed.2024.155684. Epub 2024 May 17. PMID: 38788391.
99. Kniss DA, Summerfield TL. Discovery of HeLa cell contamination in HES cells: call for cell line authentication in reproductive biology research. Reprod Sci. 2014;21(8):1015–9. https://doi.org/10.1177/1933719114522518. Epub 2014 Feb 11. PMID: 24520087; PMCID: PMC4126216.
100. Masadeh M, Mhaidat N, Alzoubi K, et al. Antibacterial activity of statins: a comparative study of Atorvastatin, Simvastatin, and Rosuvastatin. Ann Clin Microbiol Antimicrob. 2012;11:13. https://doi.org/10.1186/1476-0711-1113.
101. Mugford ST, Osbourn A. Saponin synthesis and function. Isoprenoid Synth Plants Microorg. 2012;22:405–24. https://doi.org/10.1007/9781-4614-4063-5_28. PMCID: PMC7121976.
102. Timilsena YP, Phosanam A, Stockmann R. Perspectives on saponins: food functionality and applications. Int J Mol Sci. 2023;24(17):13538. https://doi.org/10.3390/ijms241713538. PMID: 37686341; PMCID: PMC10487995.
103. Juang YP, Liang PH. Biological and pharmacological effects of synthetic saponins. Molecules. 2020;25(21):4974. https://doi.org/10.3390/molecules25214974. PMID: 33121124; PMCID: PMC7663351.
104. Petroski W, Minich DM. Is there such a thing as "anti-nutrients"? A narrative review of perceived problematic plant compounds. Nutrients. 2020;12(10):2929. https://doi.org/10.3390/nu12102929. PMID: 32987890; PMCID: PMC7600777.
105. Haskell-Ramsay CF, Schmitt J, Actis-Goretta L. The impact of epicatechin on human cognition: the role of cerebral blood flow. Nutrients. 2018;10(8):986. https://doi.org/10.3390/nu10080986. PMID: 30060538; PMCID: PMC6115745.
106. Yamazaki KG, Romero-Perez D, Barraza-Hidalgo M, Cruz M, Rivas M, Cortez-Gomez B, Ceballos G, Villarreal F. Short- and long-term effects of (-)epicatechin on myocardial ischemia-reperfusion injury. Am J Physiol Heart Circ Physiol. 2008;295(2):H761–7. https://doi.org/10.1152/ajpheart.00413.2008. Epub 2008 Jun 20. PMID: 18567705; PMCID: PMC2519218.

Chapter 7
Taste and Other Receptors

Contents

7.1 Recipe, Porotos Granados 198
7.2 Toxins and Taste Receptors—Structure and Function 201
7.3 Taste and Odour Receptors 203
7.4 Natural and Synthetic Organoleptically Active Compounds 205
7.5 Bitter Taste and Extra-Oral Taste Receptors 208
7.6 Synergy Between Taste Molecules 210
7.7 The Essence of Colour Chemistry 211
7.8 Colour in Food and Art—Additive and Subtractive Colours. Light and Shade 213
7.9 Natural and Synthetic Colours. Chromophores and Chemical Structures 217
References 218

Abstract Flavour development and detection is, in a chemical sense, dependent upon the interactions of natural products in foods and, in a biological system, reliant upon the interaction with taste receptors. In some cases (for example, capsaicin) the taste molecule is present and is liberated to have an immediate effect. In others a chemical (for example Maillard) or a biological (the 'mustard bomb') reaction is required. The correlation between taste, aversion and toxicity has been proposed however it is difficult to determine if there is a quantitative structure activity relationship (QSAR) or not. The response to many xenobiotics appears to be mediated by G-Protein Coupled Receptors (GPCRs) including those responsible for the perception of taste and odour. Some interest has been associated with extra-oral taste receptors and their potential role in mediating non-taste responses. One of the other receptor-mediated responses is vision and the appreciation of colour is an important aspect of foods.

Keywords Taste · Odour · Colour · QSAR · GPCR

B. Hanley, *Secondary Natural Products in Foods and Food Systems*,
https://doi.org/10.1007/978-3-032-00292-1_7

7.1 Recipe, Porotos Granados

Porotos Granados is a South American recipe in which vegetables and pulses are cooked together slowly. The concept of cooking things slowly is very popular in meat cooking where cheap cuts of meat can be cooked to succulence. In a vegetarian dish, long cooking can ensure that the vegetable protein is well cooked and plant cell walls are edible. It can also help in the evolution of flavours and in the elimination of some anti-nutritional factors (phytates, lectins). It demonstrates one aspect of an intersection between chemistry and cooking on a macromolecular level. On the whole, however, the worlds of cooking and chemistry are divergent. The level of detail required for chemistry is far in excess of that needed for cooking while the complexity of cooking is difficult to reduce to chemically defined molecular interactions. The study of molecules derived from cooking processes which then interact with biological systems is a rare area of overlap. Before considering this particular synergy, it is perhaps worth considering the areas of divergence between cooking and chemistry. Much of this is concerned with the scale and level of granularity required for each.

There have been significant numbers of reports, articles and even books that draw comparisons between cooking and chemistry. In some cases, these involve the detailed study of a cooking process and trying either to explain cooking in terms of science or applying some scientific techniques to culinary processes to give something that is new. Where the two areas diverge is in the level of detail and explanation required. In general, science measures things in a far more precise way and provides far more insight than is required for cooking. The level of granularity of understanding in science may be considerably greater than that required in cooking. While chemistry is concerned with interactions at a molecular level, cooking is in essence a macromolecular activity. The use of terms like ‘molecular gastronomy’ suggest a higher level of precision than is justified by the techniques that are generally employed or needed. There may be an exception in the use and understanding of taste and taste receptors, however simply combining foods because they contain similar taste molecules is more like ‘epidemiological gastronomy’ than ‘molecular gastronomy’. At the core of the divergence between chemistry and cooking is a difference in philosophical approach. Chemistry, as has been noted in previous chapters, is a reductionist and empirical science which, in the area of natural products, seeks to discover existing, novel compounds (secondary natural products) produced (actually or potentially) by nature and their interrelationships and interactions with living systems. Cooking attempts to create, on a macromolecular level, materials that are edible and nutritious and, while the products may have some novel mixtures of flavours and different ways of presenting textures, the science base is generally limited and often naïve because that is the level of input that is required. Of course, there is invariably substantial interest in how molecules respond to heating, cooling and, to an extent, changes in ionic strength, pressure and pH and the transformations of macromolecules such as starch, collagen and gelatine (amongst others) but cooking is, on the whole, more concerned with bulk properties than molecular interactions and is also less predictive and empirical. You can invoke Maillard reactions and study the mechanisms and kinetics of formation of various products but the level of detail afforded even to reactions like the formation of acrylamide from asparagine is totally out of proportion to its scientific or toxicological importance. In brief, the attempts to associate cooking with chemistry are unconvincing. The macromolecular application of chemistry principles to explain the results of cooking are of marginal interest at best.

Porotos Granados—More than the Sum of its Parts

While cooking cannot be compared directly to chemistry and most cooking involves the mixing of ingredients rather than their reaction, there are some examples where a completed recipe produces something that is distinctly different to simply the sum of its parts. This is arguably the case when emulsions are formed or when proteins are denatured (as occurs upon heating) but molecular reactions to give different products is less common. As noted above, Maillard reactions are perhaps an example of a reaction between an amine and a reducing sugar which, upon application of heat to overcome the activation energy of the reaction, may proceed to give a range of products (Fig. 7.1)***. The study of the kinetics of the Maillard reaction is interesting from a descriptive chemistry standpoint*** [1] ***but adds little, if anything to the use of chemistry in cooking.***

Put a couple of tablespoons of olive oil in a large saucepan. Heat the oil then add one chopped medium onion and two cloves of crushed chopped garlic and cook gently (stirring from time to time to stop the vegetables from sticking) for 10 min. Chopping the onion and garlic (as we saw in **Chap. 6** ***and will revisit later in this chapter) allow contact between an enzyme (alliinase) and alk(en)yl-cysteine sulphoxides to give a range of compounds of varying degrees of volatility. Add a teaspoon of paprika and a quarter teaspoon of dried oregano. Stir then add about 100g (or more if you like) of fresh or canned beans (borlottis are good but pinto or cannellini beans will be fine). Add about 250 mls of vegetable stock or water. You can use dried beans but they will require extensive cooking both to soften them and to break down some of the bitter components*** **(see Sect.** 6.6)***. After simmering for 10 minutes, add about 1 kg of Crown Prince squash (peeled, deseeded and cut into chunks of about 2–3 cm). Boil gently for 10 minutes then add trimmed French beans (cut into 2 cm pieces) and either a small tin of corn kernels or nibs from two cobs of corn. Add salt and pepper to taste quite a bit of both will probably be required) then cover and simmer gently for about 30 minutes. Add chopped spinach or chard (about 250 g) and cook for a further 20 min. If necessary add a little more boiling water to prevent the dish from getting dry. Serve in a bowl with crusty bread. It will keep overnight (once cooled) in a refrigerator and can be reheated.***

This recipe makes use of long and fairly slow cooking. This helps both with the generation of flavours and to develop the texture of the dish. It is possible to 'upgrade' the heat by adding a little cayenne pepper alongside the paprika. The nature

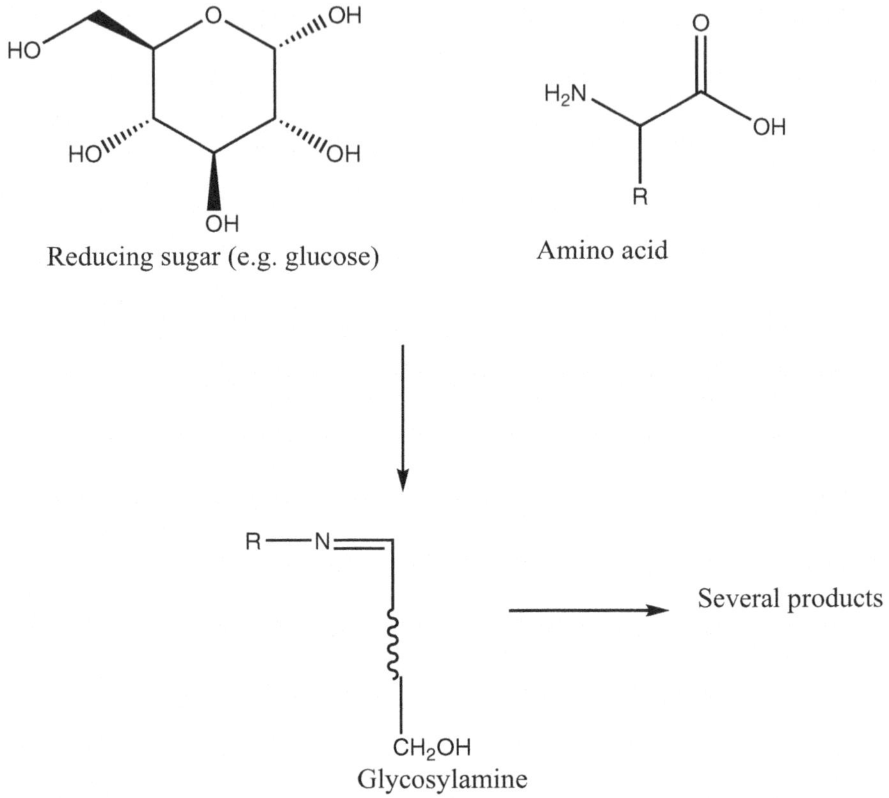

Fig. 7.1 The Maillard reaction

of this recipe means that the flavour profile results from the mixture of the individual components (the heat of paprika (or cayenne), the aromatic flavour imparted by oregano) and the chemical reactions between the ingredients (particularly onion and garlic). Chopping enables *Allium* flavour development to take place while gentle frying in oil affects the nature of the chemicals that are formed from the initially reactive thiosulphinates (Fig. 7.2).

It is possible to change the flavour profile by leaving the chopped onion and garlic to rest before frying. In addition to impacting upon the types of compounds that are formed by solubilising some of them thereby shifting reaction profiles, frying also inactivates the alliinase enzyme thereby stopping the reaction from forming more thiosulphinates de novo and adding to the supply of reactive S compounds. Changes in the types and amounts of compounds formed can also be enabled by slicing rather than crushing the garlic or by mincing (or grating) some or all the onion. In addition to containing different proportions of the flavour precursors (S-alk(en)yl-L-cysteine sulfoxides), onion and garlic also contain polymorphic forms of the alliinase enzyme

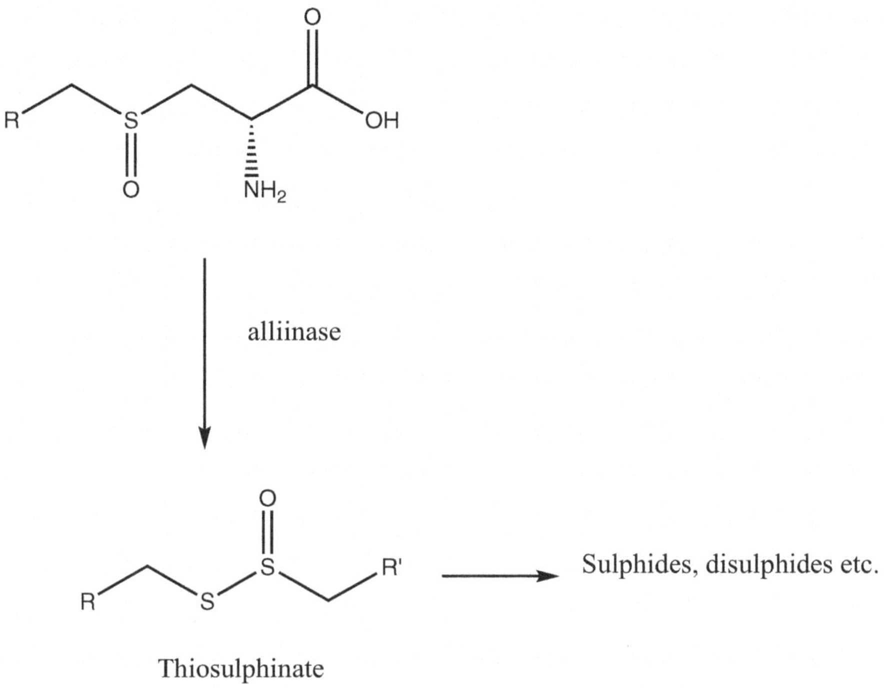

Fig. 7.2 Formation of thiol compounds from Thiosulphinates

with different specificities [2] and this may affect the flavour profiles formed [3] since onion alliinase can cleave cysteine sulphoxides from garlic and vice versa.

7.2 Toxins and Taste Receptors—Structure and Function

The supra-nutritional importance of secondary natural products and their generation in edible plants has been the subject of considerable investigation [4]. In some cases, in nature, there is direct competition for resources and the development of specific toxic entities may give one species a growth advantage over others—particularly if the toxic entity specifically targets the competing species and does not affect the producing species. This is perhaps most clearly seen in the synthesis of anti-microbial compounds which are molecularly targeted to interfere with a specific aspect of a competing species (for example, the formation of the bacterial cell wall as is the case for the bacteriastatic activity of penicillins and cephalosporins). Other products are toxic to potential invading species and can kill microbial species rather than just affecting their growth. Some plant secondary natural products fall into this category. These include bacteriocins, anti-microbial peptides which specifically permeabilise bacterial cell walls and thereby compromise their integrity [5] and other natural products [6]. Much of the interest in these compounds is driven by pharmacological investigations since they may offer an alternative mechanism of toxicity that is not affected by the development of anti-microbial resistance. Recent work on synergies between bioactive species with the same overall biological target but different modes of action, may increase research into the potentially synergistic functional properties of mixtures of plant compounds from a single source rather than the reductionist chemistry approach of separating and purifying individual compounds. In the case of plants of the genus *Brassica*, a range of pathways are employed to protect against insect and pathogen attack [7]. The levels of plant defensive compounds such as isothiocyanates (and their precursors, glucosinolates) may affect the degree of resistance to predation by insects and insect damage may affect the level of expression of the secondary natural product (and related) biosynthetic pathways [8].

A further role of secondary natural products is to act as feeding deterrence molecules to inhibit predation. A number of classes of compounds exhibit this type of activity and their mechanism of action may include inhibition of food (energy) intake or direct repellent effects [9]. Since they are stationary and cannot physically move away from an attacking species, many plant natural products are synthesised and do not primarily cause pest mortality, but they repel or deter insects or other feeding species. A proportion of these substances operate by binding to a bitter taste receptor and, through signal transduction, cause aversion [10]. In order for a plant natural product to operate effectively as an anti-feedant compound, it must be produced by the plant at a biologically relevant level and the appropriate receptor to which it binds must exist in the predatory species. It has been suggested that the development of bitter taste receptors (about 430 million years ago) occurred at about the same time as a significant explosion in the numbers of plant species and, presumably, a concomitant increase in the numbers of secondary natural products with a significant taste profile. The presence of extra-oral taste receptors with different functions suggest alternative theories may also be tenable (see below). In the case of humans, much of the taste aversion is mediated by one of the 25 known orally situated bitter receptors. It is known, however, that many bitter receptors have an extra-oral role and, in addition, natural products have a number of physiological roles so the simple 'one receptor, one ligand' model does not apply. For example, one bitter receptor in particular (T2R38) responds to a number of microbial metabolites [11] and this may be related to the extra-oral activity of bitter receptor-mediated activity (for example, modulation of the immune system and resistance to infection [12]). While a number of bitter compounds are agonists for a single bitter receptor, some of the bitter receptors have a large range of agonists [13] and the binding of some pharmaceuticals to bitter taste receptors may mediate nontarget biological activity [14]. The presence of extra-oral bitter (and sweet) taste receptors seems able to mediate a number of effects—including, importantly, pathways associated with the immune response [12]. In addition, the binding of bitter compounds to T2R bitter receptors can lead to upregulation of transporters such as ABCB1 and the active transport and excretion of toxins [15].

Although other activities have been noted (see above) taste receptors are G-Protein Coupled Receptors (GPCRs) and are primarily responsible for mediating the taste response in gustatory cells. In the case of sweet, bitter and umami taste, the ligands bind to the receptors and, through a transmembrane mechanism and a similar pathway, cause release of calcium ions in the cell [16] and, in addition, activate GPCR pathways such as Gβ3, Gγ13 and α-gustducin. In the intestine, some bitter receptors can also bind a number of microbial cell metabolites and bind microbial quorum sensing molecules and the increase in calcium ions leads to stimulation of production of IL-25, an important mediator of the type 2 immune response [17]. A detailed description of the functional properties of various taste receptors and the mechanism of the transduction effects in different cell types is beyond the scope of this chapter however the expression of taste receptors in non-taste receptor type cells is likely an important aspect of their evolution. If this is the case then, while modulation of the taste response in the oral cavity and taste aversion is the most physiologically immediate function of taste receptors, it may only be a minor function of receptor mediated responses by taste-receptors situated elsewhere in an organism. A more detailed description of the extra-oral role, specifically of bitter receptors, will be presented in Sect. 7.5.

7.3 Taste and Odour Receptors

While both taste and odour are organoleptically linked [18], the mechanisms by which they are detected at a receptor level, are very different. In order to understand this, it is relevant to consider how taste and odour receptors work on a molecular level.

Taste receptors are divided into different types and can be classified according to the type of response they elicit. Sour taste is produced by intracellular acidification. Taste bud cells are type III, neuron-like cells and acidification results in depolarisation and this leads to a calcium ion response. Salty taste is a function of the presence of sodium however other similar ions such as potassium can also be used to introduce saltiness. Monosodium glutamate, nucleotide phosphates and certain peptides can enhance salty taste. The receptor for salt is a sodium ion channel on the apical membrane of some taste cells. While, as noted above, salty taste can be enhanced by some nonintrinsically salty compounds, the sodium channel is also permeable to protons so acidic stimuli can reduce saltiness [19].

The remaining tastes (sweet, bitter and umami) are regulated by GPCRs. The first taste receptors that were characterised were found using the same cDNA subtractive library screen that had previously identified olfactory receptors (see below). The first two taste receptors were named TR1 and TR2 (later renamed as Tas1r1(T1R1) and Tas1r2 (T1R2).

The subsequent discovery of a further family of related GPCRs resulted in the characterisation of Tas2rs (T2Rs). Eventually a total of 3 T1Rs and 25 T2Rs were identified in humans. Although activated by different types of compounds (sweet and bitter), T1Rs and T2Rs share similar signalling pathways [16]. In summary, sweet, bitter, and umami tastants trigger the intracellular increase in calcium ions in particular taste receptor cells. While sweet and bitter tastes are mediated by binding to one type of receptor, umami taste is realised through the binding of molecules to the hererodimeric T1R1 and T1R3 GPCR [20]. The different taste sensations are a consequence of the type of taste cells that the receptors are associated with. While salt and sour tastes are recognised by Type III taste receptor cells, sweet, bitter and umami receptors are incorporated in Type II taste receptor cells. Analysis of the binding (and the consequences) of different molecules to receptors has been studied [21]. While much of the interest in extra-oral taste receptors has been directed towards bitter receptors (Sect. 7.5), the sweet receptor, in particular, is also present at a number of extra-oral sites and appears to have distinct physiological roles. In addition to modulating the sensation of taste, sweet and umami receptors are implicated in the assessment of the energy content of food. Indeed sweet taste receptors are implicated in the glucose response [22], in the insulin response to fructose in beta cells [23] and both sweet and the dimeric umami receptors have been implicated in nutrient sensing in the cardiovascular system [24]. A link between the binding of allyl isothiocyanate to a bitter receptor (T2R38) and an immune response has also been noted [25] and, given that polymorphisms in the T2R38 bitter receptor has been identified and correlated with different activities including susceptibility to respiratory disease [26], the possibility of a link between certain dietary components and immune response (and hence the variable susceptibility of individuals to diet-mediated immune disease) may be postulated.

Olfactory receptors are also GPCRs and, just as taste receptors have been found in extra-oral sites, olfactory receptors have been found in a number of extra-nasal sites [27]. Olfactory receptors were originally cloned and identified as a multi-gene family [28] and this suggested that there was a molecular basis for odour recognition. Humans have about 400 olfactory receptors and these, like other GPCRs have a seven transmembrane structure. Olfactory receptors are divided into Class I and Class II receptors and mammals have both types. Although olfactory receptors are represented by a large number of distinct different members of the class, certain motifs occur which are important for the functional operation of the receptor and these are critical for receptor function, ligand binding, and signal transduction. The detailed analysis of some of these conserved motifs has been carried out [29]. Rather than binding in a 'lock and key' conformation to a single olfactory receptor, odour molecules can bind to a range of receptors with variable affinity. Similarly olfactory receptors are not limited to binding to a single odour molecule and this flexibility helps to explain the high discriminatory capability of the sense of smell. As noted previously, olfactory receptors are not limited in location to the nasal cavity. For example, activation of the olfactory receptor 544 in the gut by an agonist has been found to reduce adiposity in mice fed a high-fat diet (HFD) by regulating the preference for fats [30]. Additionally, olfactory receptors have been found to play a role in glucose metabolism and therefore may be manipulated to affect obesity and diabetes [31]. The concept of modulation of metabolic and other imbalances by consideration of the binding of molecules to a network of receptors rather than simply to a single target will be greatly assisted by the development of AI approaches (including machine learning) and will be a significant departure from traditional medicinal chemistry.

The basic paradigm of receptors binding a variety of molecules may be relevant to the development of models to investigate the effects of multiple mixtures of bioactive compounds—such as are found in foods.

The evolution of the development of olfactory and taste receptors remains unclear. The widespread distribution of both types of GPCRs might suggest a common lineage while the signalling pathways between them are closely related—even though their functional properties are widely divergent—and this may also suggest some degree of commonality in their lineage. Since GPCRs are the most abundant class of mammalian receptors, they have been extensively studied both in terms of their function and their origins. There have been a number of investigations into the evolutionary development of GPCRs with some specific examples being considered [32] and some interesting relationships have been disclosed [33]. The construction of phylogenetic trees suggests, for example, that while some families of GPCRs are somewhat diverse, others (including olfactory receptors) can be considered to be a single group due to a negative selective pressure [34].

In essence both olfactory detection and taste are ways of sensing different substances through particular sets of properties of biological systems—even if the consequences of some aspects of the receptor binding is unclear. The concept of receptors having multiple uses serves to emphasise the complexity (and non-reductionist nature) of living systems. The target compounds for the receptors (agonists or antagonists) are present extracellularly and since this interaction regulates the response to extracellular events, it offers both a growth and survival advantage. From a secondary natural product standpoint, there is significance in the possibility that GPCRs that are in locations other than sensory cells may have a different biological significance in addition to simply chemosensing and that natural products that interact with these GPCRs may have an undiscovered biological profile (particularly in relation to immune response). This has two immediate consequences. The first is in the design of molecules with modified (particularly enhanced) organoleptic properties and the second is that some of these molecules may have unintended extrasensory properties.

This will be explored in Sect. 7.4 in which the properties of some natural and synthetic organoleptically active compounds will be examined.

7.4 Natural and Synthetic Organoleptically Active Compounds

The development of taste screening systems to detect novel natural and synthetic taste molecules has led to an increasing number of sensitive methods being operationalised. While a range of non-nutritive sweeteners have been developed including saccharin, aspartame and others. While there are commercial (cost) drivers behind the development of artificial (generally enhanced) sweeteners and the objectives of reducing calorie intake and improving oral health, recent studies implication them in the development of type 2 diabetes and other metabolic diseases has led to some concerns being expressed [35]. The sweet taste receptor (a dimer formed between of the T1R2 and T1R3 receptors) can detect both natural (sugar-based) and non-nutritive sweeteners [36]. The molecular structure of the receptor dimer has been solved by cryo electron microscopy [37]. The T1R2 receptor binds the sweetener molecule while the T1R3 receptor helps to assemble and stabilise the receptor complex. This, combined with the presence of the T1R2/T1R3 dimer in extra-oral organs suggests that nonnutritive sweeteners may be biologically active in addition to their sensory properties. Furthermore, it has been suggested that non-nutritive sweeteners may impact on the gut microbiome and encourage the growth of bacteria associated with an obese phenotype [38]. Since non-nutritive sweeteners do not contribute to energy intake, it is more likely that any effect is mediated at the receptor level. Although only transiently present in the oral cavity, an effect on both the oral and gut microbiome of non-nutritive sweeteners has been suggested [39]. Another area of interest is the development of sweet-tasting proteins (miraculin, monellin, thaumatin, mabinlin, pentadin, curculin, brazzein, and neoculin). These also bind to and activate the sweet taste receptor T1R2/T1R3. Some studies have been carried out to determine the more precise molecular nature of the interaction between 'sweet proteins' and the sweet receptor and these have provided some insights [40]. Biomimetic taste sensors have also been developed comprising a component that can bind to specific taste substances and a secondary element that converts the initial signal from the detector into something that can be more easily measured than a signal transduction [41].

Other compounds have been developed that can either replace the natural ligand for all of the other four tastes or can enhance or diminish the effects [42]. While the use of the term 'umami' is fairly common when referring to taste enhancers, flavour enhancers are described collectively by the more generic, multi-faceted descriptor 'kokumi'—a rich, long-lasting mouthfeel [43]. It has been suggested that sweet taste enhancers operate by an allosteric mechanism that is enabled by binding to the sweet receptor [44]. This raises the possibility of using less of a caloric sweetener to achieve the same sweetness effect and it may also overcome some of the safety issues that sometimes accompany the use of non-caloric sweeteners [45]. Consumer reaction to 'natural' rather than 'artificial' sweeteners may also contribute to research on the development of sweet taste enhancers [46]. There are a range of structures associated with bitterness. For example the NH–C=S moiety occurs both in natural and synthetic bitter agents (Fig. 7.3).

The most bitter known compound, however is denatonium (Fig. 7.4).

In addition to binding to T2R receptors in the oral cavity, denatonium also exhibits a range of physiological effects associated with bitter taste receptors in a range of cell types [47]. These are further discussed in Sect. 7.5.

While screening using biomimetic or receptor-based systems has helped in the discovery and development of new tastants, the use of structure/activity relationships has been, perhaps, a more traditional chemistry approach [48]. This is further described in the use of QSAR techniques to predict sweet and non-sweet taste molecules [49]. The approach has also been used to identify molecules with the potential to have umami characteristics through a combination of sweet and bitter

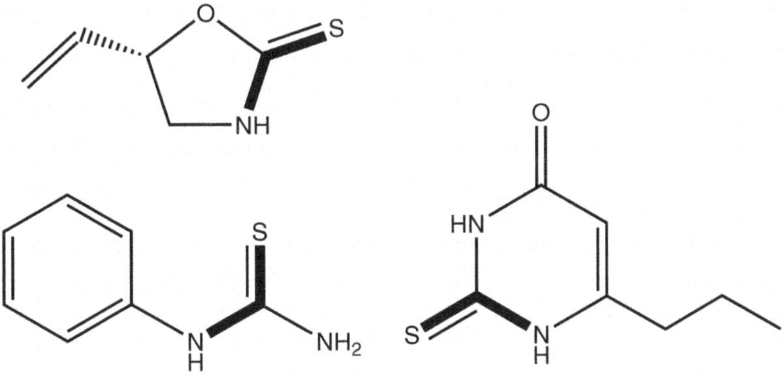

Fig. 7.3 Structures of some natural and synthetic bitter compounds

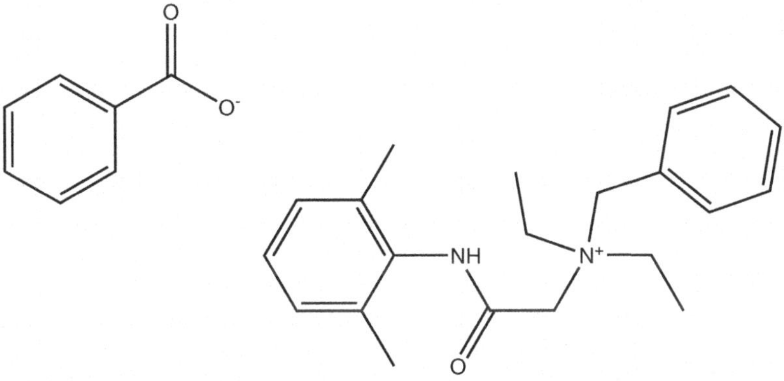

Fig. 7.4 Structure of denatonium

taste characteristics [50]. A combination of data and structure-driven approaches has been suggested to strengthen the predictive capability of in silico methods to predict the sweetness of molecules [51]. The alternative to the use of structure/activity relationships is the directly empirical approach of making use of receptor-based assays and the displacement of one molecule by another. This is somewhat similar to the eighteenth century concept of 'affinities' which was tabulated by Geoffroy in 1718 and was used to explain the 'displacement' of one element by another of higher affinity [52]. This approach, if based purely on displacement and not on the biological consequences of this action, may fail to distinguish between agonists and antagonists and modulation of taste by the antagonistic activity of small molecules is known [53].

The development of the study of odour compounds is intrinsically more complex due to the large range of olfactory receptors that exist and the complex way that they interact with odour molecules (see Sect. 7.3). In addition, the vibrational theory of odour detection has further complicated the more familiar receptor, lock and key model [54]. While there has been some success in the use of structure/activity relationships to predict the odour characteristics within some classes of compounds (for example, homologous carboxylic acids [55]), these approaches are somewhat limited in application. The use of deep learning techniques has been suggested [56] and this type of advanced data analysis my prove fruitful however complex mixtures will likely remain problematic. While some of the characteristics of odour molecules were able to be computed (molecular charge, weight, composition, shape, surface interaction, and hydrophobic interaction were closely related to the molecular odour characteristics), the number of different odour principles that could be evaluated was limited. In addition, chirality is also an important factor in the detection of some odour molecules [57, 58] and this clearly adds a further level of complexity. The use of cryo electron microscopy has helped to define how one particular odorant (propionic acid) binds to one specific olfactory receptor (OR51E2) and mutation analysis has helped to define the most important aspects of the interaction [59]. Olfactory cues and signal transduction is an important factor in the survival of insect species [60] and forms the basis for the use of odorant molecules in insect traps. The loss of the sense of smell is called anosmia and can be caused (and can result in) a number of conditions. It may be linked to a loss in the sense of taste (ageusia). Partial loss of either are called hyposmia and hypogeusia respectively. The most common causes are inflammation or infectious agents that cause conditions such as sinusitis. The link between cognition and the sense of smell and taste has been noted [61] although the extent to which this is causative or correlative remains unclear but has been postulated to be an empirically adjustable factor through olfactory training [62]. Indeed, olfactory function declines with ageing and is thought to be an early predictor of the onset of a number of neurological dysfunctional conditions including Alzheimer's and Parkinson's diseases as well as cognitive decline [63]. It is unclear if there might be a pharmacological solution to the loss of the senses of taste and olfaction however evidence is accumulating that reversing a decline in olfactory capability by training may have a positive effect on cognitive outcomes [62]. The potential connection between sinusitis and cognitive decline in the elderly has been noted and ascribed to the sub-optimal nasal microbiome [64]. While various microbiomes appear also to be involved in the onset of various conditions associated with taste and smell (and not solely in the oral cavity), it is unclear in which instances a chain of causation can be demonstrated and which are associative in nature [65].

7.5 Bitter Taste and Extra-Oral Taste Receptors

Bitter taste receptors have been mentioned previously and some of their characteristics described. While other taste receptors have been found to have an extra-oral role, the number of different bitter receptors and their capacity to bind a large range of secondary natural products means that particular interest has been generated in this particular class of GPCRs. There are examples of chemosensory cells in a range of organs—in particular in airway mucosae where they are involved in the response to invasive species and in the activation of innate immunity. In addition, the specific bitter receptor T2R38, is found in cancer cells [66] and both the bitter compound phenyl thiourea and the bacterial quorum sensing molecule, N-acetyl-dodecanoyl homoserine (AHL-12) are ligands for the receptor (see below for a summary of some of the biological consequences of these). Activation of T2R38 also leads to upregulation of the ABCB1 transporter which is involved in the excretion of xenobiotics from cells. Another function of T2R38 after binding the quorum sensing molecule, AHL-12, is involvement in biofilm production [67]. Detection of AHL-12 via the bitter receptor has been hypothesised to upregulate phagocytosis and other parts of the cellular response to invading species. Another bitter receptor, T2R14, detects a different type of quorum sensing molecule and mediates innate immune responses [68]. While the extra-oral expression of bitter taste receptors and the consequences have been the subject of increasing studies, the regulation of bitter receptor sub-type expression has been linked to specific dietary components and to the gut microbiota [69]. They are also implicated in pathogen response and as sensors of luminal content thereby regulating gut progression through non-inflammatory states and immune response [70]. Two bitter taste receptors, in particular (bitter taste receptor 138 (T2R138) which is selectively activated by isothiocyanates, and taste receptor 108 (T2R108)) appear to be regulated by luminal content but to different extents in different parts of the GI system [71]. Bitter taste receptors are also upregulated in severe asthmatics [72]. The role of other dietary constituents may be more complex. For example, cholesterol appears to bind to an orthosteric activating site on the bitter receptor T2R14 [73]. The receptor T2R14, indeed binds a large number of ligands at different sites and its gustatory activity may offer an opportunity for the rational design of activators and inhibitors. It may also facilitate modulation of the immune response to pathogenic bacteria and thereby facilitate the early stage response to pathogens. Since quorum sensing is such an intrinsic response by bacteria, it is less likely that bacterial resistance to bitter-cell mediated responses leading to antimicrobial resistance (AMR), will occur.

Extra-oral bitter taste receptors are regularly reviewed and the principal characteristics outlined [74]. Over 1000 bitter compounds are known and the extra-oral biological properties of some of these have been mapped. It is important to note that not all of the physiological properties associated with bitter compounds is necessarily causally linked to binding to T2Rs. Equally well, however, a number of bitter tasting compounds (e.g. quinine, chlorpheniramine etc.) have pharmacological activity and may, through also binding to bitter receptors at extra-oral sites, have additional pharmacological properties. Bitter receptors also exhibit polymorphisms and while this has been the subject of taste-related study and has defined a group of 'super bitter tasters' [75], it is also possible that polymorphisms in extra-oral bitter receptors may result, for example in immune response variation between individuals and, as noted previously, may provide a link between the diet of individuals and clinical outcomes based on different polymorphisms. This has already been suggested by population studies [76]. In addition, polymorphisms in T2R48 have been implicated in the variation in immune responses [77].

In a number of cases, as has been described, the effect of bitter receptor-mediated responses goes beyond sensory evaluation [78]—particularly in extra-oral receptor sites and responses. Further putative disease links continue to be suggested. For example, prevention of biofilm formation by the brain microbiome may be a modulating influence on the development of Alzheimer's and other inflammation-related neurocognitive conditions [79]. The use of GPCRs as a primary target for medicinal chemistry and the development of more precise screening tools [80] suggests that the biological sitespecific role of taste receptors may provide a useful investigative paradigm for drug discovery.

7.6 Synergy Between Taste Molecules

The study of synergy between taste molecules has previously been mentioned (Sect. 7.5). Synergy can be considered either as an extra-receptor effect (in particular the role of umami) or as an intra-receptor effect (the binding of molecules that induce an allosteric shift in a receptor which either enhances the receptor output and/or increases the number of molecules that are able to bind to the active site. This allows for effects to be proposed to be divided into three classes. Those which occur within a single receptor (allosteric), within a functional set of receptors (synergistic) or a crossfunctional effect (enhancement of effects on different receptors by the umami receptor or cross-talk between odour and taste receptors). Synergy between receptors and the interaction between odour and taste receptors and the effect of umami has been mentioned elsewhere so examples of allosteric effects and of the interaction between bitterness and odour will be mentioned but other aspects of synergy between taste molecules will not be covered at this time.

Both glutamate and ribonucleotides can enhance the umami receptor itself by binding specifically to the receptor [81]. This allosteric effect has been suggested to be the underlying molecular mechanism for the taste pairing of oysters and champagne [82]. Champagne is a source of glutamate while oysters contain both free glutamate and 5′nucleotides. Both will bind in an allosteric fashion to the umami receptor and enhance its effect. These are 'within receptor' mechanisms even though they are umamimediated effects.

In the case of the sweet receptor, multiple allosteric binding sites have been found [83] and these have been used to investigate synergy and optimisation between different sweetener molecules [84]. In addition, some binary mixtures have been found to produce enhanced sweetness as a consequence of inhibition of a specific bitter receptor [85].

While enhancement of sweet and umami tastes have practical applications in the development of new products with enhanced flavour profiles, bitterness is more commonly associated with aversion and one of the targets of research into bitterness is the reduction of the response (through the use of antagonists or de-sensitising strategies) rather than its enhancement through the mechanisms noted above. The bitterness of many bioactive compounds presents practical problems in the oral acceptability of drugs. As noted previously, binding to a bitter receptor may activate the ABCB1 transporter system which can increase the excretion of xenobiotics (including drugs) from the cell. Quantitative methods for assessing the bitterness of molecules have been reviewed [86] and the use of machine learning to predict bitterness has been suggested [87]. Different strategies have been adopted to enhance the consumption of beneficial bitter compounds and thereby increase their efficacy. These include the development of specific bitterness masking agents [88]. While bitter receptor antagonists seem often to be amino acid derived [89], some secondary natural products have been found to be bitter receptor antagonists [90]. The development of bitter taste masking compounds to facilitate the consumption of bitter drugs, based on bitter taste receptor polymorphisms has been investigated [91]. The link between bitterness and odour has also been investigated [92] and this may also be a source of synergy therefore expanding the concept beyond a single receptor (allosteric), a single set of receptors (synergistic) to a cross-functional (and cross-class) receptor effect.

7.7 The Essence of Colour Chemistry

The colour of foods is an important cue for other organisms to assess ripeness, the property of being edible or attractiveness to, for example, pollinators but colour can also denote functional properties (for example chlorophyll). As is the case with odour and taste, the secondary natural products that are involved in colour generation can be complex. The study of some of this complexity also reflects the types of techniques used for the characterisation of food-related secondary natural products. The science of taste and odour perception has, as noted previously, some degree of chemical logic and can, in a reductionist way, be considered in terms of molecular level receptor binding and signal transduction. In the case of another of the senses, vision, the understanding of molecules in terms of their contribution to colour can range from quantum effects, through wave/particle duality to molecular vibrations and the interaction with living systems on a receptor level. A comprehensive consideration of this whole spectrum of effects would be a considerable undertaking however some of the salient points will help to contextualise the role of secondary natural products in understanding the chemistry of colour. As chemistry is the most empirical of sciences, a brief discussion of the essence of colour chemistry in biological systems will be followed by examples of colour in food and art and, finally, the molecular level interactions that define colour in the context of chemistry.

The story of colour can be divided into two aspects—the perceived effect and the mechanistic basis of that effect. In that sense it has some similarity to taste and smell both of which can initially be perceived, and this perception is subjective and so is not affected by a greater or lesser knowledge of the underpinning science. Newton, in a recognisably empirical and reductionist way, suggested that while light contained all colours and these components could be revealed using a prism. By adding all the component colours together, white light could be recreated. It is noteworthy that this

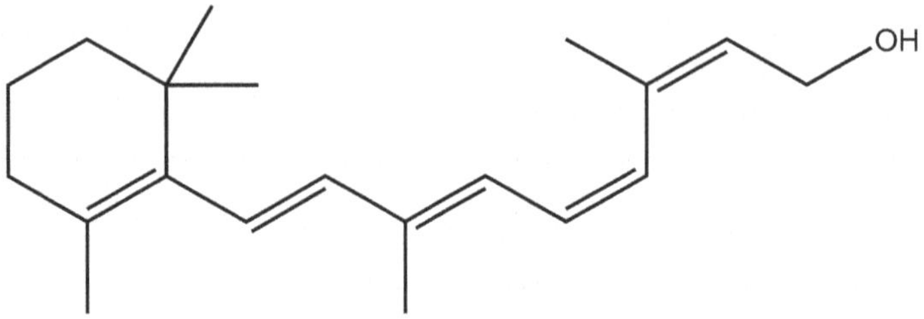

Fig. 7.5 Structure of 11-cis-retinal

additive aspect of light is in contrast to the subtractive aspect of dyes and pigments. In the case of a prism, the wavelengths of the component colours are added to eventually give white light (which contains all visible wavelengths). In the case of dyes and pigments, colours are created by absorption (subtraction) of certain wavelengths and reflecting the others therefore eventually all wavelengths are absorbed. The colour of an object is the light it reflects back to us minus the light it absorbs. The Newtonian rationalisation of colour was not the only theory being pursued. Goethe, for example was one of the most significant 'anti-Newton' voices and wrote a book containing his theories about colour (Zur Farbenlehre). He thought (as was commonly believed preNewton) that colour came from the interaction of light and dark and that dark was not simply the absence of light.

The perception of colour is governed by photoreceptors. Like taste and odour receptors, they can distinguish between effects. Photoreceptors are a type of epithelial cell that can distinguish colours and also light and dark. The biological molecule than enables photoreceptors to function is rhodopsin (or a related structure). Rhodopsin is a GPCR and is, therefore, analogous to other sensory receptors. Indeed olfactory receptors have been described as 'rhodopsin-like' GPCRs [93]. Rhodopsin has a bound chromophore (11-cis-retinal (Fig. 7.5)) [94].

While photosensitivity mediated by photoreceptors does not appear to occur as extensively as is the case, for example, with the bitter receptor (however see below), non-visual light-mediated effects have been observed [95]. Rhodopsin has been used as a model GPCR for investigations into the binding of agonists and the study of activation of the class of receptors [96]. In addition rhodopsins (and related structures) have been found in a variety of microbial species and in various tissues [97]. This may suggest a range of functions that are activated by light which may be analogous to the extra-oral effects noted for taste GPCRs. For example bacteriorhodopsin can function as a proton pump and other functions of microbial and animal rhodopsins have been noted [98]. The study of rhodopsin antagonists has suggested that such small molecules may, through interaction with rhodopsin and disruption of protein–protein interactions, have an impact on other cellular mechanisms [99]. Some secondary natural products (e.g. anthocyanins) appear to affect rhodopsin, possibly by allosteric modulation. Investigations of different types of rhodopsins and related receptors may offer a range of potential medicinal opportunities.

7.8 Colour in Food and Art—Additive and Subtractive Colours. Light and Shade

The story of the chemistry of colour has been significant in both food and art but the link between them has not always been clear. While chemists have used light in the form of spectroscopy to probe chemical structures that are present as secondary natural products, art has used natural product and, later, synthetic chemistry to represent the visible world—both real and imagined. Dyes and pigments make use of the subtractive nature of colours and colour mixtures to create these representations. The natural pigments present in foods have both been attractive and repulsive. In some cases colour is a warning of the presence of toxins while in others it has been a visible demonstration of ripeness. In addition, colour can affect mood and a visual key to actions. Unfortunately, there is a considerable amount of over interpretation in the application of colour theory to psychology and the equivalent of Double Blind Placebo controlled interventions are rare.

From a molecular chemistry standpoint, colour is associated with the movement of electrons. In organic compounds, this is associated with unsaturation. For conjugated, unsaturated ring systems, radiation in the visible part of the spectrum has sufficient energy to promote electrons to an excited state and, as a result, the corresponding compounds are coloured. For example, tetracene (Fig. 7.6) absorbs light from the blue end of the spectrum and so forms bright orange crystals.

Similarly, pentacene (Fig. 7.7) absorbs light from the red end of the spectrum and so is deep purple.

Acyclic conjugated molecules can also be coloured. For example, beta-carotene has 11 conjugated double bonds and will absorb light with wavelengths in the lower energy blue and green portions of the spectrum therefore it will appear as orange and red. Colour can also be a consequence of changes in an inorganic substance—once

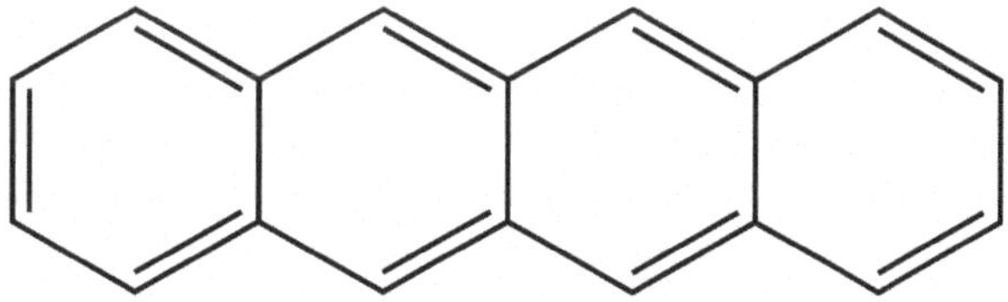

Fig. 7.6 Structure of tetracene

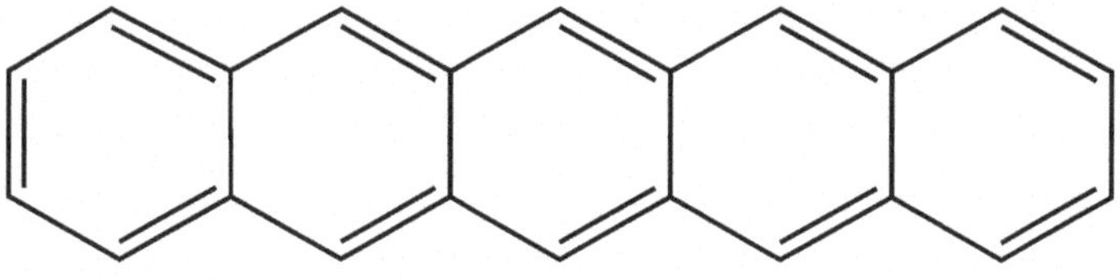

Fig. 7.7 Structure of pentacene

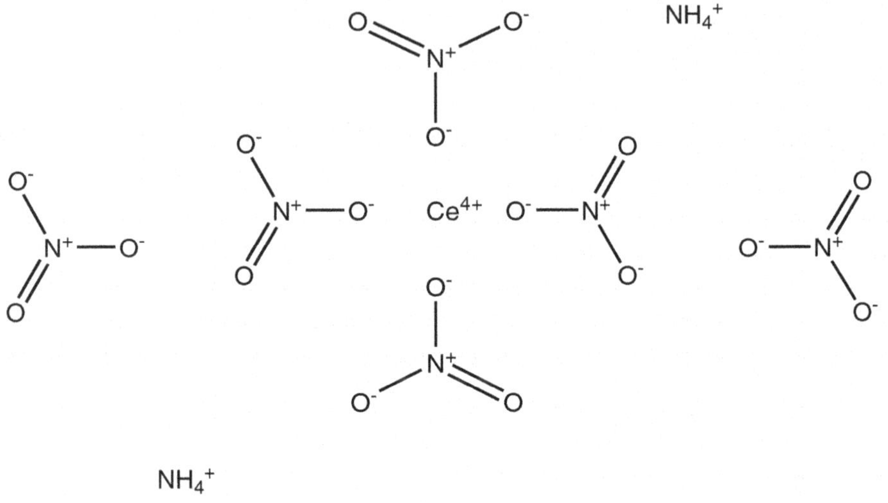

Fig. 7.8 Ceric ammonium nitrate

again this is a consequence of the absorption of energy by electrons. For example, ceric ammonium nitrate is yellow (Fig. 7.8).

The nature of the functional groups attached to the inorganic complex can affect the electron energy levels so if one of the nitrate groups in ceric ammonium nitrate is replaced with an alcohol, the colour of the complex changes from yellow to red. This effect can, of course, be used to detect alcohols. Colorimetric detection methods for a range of functional groups were developed and many of these are still in use due to their simplicity and specificity. More recently, colorimetric bioanalytical assays have been developed including the use of aptamers [100]. One important principle of colour concerns the extreme situations alluded to previously. If all the wavelengths of visible light are reflected back to the observer then the object will be white. If all the wavelengths of light are absorbed then the object will appear to be black. If all the wavelengths of light pass through an object unchanged then that object will be colourless (for example, glass). The distinction between white, black and colourless is important both chemically and philosophically.

Prior to the advent of synthetic dyes in the nineteenth century, colours were obtained from natural sources. The most accessible natural source were inorganic pigments based upon earth colours. Yellow and brown ochres together with chalk (white) and black (from the burning of animal fats) were among the first colours used in, for example, cave paintings such as those found at Lascaux in France [101]. In addition, red was represented by iron oxides. As with living systems, the cave paintings are in a state of continual change and it is not possible to ascertain with any degree of certainty what the paintings looked like at the moment they were completed. Equally imposition of 'homeostasis' in such a multi-factorial system will be unsuccessful. Looking beyond the simple representational depiction of humans and animals, it has been suggested that the cave paintings represent the first recorded attempt to explain the anatomy and physiology of animals [102] rather than simply being a representational record of what was present. Following the use of earth colours, techniques were developed to produce stronger, mineral colours such as malachite. One of the most appreciated colours was cinnabar (mercury sulphide) which is found naturally in volcanic eruptions and is deep red (this intensity of colour gives it a distinct advantage over the more muted tone of iron oxides). Cinnabar was the most commonly used red pigment used until the advent of cadmium red. The popularity of cinnabar (and vermilion—which is chemically identical to it) was a consequence of the strength of the colour and also its availability. Alchemists were interested in the properties of combinations (not simply mixtures but, rather, 'transformed' combinations) of mercury and sulphur. Indeed there is evidence that suggests cinnabar might be considered to be the first 'synthetic' colour [103]. The link between Kant and chemistry has been examined through cinnabar [104]. And the oft-quoted view of Kant that chemistry is not a true science, unlike mathematics. While the transformative aspect of chemistry (and alchemy) is certainly visible, the failure of alchemy to create gold from base metal is seen as a critical failure of methodology and thinking. At the same time, creation of cinnabar is a significant transformative achievement. The question of how chemistry is constituted remains. If chemists pin their hopes on chemistry as a transformative endeavour, then a failure in transformative capability does, in a Popperian sense, condemn the whole of chemistry. As natural product chemists are all too aware, synthesis of any compound (real or imagined) is not a given and the history of the development of colours is an appropriate example of this.

Following the development of earth-derived colours (and cinnabar/vermilion) and the development of mineral colours (including, perhaps most significantly, lapis lazuli), the use of living systems as sources of colour was the next step in the use of colour was the isolation of natural colours from living systems. One of the most significant sources of natural colours are lichen [105]. The chemical constituents of lichens has been reviewed and the colours derived from them have, in some cases, been chemically characterised [106]. A second major natural pigment is Tyrian Purple which is derived from a particular body part of a sea snail and was produced as long ago as the late Bronze Age [107]. The compounds responsible for the colour Tyrian Purple are mono and dibromo indigo, 6,6′-dibromoindirubin and indigo (Fig. 7.9).

The colour hue depends on the relative proportions of these compounds that are present. A number of organic dyes were isolated by extraction from natural sources. While British chemists hold onto the idea that the first synthetic dye was that created by Perkin in 1856, it is worth noting that both Prussian blue (1704) [108] and synthetic ultramarine (1828) [109] predate Perkin's mauve. It is a well-known story that Perkin had been tasked with synthesising quinine (at that time an important anti-malarial compound isolated from the bark of the *Cinchona* tree whose supply was restricted by the ownership of plantations in Asia by the Dutch). At that time, there was little or no structural information available about how organic molecules were assembled. Perkin relied upon information on the proportions of C, H, N and O atoms obtained from the empirical formula. This enabled him to postulate that combination of two

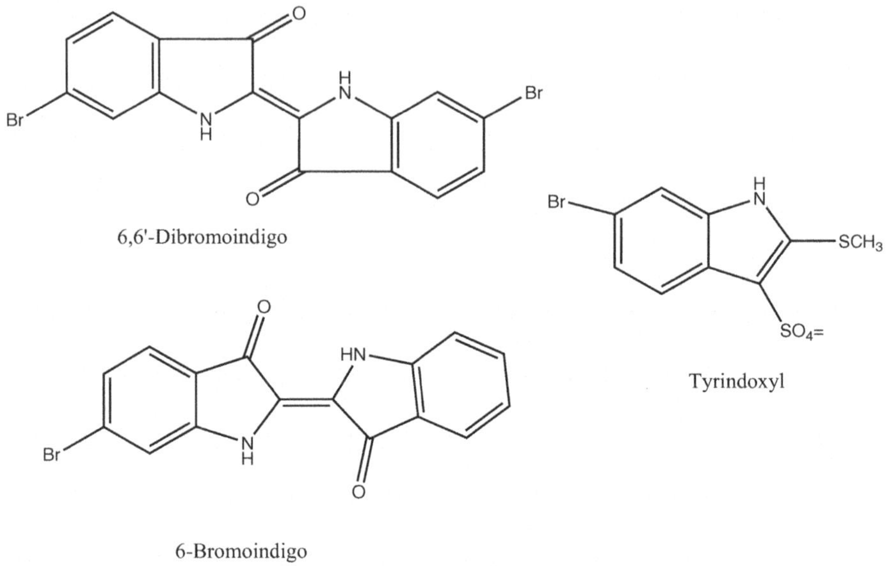

Fig. 7.9 Chemical constituents of tyrian purple

molecules of allyl toluidine ($C_{10}H_{13}N$) plus 3 O atoms from potassium dichromate could be combined to give quinine ($C_{20}H_{24}N_2O_2$) + H_2O (Fig. 7.10).

What Perkin accidently discovered was the first aniline dye and it was the first 'new to nature' synthetic dye that was produced commercially. The dyestuff industry became very important—particularly in France and Germany however it was unclear for many years exactly what Perkin had synthesised. This is partially due to Perkin changing the ratios of the starting materials as he scaled up his dye production process—a detail he did not disclose so as to keep the information secret from his continental competitors [110]. The combination of the invention of new dyes, the use of the metal tube to store dye preparations and the spread of the railways all contributed to the development of the Impressionist school of painting and the trend towards painting en plein air. One of the biggest drawbacks to aniline dyes was their lack of lightfastness. They quickly faded upon exposure to sunlight. Other pigments are also known to fade and, in the case of one colour favoured by impressionists, cadmium yellow, has been examined using a non-invasive, non-destructive technique [111].

Light and shade were important concepts in painting and as the nineteenth and twentieth centuries progressed, new synthetic paints were developed than enabled artists to develop new ways of creating effects. One of these was translucence which is a sub-surface effect. Translucency as an optical phenomenon has been investigated [112] but, although it is relatively simple to describe, the factors that affect it are more complex. Another phenomenon, which is a function of the wave nature of light, is iridescence. This phenomenon is found in nature and is an interface effect [113]. The visible sign of iridescence is a change in colour depending on the viewing

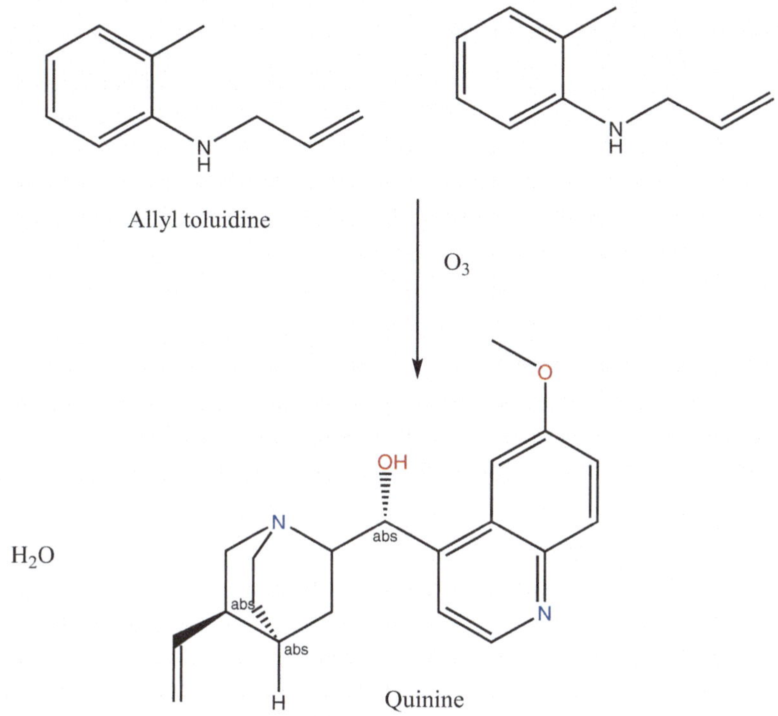

Fig. 7.10 Perkin's putative synthesis of quinine

angle of the observation system. The cause of iridescence is the scattering of light by a semi-transparent surface and sub-surface structures [114]. It is possible to use nanoparticles to produce iridescence. The essence of iridescence in nature is the presence of alternate layers of high and low refractive index dielectric materials [115]. Iridescence can be considered a structural colour rather than a pigment. This means that the study of iridescence falls within the remit of surface and interface science rather than natural product chemistry. Interference at a quantum level has been hypothesised [116]. One of the possible suggested uses of iridescence in nature is in improving the efficiency of photosynthesis and this may be an example of a quantum effect mediated by the quantum wave properties of iridescence [117].

7.9 Natural and Synthetic Colours. Chromophores and Chemical Structures

Natural and synthetic colours can be examined from the standpoint of structure/activity relationships. As noted previously, prediction of colours based upon some structural aspects of molecules is possible—at least on a comparative basis. An AI platform has been developed, based on data mining, for predicting colour based on molecular structure [118]. The data handling system was extended to also predict odour. The researchers found that 24 physicochemical features were sufficient for the accurate prediction of colour, while 39 physicochemical features were sufficient for the accurate prediction of odour. One factor which can affect the colour of individual pigments is the crystal structure and methods have been developed to determine the crystal structure of pigments from X-ray powder diffraction patterns [119]. One of the most important aspects of the measurement of colour is, as noted previously in the case of the deterioration of cadmium yellow, the effect on particular works of art and a range of techniques have been developed and employed to assess early stages in the loss of chromatic integrity [120]. Colour is not simply a visual cue. A number of coloured natural products that may provide a link between food components and health have also been suggested to have potential as beneficial xenobiotics [121] and in many cases do not face the regulatory challenges of synthetic compounds. In addition, nutritionists have associated multi-coloured vegetables with health benefits [122].

Taste, odour and colour are basic qualities associated with secondary natural products—particularly those associated with foods and food systems. The development of odour and colour, in particular, are important signifiers of ripeness and suitability for consumption. While there has been some progress in the development of predictive models (indeed some based upon quantum mechanical calculations from first principles for both colour—[123] and odour [124]). In the case of the latter, a quantum approach provides evidence for a vibrational theory of odour perception rather than a simple, lock and key receptor approach. Such multivariate approaches may help in the development of better hypotheses to explain complex biological systems.

References

1. van Boekel MA. Kinetic aspects of the Maillard reaction: a critical review. Nahrung. 2001;45(3):150–9. https://doi.org/10.1002/1521-3803(20010601)45:3%3c150::AID-FOOD150%3e3.0.CO;2-9. PMID: 11455780.
2. Ovesná J, Mitrová K, Kučera L. Garlic (A. sativum L.) alliinase gene family polymorphism reflects bolting types and cysteine sulphoxides content. BMC Genet. 2015;16:53. https://doi.org/10.1186/s12863-015-0214-z. PMID: 25997498; PMCID: PMC4440563.
3. Nock LP, Mazelis M. The C-S lyases of higher plants: direct comparison of the physical properties of homogeneous alliin lyase of garlic (Allium sativum) and onion (Allium cepa) Plant Physiol. 1987;85:1079–83. https://doi.org/10.1104/pp.85.4.1079.
4. Seca AML, Trendafilova A. Secondary metabolites in edible species: looking beyond nutritional value. Foods. 2021;10(5):1131. https://doi.org/10.3390/foods10051131. PMID: 34069570; PMCID: PMC8161122.
5. Darbandi A, Asadi A, Mahdizade Ari M, Ohadi E, Talebi M, Halaj Zadeh M, Darb Emamie A, Ghanavati R, Kakanj M. Bacteriocins: properties and potential use as antimicrobials. J Clin Lab Anal. 2022;36(1):e24093. https://doi.org/10.1002/jcla.24093. Epub 2021 Dec 1. PMID: 34851542; PMCID: PMC8761470.

6. Stan D, Enciu AM, Mateescu AL, Ion AC, Brezeanu AC, Stan D, Tanase C. Natural compounds with antimicrobial and antiviral effect and nanocarriers used for their transportation. Front Pharmacol. 2021;6(12): 723233. https://doi.org/10.3389/fphar.2021.723233. PMID: 34552489; PMCID: PMC8450524.
7. Mukarram AJ, Ojo M, Dawam J, Riyazuddin N, Ghramh R, Khan HA, Chen KA, Kurjak R, Bayram A. Harnessing phytohormones: advancing plant growth and defence strategies for sustainable agriculture. Physiol Plant. 2024;176(3):e14307. https://doi.org/10.1111/ppl.14307. PMID: 38705723.
8. Loitongbam A, Samal NK, Kumar NR, et al. Differential regulation of glucosinolate-myrosinase mediated defense determines host-aphid interaction in Indian mustard Brassica juncea L. Mol Biol Rep. 2024;51:1079. https://doi.org/10.1007/s11033-024-10002-z.
9. Pavela R, Kovaříková K, Novák M. Botanical antifeedants: an alternative approach to pest control. Insects. 2025;16:136. https://doi.org/10.3390/insects16020136.
10. Wooding SP, Ramirez VA, Behrens M. Bitter taste receptors: Genes, evolution and health. Evol Med Public Health. 2021;9(1):431–47. https://doi.org/10.1093/emph/eoab031. PMID: 35154779; PMCID: PMC8830313.
11. Verbeurgt C, Veithen A, Carlot S, et al. The human bitter taste receptor T2R38 is broadly tuned for bacterial compounds. PLoS ONE. 2017;12(9): e0181302.
12. Harmon CP, Deng D, Breslin PAS. Bitter taste receptors (T2Rs) are sentinels that coordinate metabolic and immunological defense responses. Curr Opin Physiol. 2021;20:70–6. https://doi.org/10.1016/j.cophys.2021.01.006. Epub 2021 Jan 12. PMID: 33738371; PMCID: PMC7963268.
13. Meyerhof W, Batram C, Kuhn C, et al. The molecular receptive ranges of human TAS2R bitter taste receptors. Chem Senses. 2010;35(2):157–70.
14. Clark AA, Liggett SB, Munger SD. Extraoral bitter taste receptors as mediators of off-target drug effects. FASEB J. 2012;26(12):4827–31.
15. Mori S, Nakamura N, Fuchigami A, Yoshimoto S, Sakakibara M, Ozawa T, Aoki J, Inoue A, Sumida H, Ando H, Nakamura M. Intracellular TAS2Rs act as a gatekeeper for the excretion of harmful substances via ABCB1 in keratinocytes. FASEB Bioadv. 2024;6(10):424–41. https://doi.org/10.1096/fba.2024-00074. PMID: 39372126; PMCID: PMC11452442.
16. Zhang Y, Hoon MA, Chandrashekar J, Mueller KL, Cook B, Wu D, Zuker CS, Ryba NJ. Coding of sweet, bitter, and umami tastes: different receptor cells sharing similar signaling pathways. Cell. 2003;112:293–301. https://doi.org/10.1016/S00928674(03)00071-0.
17. Jiang W, Wang Z, Zhang J, Li M. Interleukin 25 and its biological features and function in intestinal diseases. Cent Eur J Immunol. 2022;47(4):362–372. https://doi.org/10.5114/ceji.2022.124416. Epub 2023 Jan 31. PMID: 36817397; PMCID: PMC9901255.
18. Small DM, Prescott J. Odor/taste integration and the perception of flavor. Exp Brain Res. 2005;166(3–4):345–57. https://doi.org/10.1007/s00221-005-2376-9. Epub 2005 Jul 19. PMID: 16028032.
19. Simon SA. Interactions between salt and acid stimuli: a lesson in gustation from simultaneous epithelial and neural recordings. J Gen Physiol. 2002;120(6):787–91. https://doi.org/10.1085/jgp.20028735. PMID: 12451049; PMCID: PMC2229563.
20. Servant G, Frerot E. Pharmacology of the Umami taste receptor. Handb Exp Pharmacol. 2022;275:109–36. https://doi.org/10.1007/164_2021_439. PMID: 33580387.
21. Diepeveen J, Moerdijk-Poortvliet TCW, van der Leij FR. Molecular insights into human taste perception and umami tastants: a review. J Food Sci. 2022;87(4):1449–65. https://doi.org/10.1111/1750-3841.16101. Epub 2022 Mar 17. PMID: 35301715; PMCID: PMC9314127.
22. Kyriazis GA, Smith KR, Tyrberg B, Hussain T, Pratley RE. Sweet taste receptors regulate basal insulin secretion and contribute to compensatory insulin hypersecretion during the development of diabetes in male mice. Endocrinology. 2014;155(6):2112–21. https://doi.org/10.1210/en.2013-2015. Epub 2014 Apr 8. PMID: 24712876; PMCID: PMC4020927.
23. Kyriazis GA, Soundarapandian MM, Tyrberg B. Sweet taste receptor signaling in beta cells mediates fructose-induced potentiation of glucose-stimulated insulin secretion. Proc Natl Acad Sci U S A. 2012;109(8):E524–32. https://doi.org/10.1073/pnas.1115183109. Epub 2012 Feb 6. PMID: 22315413; PMCID: PMC3286985.

24. Foster SR, Porrello ER, Purdue B, Chan HW, Voigt A, Frenzel S, Hannan RD, Moritz KM, Simmons DG, Molenaar P, Roura E, Boehm U, Meyerhof W, Thomas WG. Expression, regulation and putative nutrient-sensing function of taste GPCRs in the heart. PLoS ONE. 2013;8(5): e64579. https://doi.org/10.1371/journal.pone.0064579. PMID: 23696900; PMCID: PMC3655793.
25. Tran HTT, Stetter R, Herz C, Spöttel J, Krell M, Hanschen FS, et al. Allyl isothiocyanate: a Tas2r38 receptor-dependent immune modulator at the interface between personalized medicine and nutrition. Front Immunol. 2021;12: 669005. https://doi.org/10.3389/fimmu.2021.669005.
26. Cohen NA. The genetics of the bitter taste receptor T2R38 in upper airway innate immunity and implications for chronic rhinosinusitis. Laryngoscope. 2017;127(1):44–51. https://doi.org/10.1002/lary.26198. Epub 2016 Sep 21. PMID: 27650657; PMCID: PMC5177547.
27. Wu C, et al. The structure and function of olfactory receptors. Trends Pharmacol Sci. 45(3):268–80.
28. Buck L, Axel R. A novel multigene family may encode odorant receptors: a molecular basis for odor recognition. Cell. 1991;65(1):175–87. https://doi.org/10.1016/0092-8674(91)90418-x. PMID: 1840504.
29. Ryu SE, Bae J, Shim T, et al. Conserved pattern-based classification of human odorant receptor multigene family. Sci Rep. 2024;14:27271. https://doi.org/10.1038/s41598-024-79183-8.
30. Wu C, Jeong MY, Kim JY, Lee G, Kim JS, Cheong YE, Kang H, Cho CH, Kim J, Park MK, Shin YK, Kim KH, Seol GH, Koo SH, Ko G, Lee SJ. Activation of ectopic olfactory receptor 544 induces GLP-1 secretion and regulates gut inflammation. Gut Microbes. 2021;13(1):1987782. https://doi.org/10.1080/19490976.2021.1987782. PMID: 34674602; PMCID: PMC8632334.
31. Yang Z, Cheng J, Shang P, Sun JP, Yu X. Emerging roles of olfactory receptors in glucose metabolism. Trends Cell Biol. 2023;33(6):463–76. https://doi.org/10.1016/j.tcb.2022.09.005. Epub 2022 Oct 10. PMID: 36229334.
32. Pasquier J, Kamech N, Lafont AG, Vaudry H, Rousseau K, Dufour S. Molecular evolution of GPCRs: Kisspeptin/kisspeptin receptors. J Mol Endocrinol. 2014;52(3):T101–17. https://doi.org/10.1530/JME-13-0224. Epub 2014 Feb 27. PMID: 24577719.
33. Vaudry H. Molecular evolution of GPCRS: what we know and what the future holds. J Mol Endocrinol. 2014;52(3):E1–2.
34. Hu GM, Mai TL, Chen CM. Visualizing the GPCR network: classification and evolution. Sci Rep. 2017;7:15495. https://doi.org/10.1038/s41598-01715707-9.
35. Liauchonak I, Qorri B, Dawoud F, Riat Y, Szewczuk MR. Non-nutritive sweeteners and their implications on the development of metabolic syndrome. Nutrients. 2019;11(3):644. https://doi.org/10.3390/nu11030644. PMID: 30884834; PMCID: PMC6471792.
36. Laffitte A, Neiers F, Briand L. Functional roles of the sweet taste receptor in oral and extraoral tissues. Curr Opin Clin Nutr Metab Care. 2014;17(4):379–85. https://doi.org/10.1097/MCO.0000000000000058. PMID: 24763065; PMCID: PMC4059820.
37. Juen Z, et al. The structure of human sweetness. Cell. 2025. https://doi.org/10.1016/j.cell.2025.04.021.
38. Suez J, Korem T, Zeevi D, Zilberman-Schapira G, Thaiss CA, Maza O, Israeli D, Zmora N, Gilad S, Weinberger A, et al. Artificial sweeteners induce glucose intolerance by altering the gut microbiota. Nature. 2014;514:181–6. https://doi.org/10.1038/nature13793.
39. Conz A, Salmona M, Diomede L. Effect of non-nutritive sweeteners on the gut microbiota. Nutrients. 2023;15(8):1869. https://doi.org/10.3390/nu15081869. PMID: 37111090; PMCID: PMC10144565.
40. Zhao X, Wang C, Zheng Y, Liu B. New insight into the structure-activity relationship of sweet-tasting proteins: protein sector and its role for sweet properties. Front Nutr. 2021;18(8): 691368. https://doi.org/10.3389/fnut.2021.691368. PMID: 34222309; PMCID: PMC8249704.
41. Kang J, et al. Recent advances in biomimetic taste-based biosensors and their applications. Sens Diagnostics. 2025;4:24–34.

42. Vasilaki A, Panagiotopoulou E, Koupantsis T, Katsanidis E, Mourtzinos I. Recent insights in flavor-enhancers: definition, mechanism of action, taste-enhancing ingredients, analytical techniques and the potential of utilization. Crit Rev Food Sci Nutr. 2022;62(32):9036–52. https://doi.org/10.1080/10408398.2021.1939264. Epub 2021 Jun 18. PMID: 34142890.
43. Li Q, Zhang L, Lametsch R. Current progress in kokumi-active peptides, evaluation and preparation methods: a review. Crit Rev Food Sci Nutr. 2022;62(5):1230–41. https://doi.org/10.1080/10408398.2020.1837726. Epub 2020 Oct 26. PMID: 33103468.
44. Li X, Servant G, Tachdjian C. The discovery and mechanism of sweet taste enhancers. Biomol Concepts. 2011;2(4):327–32. https://doi.org/10.1515/bmc.2011.021. PMID: 25962040.
45. Steindler DA. Sugar substitutes and taste enhancers need more science, sensitivity- and allergy-guided labeling. NPJ Sci Food. 2023;7(1):64. https://doi.org/10.1038/s41538-023-00240-z. PMID: 38071344; PMCID: PMC10710428.
46. Jurcevic Zidar B, Knezovic Z, Pribisalic A, Luetic S, Jurcic K, Knezovic N, Sutlovic D. Consumer perceptions of artificial sweeteners in food products, consumption frequency, and body mass index: a multivariate analysis. Nutrients. 2025;17(5):814. https://doi.org/10.3390/nu17050814. PMID: 40077684; PMCID: PMC11902076.
47. Hamdard E, Lv Z, Jiang J, Wei Q, Shi Z, Malyar RM, Yu D, Shi F. Responsiveness expressions of bitter taste receptors against denatonium benzoate and genistein in the heart, spleen, lung, kidney, and Bursa Fabricius of Chinese fast yellow chicken. Animals (Basel). 2019;9(8):532. https://doi.org/10.3390/ani9080532. PMID: 31390726; PMCID: PMC6719124.
48. Birch GG. Structural relationships of sugars to taste. CRC Crit Rev Food Sci Nutr. 1976;8(1):57–95. https://doi.org/10.1080/10408397609527217. PMID: 801349.
49. Rojas C, Ballabio D, Consonni V, et al. Quantitative structure–activity relationships to predict sweet and non-sweet tastes. Theor Chem Acc. 2016;135:66. https://doi.org/10.1007/s00214-016-1812-1.
50. Xiu H, Liu Y, Yang H, Ren H, Luo B, Wang Z, Shao H, Wang F, Zhang J, Wang Y. Identification of novel umami molecules via QSAR models and molecular docking. Food Funct. 2022;13(14):7529–39. https://doi.org/10.1039/d2fo00544a. PMID: 35765918.
51. Goel A, Gajula K, Gupta R, Rai B. In-silico screening of database for finding potential sweet molecules: a combined data and structure based modeling approach. Food Chem. 2021;343:128538. https://doi.org/10.1016/j.foodchem.2020.128538. Epub 2020 Nov 4. PMID: 33183872.
52. Nikolaychuk PA. How the concept of chemical affinity evolved and how it contributed to chemical science. ChemTexts. 2024;10:2. https://doi.org/10.1007/s40828-023-00185-6).
53. Slack JP, Brockhoff A, Batram C, Menzel S, Sonnabend C, Born S, Galindo MM, Kohl S, Thalmann S, Ostopovici-Halip L, Simons CT, Ungureanu I, Duineveld K, Bologa CG, Behrens M, Furrer S, Oprea TI, Meyerhof W. Modulation of bitter taste perception by a small molecule hTAS2R antagonist. Curr Biol. 2010;20(12):1104–9. https://doi.org/10.1016/j.cub.2010.04.043. Epub 2010 May 27. PMID: 20537538; PMCID: PMC2925244.
54. Ameta D, Behera L, Chakraborty A, et al. Predicting odor from vibrational spectra: a data-driven approach. Sci Rep. 2024;14:20321. https://doi.org/10.1038/s41598-024-70696-w.
55. Cometto-Muñiz JE, Abraham MH. Structure-activity relationships on the odour detectability of homologous carboxylic acids by humans. Exp Brain Res. 2010;207(1–2):75–84. https://doi.org/10.1007/s00221-010-2430-0. Epub 2010 Oct 8. PMID: 20931179; PMCID: PMC2964470.
56. Bo W, Yu Y, He R, Qin D, Zheng X, Wang Y, Ding B, Liang G. Insight into the structure-odor relationship of molecules: a computational study based on deep learning. Foods. 2022;11(14):2033. https://doi.org/10.3390/foods11142033. PMID: 35885276; PMCID: PMC9320518.
57. Kirsch F, Buettner A. Odor qualities and thresholds of physiological metabolites of 1,8-cineole as an example for structure-activity relationships considering chirality aspects. Chem Biodivers. 2013;10(9):1683–95. https://doi.org/10.1002/cbdv.201300097. PMID: 24078601.
58. Engel KH. Chirality: an important phenomenon regarding biosynthesis, perception, and authenticity of flavor compounds. J Agric Food Chem. 2020;68(38):10265–74. https://doi.org/10.1021/acs.jafc.0c01512. Epub 2020 Apr 9. PMID: 32223150.

59. Billesbølle CB, de March CA, van der Velden WJC, et al. Structural basis of odorant recognition by a human odorant receptor. Nature. 2023;615:742–9. https://doi.org/10.1038/s41586-023-05798-y.
60. Fleischer J, Pregitzer P, Breer H, Krieger J. Access to the odor world: olfactory receptors and their role for signal transduction in insects. Cell Mol Life Sci. 2018;75(3):485–508. https://doi.org/10.1007/s00018-017-2627-5. Epub 2017 Aug 21. PMID: 28828501; PMCID: PMC11105692.
61. Kostka JK, Bitzenhofer SH. How the sense of smell influences cognition throughout life. Neuroforum. 2022;28(3):177–85. https://doi.org/10.1515/nf-20220007. Epub 2022 Jun 6. PMID: 36067120; PMCID: PMC9380998.
62. Vance DE, Del Bene VA, Kamath V, Frank JS, Billings R, Cho DY, Byun JY, Jacob A, Anderson JN, Visscher K, Triebel K, Martin KM, Li W, Puga F, Fazeli PL. Does olfactory training improve brain function and cognition? A systematic review. Neuropsychol Rev. 2024;34(1):155–91. https://doi.org/10.1007/s11065-022-09573-0. Epub 2023 Feb 2. PMID: 36725781; PMCID: PMC9891899.
63. Tian Q, An Y, Kitner-Triolo MH, Davatzikos C, Studenski SA, Ferrucci L, Resnick SM. Associations of olfaction with longitudinal trajectories of brain volumes and neuropsychological function in older adults. Neurology. 2023;100(9):e964–74. https://doi.org/10.1212/WNL.0000000000201646. Epub 2022 Dec 2. PMID: 36460474; PMCID: PMC9990434.
64. Harrass S, Yi C, Chen H. Chronic rhinosinusitis and Alzheimer's disease-a possible role for the nasal microbiome in causing neurodegeneration in the elderly. Int J Mol Sci. 2021;22(20):11207. https://doi.org/10.3390/ijms222011207. PMID: 34681867; PMCID: PMC8541405.
65. Cattaneo C, Gargari G, Koirala R, Laureati M, Riso P, Guglielmetti S, Pagliarini E. New insights into the relationship between taste perception and oral microbiota composition. Sci Rep. 2019;9(1):3549. https://doi.org/10.1038/s41598-019-40374-3. PMID: 30837660; PMCID: PMC6401163.
66. Gaida MM, Mayer C, Dapunt U, Stegmaier S, Schirmacher P, Wabnitz GH, Hänsch GM. Expression of the bitter receptor T2R38 in pancreatic cancer: localization in lipid droplets and activation by a bacteria-derived quorumsensing molecule. Oncotarget. 2016;7(11):12623–32. https://doi.org/10.18632/oncotarget.7206. PMID: 26862855; PMCID: PMC4914309.
67. Gaida MM, Dapunt U, Hänsch GM. Sensing developing biofilms: the bitter receptor T2R38 on myeloid cells. Pathogens Dis. 2016;74(3):ftw004, https://doi.org/10.1093/femspd/ftw004.
68. Medapati MR, Singh N, Bhagirath AY, Duan K, Triggs-Raine B, Batista EL Jr, Chelikani P. Bitter taste receptor T2R14 detects quorum sensing molecules from cariogenic Streptococcus mutans and mediates innate immune responses in gingival epithelial cells. FASEB J. 2021;35(3): e21375. https://doi.org/10.1096/fj.202000208R. PMID: 33559200.
69. Caremoli F, Huynh J, Lagishetty V, Markovic D, Braun J, Dong TS, Jacobs JP, Sternini C. Microbiota-dependent upregulation of bitter taste receptor subtypes in the mouse large intestine in high-fat diet-induced obesity. Nutrients. 2023;15(19):4145. https://doi.org/10.3390/nu15194145. PMID: 37836428; PMCID: PMC10574285.
70. Sternini C, Rozengurt E. Bitter taste receptors as sensors of gut luminal contents. Nat Rev Gastroenterol Hepatol. 2025;22:39–53. https://doi.org/10.1038/s41575-024-01005-z.
71. Vegezzi G, Anselmi L, Huynh J, Barocelli E, Rozengurt E, Raybould H, Sternini C. Diet-induced regulation of bitter taste receptor subtypes in the mouse gastrointestinal tract. PLoS ONE. 2014;9(9): e107732. https://doi.org/10.1371/journal.pone.0107732. PMID: 25238152; PMCID: PMC4169553.
72. Orsmark-Pietras C, James A, Konradsen JR, Nordlund B, Söderhäll C, Pulkkinen V, Pedroletti C, Daham K, Kupczyk M, Dahlén B, Kere J, Dahlén SE, Hedlin G, Melén E. Transcriptome analysis reveals upregulation of bitter taste receptors in severe asthmatics. Eur Respir J. 2013;42(1):65–78. https://doi.org/10.1183/09031936.00077712. Epub 2012 Dec 6. PMID: 23222870.
73. Hu X, Ao W, Gao M, Wu L, Pei Y, Liu S, Wu Y, Zhao F, Sun Q, Liu J, Jiang L, Wang X, Li Y, Tan Q, Cheng J, Yang F, Yang C, Sun J, Hua T, Liu ZJ. Bitter taste TAS2R14 activation

by intracellular tastants and cholesterol. Nature. 2024;631(8020):459–66. https://doi.org/10.1038/s41586-024-07569-9. Epub 2024 May 22. PMID: 38776963.

74. Tuzim K, Korolczuk A. An update on extra-oral bitter taste receptors. J Transl Med. 2021;19:440. https://doi.org/10.1186/s12967-021-03067-y.
75. Risso DS, Mezzavilla M, Pagani L, et al. Global diversity in the TAS2R38 bitter taste receptor: revisiting a classic evolutionary PROPosal. Sci Rep. 2016;6(1):1–9.
76. Triantafillou V, Workman AD, Kohanski MA, Cohen NA. Taste receptor polymorphisms and immune response: a review of receptor genotypic-phenotypic variations and their relevance to chronic rhinosinusitis. Front Cell Infect Microbiol. 2018;7(8):64. https://doi.org/10.3389/fcimb.2018.00064. PMID: 29564227; PMCID: PMC5845873.
77. Liszt KI, Wang Q, Farhadipour M, Segers A, Thijs T, Nys L, Deleus E, Van der Schueren B, Gerner C, Neuditschko B, Ceulemans LJ, Lannoo M, Tack J, Depoortere I. Human intestinal bitter taste receptors regulate innate immune responses and metabolic regulators in obesity. J Clin Invest. 2022;132(3): e144828. https://doi.org/10.1172/JCI144828. PMID: 34784295; PMCID: PMC8803326.
78. Harris JC, Lee RJ, Carey RM. Extragustatory bitter taste receptors in head and neck health and disease. J Mol Med. 2024;102:1413–24. https://doi.org/10.1007/s00109-024-02490-0.
79. Hanley AB. Alzheimer's disease—a clear and future epidemic. Sci Parliament. 2021;77(1):27–9.
80. Zeghal M, Laroche G, Freitas JD, Wang R, Giguère PM. Profiling of basal and ligand-dependent GPCR activities by means of a polyvalent cell-based high throughput platform. Nat Commun. 2023;14(1):3684. https://doi.org/10.1038/s41467023-39132-x. PMID: 37407564; PMCID: PMC10322906.
81. Zhang F, Klebansky B, Fine RM, Xu H, Pronin A, Liu H, Tachdjian C, Li X. Molecular mechanism for the umami taste synergism. Proc Natl Acad Sci U S A. 2008;105(52):20930–4. https://doi.org/10.1073/pnas.0810174106. Epub 2008 Dec 22. PMID: 19104071; PMCID: PMC2606899.
82. Schmidt CV, Olsen K, Mouritsen OG. Umami synergy as the scientific principle behind taste-pairing champagne and oysters. Sci Rep. 2020;10(1):20077. https://doi.org/10.1038/s41598-020-77107-w. PMID: 33208820; PMCID: PMC7676262.
83. Servant G, Kenakin T, Zhang L, Williams M, Servant N. The function and allosteric control of the human sweet taste receptor. Adv Pharmacol. 2020;88:59–82. https://doi.org/10.1016/bs.apha.2020.01.002. Epub 2020 Feb 10. PMID: 32416872.
84. Belloir C, Jeannin M, Karolkowski A, Scott C, Briand L. A receptor-based assay to study the sweet and bitter tastes of sweeteners and binary sweet blends: the SWEET project. Chem Senses. 2024;49:bjae041, https://doi.org/10.1093/chemse/bjae041.
85. Munger SD. A bitter tale of sweet synergy. Cell Chem Biol. 2017;24(10):1191–2. https://doi.org/10.1016/j.chembiol.2017.10.003. PMID: 29053947.
86. Wang P, Li H, Wang Y, Dong F, Li H, Gui X, Ren Y, Gao X, Li X, Liu R. One of the major challenges of masking the bitter taste in medications: an overview of quantitative methods for bitterness. Front Chem. 2024;14(12):1449536. https://doi.org/10.3389/fchem.2024.1449536. PMID: 39206439; PMCID: PMC11349634.
87. Margulis E, Dagan-Wiener A, Ives RS, Jaffari S, Siems K, Niv MY. Intense bitterness of molecules: machine learning for expediting drug discovery. Comput Struct Biotechnol J. 2020;25(19):568–76. https://doi.org/10.1016/j.csbj.2020.12.030. PMID: 33510862; PMCID: PMC7807207.
88. Ke X, Ma H, Yang J, Qiu M, Wang J, Han L, Zhang D. New strategies for identifying and masking the bitter taste in traditional herbal medicines: the example of Huanglian Jiedu Decoction. Front Pharmacol. 2022;17(13): 843821. https://doi.org/10.3389/fphar.2022.843821. PMID: 36060004; PMCID: PMC9431955.
89. Pydi SP, Sobotkiewicz T, Billakanti R, Bhullar RP, Loewen MC, Chelikani P. Amino acid derivatives as bitter taste receptor (T2R) blockers. J Biol Chem. 2014;289(36):25054–66. https://doi.org/10.1074/jbc.M114.576975. Epub 2014 Jul 24. PMID: 25059668; PMCID: PMC4155672.

90. Roland WS, Gouka RJ, Gruppen H, Driesse M, van Buren L, Smit G, Vincken JP. 6methoxyflavanones as bitter taste receptor blockers for hTAS2R39. PLoS ONE. 2014;9(4): e94451. https://doi.org/10.1371/journal.pone.0094451. PMID: 24722342; PMCID: PMC3983201.
91. Nguyen H, Lin C, Bell K, Huang A, Hannum M, Ramirez V, Christensen C, Rawson NE, Colquitt L, Domanico P, Sasimovich I, Herriman R, Joseph P, Braimah O, Reed DR. Worldwide study of the taste of bitter medicines and their modifiers. bioRxiv [Preprint]. 2024 Apr 28:2024.04.24.590957. https://doi.org/10.1101/2024.04.24.590957. Update in: Chem Senses. 2025 Feb 04:bjaf003. https://doi.org/10.1093/chemse/bjaf003. PMID: 38712219; PMCID: PMC11071635.
92. Margulis E, Lang T, Tromelin A, Ziaikin E, Behrens M, Niv MY. Bitter odorants and odorous bitters: toxicity and human TAS2R targets. J Agric Food Chem. 2023;71(23):9051–61. https://doi.org/10.1021/acs.jafc.3c00592. Epub 2023 Jun 1. PMID: 37263600; PMCID: PMC10273317.
93. Katada S, Touhara K. A molecular basis for odorant recognition: olfactory receptor pharmacology. Nihon Yakurigaku Zasshi. 2004;124(4):201–9. Japanese. https://doi.org/10.1254/fpj.124.201. PMID: 15467253.
94. Hofmann KP, Lamb TD. Rhodopsin, light-sensor of vision. Prog Retin Eye Res. 2023;93: 101116. https://doi.org/10.1016/j.preteyeres.2022.101116. Epub 2022 Oct 21. PMID: 36273969.
95. Zhao C, Bo J, Li T, Tian J, Long T, He Y, Chen S, Liu C. Blue light-driven cell cycle arrest in thyroid cancer via Retinal-OPN3 complex. Cell Commun Signal. 2024;22(1):530. https://doi.org/10.1186/s12964-024-01908-z. PMID: 39487504; PMCID: PMC11531186.
96. Deupi X, Standfuss J. Structural insights into agonist-induced activation of Gprotein-coupled receptors. Curr Opin Struct Biol. 2011;21(4):541–51. https://doi.org/10.1016/j.sbi.2011.06.002. Epub 2011 Jun 30. PMID: 21723721.
97. Kojima K, Sudo Y. Convergent evolution of animal and microbial rhodopsins. RSC Adv. 2023;13(8):5367–81. https://doi.org/10.1039/d2ra07073a. Erratum in: RSC Adv. 2023;13(11):7222–4. https://doi.org/10.1039/d3ra90016a. PMID: 36793294; PMCID: PMC9923458.
98. Ernst OP, Lodowski DT, Elstner M, Hegemann P, Brown LS, Kandori H. Microbial and animal rhodopsins: structures, functions, and molecular mechanisms. Chem Rev. 2014;114(1):126–63. https://doi.org/10.1021/cr4003769. Epub 2013 Dec 23. PMID: 24364740; PMCID: PMC3979449.
99. Yanamala N, Gardner E, Riciutti A, Klein-Seetharaman J. The cytoplasmic rhodopsin-protein interface: potential for drug discovery. Curr Drug Targets. 2012;13(1):3–14. https://doi.org/10.2174/138945012798868461. PMID: 21777183; PMCID: PMC3275648.
100. Liu R, Li J, Salena BJ, Li Y. Aptamer and DNAzyme based colorimetric biosensors for pathogen detection. Angew Chem Int Ed Engl. 2025;64(4):e202418725. https://doi.org/10.1002/anie.202418725. Epub 2024 Nov 25. PMID: 39551709; PMCID: PMC11753613.
101. Leroi-Gourhan A. The archaeology of Lascaux cave. Sci Am. 1982;246(6):104–13. http://www.jstor.org/stable/24966617.
102. Jeon JH. Lascaux cave painting: the earliest drawing of gastrointestinal anatomy and physiology? J Korean Med Sci. 2024;39(3): e34. https://doi.org/10.3346/jkms.2024.39.e34. PMID: 38258366; PMCID: PMC10803204.
103. Marchini M, Montanari G, Casali L, Martelli M, Raggetti L, Baláž M, Baláž P, Maini L. What makes every work perfect is cooking and grinding: the ancient roots of mechanochemistry. RSC Mechanochem. 2024;1: 123129. https://doi.org/10.1039/D3MR00035D.
104. Babich B. Kant between chemistry and alchemy: cinnabar, 'now red, now black.' Kant-Studien. 2023;114(4):796–813. https://doi.org/10.1515/kant-2023-2042.
105. Rather LJ, Jameel S, Ganie SA, Bhat KA. Lichen derived natural colorants: history, extraction, and applications. In: Yusuf M, editor. Handbook of renewable materials for coloration and finishing. Beverly, MA: Scrivener Publishing LLC; 2018. pp. 103–14.
106. Yousuf S, et al, Lichens: chemistry and biological activities. Stud Nat Prod Chem. 2014;43(2014):223–59.

107. Berger L, Forstenpointner G, Frühauf P, Kanz F. More than just a color: Archaeological, analytical, and procedural aspects of Late Bronze Age purpledye production at Cape Kolonna, Aegina. PLoS ONE. 2024;19(6): e0304340. https://doi.org/10.1371/journal.pone.0304340. PMID: 38865333; PMCID: PMC11168678.
108. Guichard C, Le Hô A-S, Williams H, Blue P. Chemistry, commerce, and colour in eighteenth-century paris. Art History. 2023;46(1):154–86. https://doi.org/10.1111/1467-8365.12695.
109. Mertens J. The history of artificial ultramarine (1787–1844): science industry and secrecy. Ambix. 2004;51(3):219–44. https://doi.org/10.1179/amb.2004.51.3.219.
110. Cova TFGG, Pais AACC, Seixas de Melo JS. Reconstructing the historical synthesis of mauveine from Perkin and Caro: procedure and details. Sci Rep. 2017;7:6806. https://doi.org/10.1038/s41598-017-07239-z.
111. Zhou Y, Grass D, Warren WS, Fischer MC. Non-destructive three dimensional imaging of artificially degraded CdS paints by pump-probe microscopy. J Phys: Photonics. 2024;6(2): 025013. https://doi.org/10.1088/2515-7647/ad3e65.
112. Gigilashvili D, Thomas JB, Hardeberg JY, Pedersen M. Translucency perception: a review. J Vis. 2021;21(8):4. https://doi.org/10.1167/jov.21.8.4. PMID: 34342646; PMCID: PMC8340651.
113. Doucet SM, Meadows MG. Iridescence: a functional perspective. J R Soc Interface. 2009;6(Suppl 2):S115–32. https://doi.org/10.1098/rsif.2008.0395.focus. PMID: 19336344; PMCID: PMC2706478.
114. Tilley RJD (2019) Iridescence (Goniochromism). In: Shamey R, editors. Encyclopedia of color science and technology. Springer, Berlin, Heidelberg. https://doi.org/10.1007/978-3-642-27851-8_221-2.
115. Jordan TM, Partridge JC, Roberts NW. Disordered animal multilayer reflectors and the localization of light. J R Soc Interface. 2014;11(2014):0948.
116. Chen Z, Grace IM, Woltering SL, et al. Quantum interference enhances the performance of single-molecule transistors. Nat Nanotechnol. 2024;19:986–92. https://doi.org/10.1038/s41565-024-01633-1.
117. Jacobs M, Lopez-Garcia M, Phrathep OP, et al. Photonic multilayer structure of Begonia chloroplasts enhances photosynthetic efficiency. Nature Plants. 2016;2:16162. https://doi.org/10.1038/nplants.2016.162.
118. Zhang X, Zhang K, Lin D, Zhu Y, Chen C, He L, Guo X, Chen K, Wang R, Liu Z, Wu X, Long E, Huang K, He Z, Liu X, Lin H. Artificial intelligence deciphers codes for color and odor perceptions based on large-scale chemoinformatic data. GigaScience. 2020;9(2):giaa011. https://doi.org/10.1093/gigascience/giaa011.
119. Panina N, Leusen FJ, Janssen FF, Verwer P, Meekes H, Vlieg E, Deroover G. Crystal structure prediction of organic pigments: quinacridone as an example. J Appl Crystallogr. 2007;40(Pt 01):105–14. https://doi.org/10.1107/S0021889806043767. Epub 2007 Jan 12. PMID: 19461846; PMCID: PMC2483485.
120. Miliani C, Monico L, Melo MJ, Fantacci S, Angelin EM, Romani A, Janssens K. Photochemistry of artists' dyes and pigments: towards better understanding and prevention of colour change in works of art. Angew Chem Int Ed Engl. 2018;57(25):7324–34. https://doi.org/10.1002/anie.201802801. Epub 2018 May 28. PMID: 29696761.
121. Di Salvo E, Lo Vecchio G, De Pasquale R, De Maria L, Tardugno R, Vadalà R, Cicero N. Natural pigments production and their application in food, health and other industries. Nutrients. 2023;15(8):1923. https://doi.org/10.3390/nu15081923. PMID: 37111142; PMCID: PMC10144550.
122. Petropoulos SA, Sampaio SL, Gioia FD, Tzortzakis N, Rouphael Y, Kyriacou MC, Ferreira I. Grown to be blue-antioxidant properties and health effects of colored vegetables. Part I: root vegetables. Antioxidants (Basel). 2019;8(12):617. https://doi.org/10.3390/antiox8120617. PMID: 31817206; PMCID: PMC6943509.
123. Berthier M, Provenzi E. Quantum measurement and colour perception: theory and applications. Proc R Soc A. 2022;478.

124. Bandar A. Human smell perception is governed by quantum spin-residual information. Scilight 2019;2019(30):300010.

Chapter 8
Secondary Natural Products and Food Processing

Contents

8.1 Recipe—Chicken Scallopine with White Wine, Lime and Herbs 228
8.2 Food Processing—Principles and Systems. Heating, Freezing, Fermentation and Preservation 229
8.3 Macronutrient Processing. Resistant Starch and Prebiotics 236
8.4 Processing Contaminants—Acrylamide and Heterocyclic Amines 239
8.5 Processing to Remove Natural Toxins 240
8.6 Mixtures of Bioactive Principles in Foods 241
8.7 Fermentation for Processing and Taste/Flavour. Yeast and Lactic Fermentations 242
8.8 Effect of Processing on Enzymes and Protein Structure 243
References 243

Abstract Processing (including cooking) of food is an important aspect of food—not least because of the effect on secondary natural products. Four of the most widely used (and most significant) types of processing are heating, freezing, drying and fermentation. While a primary function of each is the preservation of food and the prevention of microbial spoilage, they each also have an effect (directly or indirectly) on the secondary natural products that are present. In addition, some compounds are formed de novo as a consequence of processing and these present a challenge. In addition to the effect of processing on secondary natural products, it can also impact upon food macro and micronutrients and change the nutritional profile of foods. Processing can remove or degrade some natural toxicants. In addition, a number of natural products are subjected to enzyme-mediated modification and processing can alter enzyme activity and specificity thereby changing the profile of compounds that are present.

Keywords Freezing · Heating · Drying · Fermentation · Enzyme activity · Process contaminants

B. Hanley, *Secondary Natural Products in Foods and Food Systems*,
https://doi.org/10.1007/978-3-032-00292-1_8

8.1 Recipe—Chicken Scallopine with White Wine, Lime and Herbs

As noted in previous chapters, variations in physicochemical conditions can alter both the extent of extraction and the reactions of secondary natural products. The respective polarities of the solvent and the natural product will affect solubility while ionic strength, pH and the presence of certain functional groups and reaction conditions will determine if some reactions take place. For example, the phytochemicals present in oregano (*Origanum vulgare)* will be extracted differently into wine compared to water extraction. The volatile profile of fresh oregano is also different to that of the dried material and is also dependent upon the method of drying [1]. Similarly, if reaction conditions change then the type of chemical reaction that may take place can also change. For example, the hydrolysis of glucosinolates at neutral pH leads to formation, primarily, of isothiocyanates but at a low pH the predominant product is the corresponding nitrile [2]. Similarly crushing garlic (*Allium sativum*) at a low pH can lead to the formation of different products from the same reaction carried out at pH 7 [3]. The other factor which will play a role in the development of the taste and odour of a cooked meal is whether the vessel it is being prepared in is covered or uncovered. In the case of the former, the temperature will rise more quickly and will be maintained. Some volatile oils will condense on the lid rather than escaping into the atmosphere (see below). The amount of oil or fat present will also have an impact [4]. Finally, cooking with a lid will, in the case of stove top cooking, turn the cooking pot into a reaction vessel in which steam distillation of volatile components can occur and a proportion of these will condense on the lid and be re-introduced. This will cause a reflux and will affect the reaction that take place in the cooking vessel. Uncovered cooking will result in a greater loss of volatiles and of liquid (primarily water) and this will change the concentration of any involatile chemically reactive species left behind thereby resulting in a change in the types and rates of reactions that may occur. The Arrhenius equation has been used to model the cooking process [5] however the complexity of the system poses distinct challenges and it unclear that greater precision in measuring variables is of value.

In the case of chicken scallopine (see below) the chicken pieces are gently fried in olive oil which both seals them and causes a degree of browning (Maillard reaction). The appropriateness of sealing meat prior to longer cooking has been the subject of considerable debate however it appears to have an effect on both taste and texture [6]. This is not surprising since proteins and fats are affected by heat and the extent to which a Maillard reaction occurs is dependent upon a number of factors including the available moisture content [7] and the temperature.

Chicken Scallopine—Twice Cooked Chicken with Wine, Lime and Thyme

Changing the reaction conditions (pH, moisture content, solvent) can have an impact on the result of cooking. This is particularly the case in this recipe for chicken scallopine. To compare the effects of some of the variation, omit the lime juice (but keep the lime zest so as to retain some lime flavour) and, instead of white

wine, just add water. Cooking uncovered will affect both the temperature of the reaction and the concentration of the ingredients (and hence the reactions that can take place) but it will also affect the content of volatile oils.

Take about two tablespoons of olive oil and heat it in a heavy skillet pan. Add chicken (breast or boneless thighs cut into bite-sized pieces). Leave to cook over a low/medium heat. Do not move them around or they will stick to the bottom of the pan. In the meantime zest and juice a lime, crush and chop two garlic cloves and mix the zest with the garlic and a good pinch of dried thyme and dried oregano. Add a small pinch of dried tarragon to the herb mixture. Turn over the chicken pieces. They should have started to brown. Add the garlic, lime zest and dried herbs to the skillet and cover immediately. Cook for 10–15 s. Add the lime juice and a good glass of dry white wine. Cover and cook on a low heat for 20 min. If they start to dry out (they shouldn't if the lid is tight fitting) then add more wine. Take a teaspoon of cornflour in a small teacup. Add water (half the cup depth) and stir to distribute. Add this to the skillet and stir to distribute and thicken the liquid. Serve with sauté potatoes or pasta. As mentioned previously (Chap. 7)***, one of the effects of a lower pH (in this case, the addition of lime juice) is to lower the perception of salt taste*** [8]***, since hydrogen ions can also use the sodium channel that mediates salt taste and perception. It is also known that some Maillard products can enhance salt taste*** [9]. ***For these reasons, rather than seasoning the dish with salt that will be at a broadly equal concentration throughout, addition of surface salt by the individual is a better way of ensuring the most acceptable level of seasoning for the individual. Wine does not simply introduce ethanol to the cooking process. Carbonyl compounds (aldehydes and ketones) and phenolic secondary natural products are also present and the reactions of these can give rise to significant effects*** [10].

8.2 Food Processing—Principles and Systems. Heating, Freezing, Fermentation and Preservation

Just as different cooking processes can affect the extraction of certain organoleptically important food components, processing can also influence the health properties of individual ingredients. Food is not necessarily created primarily as a nutritious option for a consuming organism with an 'Eat Me' label.

Indeed, one of the major roles of secondary natural products is to increase the unpalatability of foods in order that the host organism will survive predation better than competing species in the ecosystem. Foods can be classified according to their macro nutrient content (protein, carbohydrate or lipid) or the extent to which they meet a requirement for an essential vitamin or mineral (micronutrient profile). The consideration of the effects of processing on foods is often described in terms either of loss of nutrients or altered bioavailability [11]. The perceived loss in nutrient value has fuelled the demonisation of processed foods (particularly ultra-processed foods) compared to 'wholefoods' [12]. In many cases there is a distinct lack of appropriate longitudinal studies and much of the cited evidence is correlative rather than demonstrating causality and fails to take confounders into account [13]. The traditional reductionist approach of nutritional sciences has encouraged focus to be placed upon single nutrients and a limited number of health outcomes. This is increasingly being seen as inadequate [14] and more complex pictures are arising where diets are considered to be more than simply a collection of nutrients. For example, the absorption of vitamin A is highly dependent upon the food source. Vitamin A equivalency of β-carotene (VEB) is defined as the amount of ingested β-carotene in μgs that is absorbed and converted into 1 μg retinol (vitamin A). Using this type of equivalence and considering conversion factors, it is found that, while raw carrots are a relatively poor source of vitamin A, dietary fat can increase the functional uptake of vitamin A [15]. In addition, the presence of dietary fibre can reduce the absorption of vitamin A precursors [16] so the simplistic assessment that suggests that processing is nutritionally deleterious and whole food is always a better nutritional option is not invariably true. Indeed, processing can contribute to the development of nutritionally improved food with an enhanced health profile [17].

In addition to enhanced bioavailability of some nutrients, a second attribute of food processing is the enhanced safety of processed foods compared to the raw material. This is not only a function of the killing of pathogens—particularly those that form biofilms and are, therefore, resistant to a number of control techniques - by various types of processing [18] but also the breakdown of toxins and antinutritional factors [19].

One of the underpinning public health principles that has grown in popularity is the reduction of sugar, salt and some types of fat as part of a strategy to reduce the levels of non-communicable diseases and while processing is commonly assumed to cause an increase in the levels of these 'deleterious' dietary constituents, this is not invariably the case and processed foods may contribute to a healthy diet [20]. It is important, therefore, to consider foods in their entirety and not just as a source of specific less nutritionally favoured components.

Four of the most commonly used processes used in the processing of food are:

- Heating
- Freezing
- Fermentation
- Drying.

The application of each of these procedures has the potential to reduce the levels of components that may give rise to adverse health effects but they may also introduce contaminants that may have deleterious consequences [21] and, in some cases, they may also adversely affect food quality. In most cases process contaminant levels can be minimised by modification of process conditions and this can help to minimise the risk to public health [22]. Quality can be maintained (or even enhanced) by understanding what the impact of each type of processing is on food quality parameters.

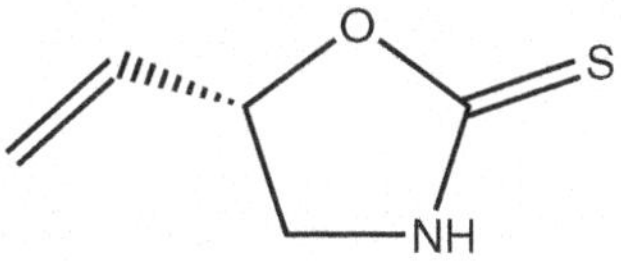

Fig. 8.1 The structure of vinyl oxazolidinethione (goitrin)

Heating is perhaps the oldest and most established form of food processing. It should be considered in the context of its effect on food components and, in particular, on secondary natural products. Foods contain a range of bioactive secondary natural products and, while there is a public perception that toxic substances purposely introduced (for example, pesticides, colours and flavourings) represent a greater risk to health, these man-made substances have a specific biological function (for example, crop protection) and are extensively tested before being given approval for use, unlike natural toxins which can have a range of biological activities and are present as part of an uncontrolled cocktail of bioactive principles [23]. The levels of synthetic compounds are fixed as part of the regulatory approval process while natural toxins may be produced in variable amounts as a response to agronomic conditions or stress [24]. Defining the impacts of non-nutritive food components (natural or synthetic) can be challenging. For example, the biological effect that has the potential to have the greatest impact on a consuming species may not be related to the primary function of the compound as produced by the host. For example, the compound vinyl-oxazolidinethione (Fig. 8.1) contains the NH–C=S moiety that enables it to bind to some polymorphisms of the bitter receptor T2R38 [25]—a structural motif shared with other bitter compounds such as 6-propyl-2-thiouracil (PROP) and phenylthiocarbamide (PTC).

Plants that contain vinyl oxazolidinethione will taste bitter and will therefore have a particular advantage in preventing predation by other species over those which do not contain this bioactive principle. Apart from bitterness, the most significant biological effect of vinyl oxazolidinethione on any consuming mammal is indicated by its more common and alternative, trivial name—goitrin—it disrupts thyroid function. A goitrogenic effect is unlikely to provide an immediate deterrence to consumption, unlike bitterness, and it is the latter rather than the former that is likely to provide the particular growth and survival result for the development of the formation of the compound by plants. It is notable that bitterness seems to be a primary plant defence. For example, many alkaloids also have bitter properties and this is more likely to be the driving force for their biosynthesis and accumulation than any other medicinal properties they may possess. Since the success of any bitterness-based strategy relies upon the consuming species having bitter receptors, it is difficult to immediately ascertain why bitter receptors may have been developed by consuming species in the first place. One part of the answer might lie in the discovery of the range of properties of extra-oral bitter receptors in other organs [26]. If bitter receptors have a range of extra-oral functions (for example upregulating the ABCB1 transport system to eliminate xenobiotics) [27] or modulating the innate immune system [28] then adapting that response by using the same receptors on epithelial cells to give an immediate sensory response to ingested, potentially harmful bioactive compounds in the form of aversion may provide the consuming species that can sense bitter taste with an advantage. This may also explain why, although being a source of beneficial pharmaceuticals, many secondary natural products have a toxicity profile—particularly at higher dose levels [29].

In addition to increasing palatability and digestibility, heat can also enhance the safety of foods. Most bacteria and viruses are destroyed by heating and, in addition, some toxins such as haemagglutinins and lysins—particularly lectins [30] are also heat sensitive. Among the most prevalent food-borne deleterious bioactive secondary natural products are mycotoxins [31]. While some of these are fairly heat labile (for example aflatoxin B1) others are much more heat stable (ochratoxin A) and may pose a threat even in processed foods [32]. It has been estimated that contamination by mycotoxins affects about 25% of the global food supply however, this represents the amount of contamination that is above the limit for regulatory action to be taken (Codex and EU). Exposure to toxins below this level should also be a matter of concern due to the development of cumulative and bioconcentration effects and the potential for synergistic effects arising from exposure to a number of toxins simultaneously [33]. The long term, chronic effects of natural food toxins has been reviewed and is of concern [34] although in many cases it remains difficult to quantify. The heat stability of natural toxins can, as indicated previously, vary however public perception (sometimes influenced by regulatory authorities) is that natural toxins present less of a risk than designed, synthetic chemicals. This is, in many (if not most) instances, a false premise. One area of particular interest in terms of the impact of heat on the biological properties of food-borne entities is the extent to which allergens may be stable to heat and other processing conditions [35]. The resistivity of allergic epitopes to disruption is dependent upon a number of factors including the structural characteristics of the allergenic protein and the nature of the processing carried out. Linear epitopes seem to be more stable than discontinuous epitopes and this is a consequence of the relative stabilities of protein primary and secondary structures [36]. Food processing has an effect on allergic response and mostly reduces allergenic potential [37] however in some cases, heating appears to enhance allergenicity [38]. The stability of the protein structure (for example, the presence of disulphide bonds) can influence stability [39]—particularly during digestion when the accessibility of cleavage sites to proteases is important. Another factor to be considered is the potential for cross-reactivity. This is a common feature of food allergens and should be considered in any assessment of potential risk [40]. While heating is often a beneficial type of food processing—even if it does not completely remove chemical or immunological hazards—there are some aspects to heating that can increase the potential toxin load. While these are considered to be 'processing contaminants' rather than natural products, they are often derived from and influenced by, natural products. For illustrative purposes, the formation of 3monochloropropane diol (3-MCPD) and acrylamide (also discussed in more detail in Sect. 8.4) will be used as examples (Fig. 8.2).

The importance of 3-MCPD is reflected in the number of research papers and regulatory assessments that have been carried out. The formation of 3-MCPD (and

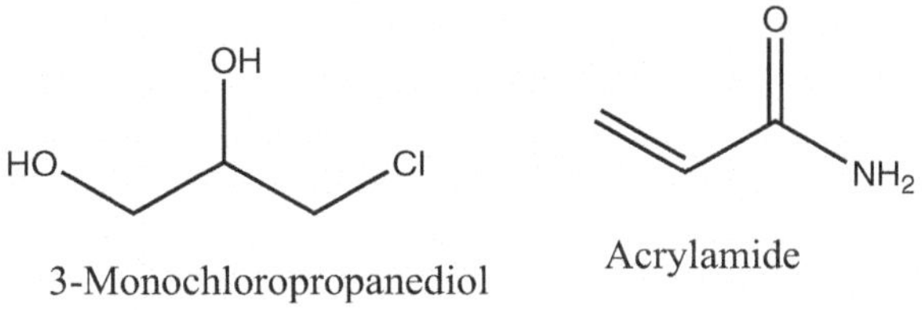

Fig. 8.2 Structures of 3-monochloropropane diol and acrylamide

a range of other chloro propanols, and the corresponding glycidyl esters) occurs when vegetable oil is heated in the presence of chloride ions. It was first detected in acid hydrolysed vegetable protein [41] and in soy sauce [42]. The occurrence of 3-MCPD and related compounds is widespread and consumption is, therefore, difficult to avoid. The first investigations of 3-MCPD concentrated on its potential as a male anti-fertility treatment [43] however its toxic effects limited any potential application. Nonetheless, since intakes of 3-MCPD are difficult to avoid, the possibility of certain diets having a negative long-term effect on male fertility cannot be ruled out. EFSA have arrived at a tolerable daily intake (TDI) of 2 μg/kg bw per day for 3-MCPD [44] and based upon intake estimates, this figure is not exceeded in adults (indeed it is estimated that typically intakes are less than 0.1% of the TDI). The classification of 3-MCPD as a possible human carcinogen by the International Agency for Research on Cancer (IARC) [45] suggests that establishing a comprehensive risk assessment is not a simple process and reduction of levels of 3-MCPD remains a food processing objective. The case of acrylamide (see Sect. 8.4) has some similarities to that of 3-MCPD. As for 3-MCPD, the compound is not a secondary natural product but is produced from the reaction between a naturally occurring compound (asparagine) and a carbohydrate (a reducing sugar) after the application of heat and with low amounts of water present. One of the features of heating food in a system with little water present is the reaction between amino acids and reducing sugars resulting in the production of a brown colour and the formation of various flavour compounds (some of which, as noted previously in Chap. 7, are flavour enhancers). This is called a Maillard reaction. It occurs at temperatures between 140 and 165 °C. It was first reported in 1912 by Louis Maillard [46]. Despite the significant efforts made to try to understand the mechanism of the Maillard reaction, the discovery that the reaction between asparagine and a dicarbonyl compound could give rise to acrylamide was unexpected although once the product had been firmly identified its genesis was confirmed [47]. Indeed, the Maillard reaction was the subject of considerable interest previously in studies of the kinetics of the reaction [48] but asparagine seems to have escaped consideration. Following the discovery of the formation of acrylamide, there have been significant efforts made to limit its presence in foods while retaining the production of beneficial flavour compounds. Unsurprisingly these efforts have included genetic modification and breeding to reduce the levels of asparagine in the raw materials (grain crops) that are the most common source of asparagine in the diet [49], removal of asparagine, once formed in the crop, by treatment with the enzyme asparaginase [50] and removal of asparagine by physicochemical means [51]. Future research is likely to be driven by regulatory considerations rather than toxicology and this may encourage a significant volume of empirical rather than hypothesis-driven science—particularly if the analytical limit of detection of acrylamide can be further reduced and thereby lead to a lowering of the regulatory limit [52].

The freezing of food was a significant advance in the preservation of safe and nutritious food. Freezing did not require the invention of domestic or commercial freezers since snow and ice could be collected during winter months and used for the storage of food. The use of rapid freezing methods to enable small ice crystals to form and therefore to minimise cellular damage to the frozen material was an important advance. A refrigeration system with no moving parts was the subject of a patent by Albert Einstein and Leo Szilard in 1927 (filed) and 1930 (granted in the US) [53]. Freezing is known to introduce a number of changes to food. Perhaps the most significant of these is the inhibition of microbial growth and the positive effects on food preservation (the kinetics of which have been studied) [54]. In addition, freezing can have an impact on food structure and morphology and hence on a number of quality characteristics [55]. One aspect of freezing that has attracted some interest is the use of freeze drying or hybrid freeze drying techniques [56] and these have been adopted to temporarily and then permanently reduce the water activity of the subject matter. In nature, freeze/thaw cycles are common and one interesting aspect of freeze/thaw cycles is the suggestion that these could facilitate a pathway from nucleotide activation to nonenzymatic RNA copying in a prebiotic environment [57]. Other side effects of natural temperature fluctuations are likely to have biological significance. The rate of freezing and subsequent thawing can affect the degree of denaturation of proteins [58] and hence enzyme activity so, in cases where enzymic hydrolysis of an organoleptically inert precursor is required for the production of flavour (in *Allium* and *Brassica* species, for example), a freeze thaw cycle may have a deleterious effect on flavour production. In addition to a potential effect on protein structural integrity, freeze/thaw cycles may also affect starch structure and properties [59]. This may impact starch digestibility [60] and will be discussed in more detail in Sect. 8.3.

The practice of fermentation has made a significant contribution to the preparation and consumption of food. The major types of fermentation will be considered in Sect. 8.7; however some mention will be made of the basic principles here to introduce the concept. Fermentation is anaerobic metabolism by a microorganism. The most common type of fermentation is carried out by yeasts [61] however it can also be employed by bacteria as a way of generating energy from a suitable carbon source. Unlike aerobic metabolism, fermentation is an inefficient process and it consumes large amounts of feedstock. The endpoint in fermentation is the production of alcohol or organic acids. While spontaneous fermentation has been used for a considerable time, the use of defined strains of microorganisms to give consistent product quality has been an ongoing aim of researchers [62]. The combination of this drive to develop reproducible fermentations with tools such as gene editing has created the means to carry out precision fermentation [63]. This has been supported by the development of omics technologies and the subsequent increased knowledge of the functional manipulation of organisms [64]. In addition to ex vivo fermentation, the gut microbiome can also be responsible for in vivo fermentation, and this has nutritional and health consequences—particularly in animal husbandry [65]. Some products from fermentation (particularly polysaccharides with probiotic properties) can influence both the composition and the metabolic balance of the microbiome [66]. This will be considered in more detail in Sect. 8.3. In addition, it is known that the proportions and types of carbon sources can affect fermentation in the gut and, in particular, the production of short chain fatty acids (SCFAs) [67]. While the production of SFCAs has a number of physiological consequences, the biological consequences of the production of gases, particularly hydrogen [68] and hydrogen sulphide [69] are of increasing interest. The importance of gut fermentation is apparent for a number of reasons apart from the utilisation of carbon sources. In addition to being a signifier of gut health, the production of different metabolites demonstrates directly the challenge associated with the systemic response to a range of immediately bioavailable compounds. In addition to the primary metabolic products of fermentation, the production of a range of metabolites of secondary natural products and secondary natural products de novo as gut microbial fermentation metabolites (for example equol) is also possible and this can have significant physiological consequences [70, 71].

Drying is one of the oldest methods of food processing. Drying is effective for food preservation by reducing the water activity and therefore the capacity for microbial growth. In addition, many enzymes also require the presence of at least a shell of water and drying can also inhibit the migration of substrates to the active site and products away from it. If drying is accompanied by freezing and sublimation then it is termed freeze-drying [72]. A range of techniques can be used for food drying [73]. In some cases, the drying process can have a deleterious effect on the food and a range of pretreatments have been advocated to minimise damage to the food. This is particularly relevant to produce such as soft fruit where the structural integrity of the material can be lost and material such as herbs and spices with a high volatile oil content [74]. As noted previously, drying of proteins can lead to a loss of functionality that is not regained upon rehydration. This is particularly the case with enzymes [75] and this is affected, in part, by the relative hydrophobicity of the enzyme structure [76]. In the case of natural products, the method used for drying can have an impact on the secondary natural products present. It has been demonstrated, for example, that the metabolomic pattern of a plant extract can be affected by the type of extraction (water or aqueous 50% ethanol) and the post extraction treatment (oven drying or freeze drying). Ethanol extraction appeared to have a protective effect even if the extracts were oven dried while freeze drying was most effective [77]. Drying may also have a positive effect upon some secondary natural products (if they are stable to the drying process itself) since some degradative (hydrolytic) chemical and enzymic reactions may be inhibited in a dried product and the natural products are retained.

8.3 Macronutrient Processing. Resistant Starch and Prebiotics

The effect of processing on some aspects of macronutrient retention and behaviour has already been mentioned. Both starch (and other carbohydrates) and proteins can undergo reactions on a macromolecular level during and after processing which can alter their structure and function.

In the case of proteins, processing can have more than one meaning. In addition to the physicochemical consequences of processing, proteins may also be subjected to biological processing. In biological systems, proteins are synthesised and, post-translationally, they can be subject to a number of modification processes including folding in a specific way to give a functional protein [78]. In addition, proteins can undergo specific side chain modifications [79]. Post-translational modification can also include cleavage of a pre-protein to give the active form. For example, in the case of HIV, a virally encoded protease carries out post-translational modification to give active proteins from a precursor polyprotein [80, 81]. The targeting of protease or cleavage sites is a recognised anti-viral strategy [82]. The possibility of modulating the posttranslational modification systems as a target for disease reduction has been noted [83]. There has been a particular focus on this as a target for cancer immunotherapy [84]. The physicochemical aspect of processing that proteins are subjected to was noted previously (including heating, freezing, pH changes, fermentation and drying). The consequences of some of these have already been mentioned and these will be briefly reprised since the specific impact on secondary natural products is not profound. Fermentation will be considered more fully in Sect. 8.7. Heating most commonly gives rise to two effects in food proteins—denaturation and coagulation [85]. Both processes have been extensively studied since they are important components of the functional and sensory properties of foods. Heating also denatures (and functionally deactivates) many enzymes—including those associated with food spoilage. Some proteins are more resistant to heat denaturation than others [86] and the discovery of extremophile organisms has enabled a number of heat stable enzymes to be discovered and used—including, most significantly, Taq polymerase [87]. While heating can cause denaturation of proteins, freezing can also disrupt protein structure and function [88] and repeated freeze/thaw cycles can be particularly deleterious to enzyme activity. The effect of pH (and ionic strength) is significant for some proteins however proteins are frequently sufficiently stable (due to intrinsic buffering capability) to deal with relatively small fluctuations in pH since such fluctuations occur as a consequence of normal enzyme activity [89]. While enzyme effectiveness can vary depending on pH, protonation of specific amino acids (for example aspartate or glutamate) may have more significant effects including inducing conformational changes [90]. The effect of ionic strength on the intramolecular interactions in proteins has been studied in detail [91]. For enzymes, it can be difficult to disentangle the effects of ionic strength per se from the effects of specific ionic species [92]. In addition to specific effects, ionic strength can affect the electrostatic interactions between an enzyme and its substrate. While not a direct consequence of ionic strength or an ion-specific effect, calcium ions are important physiological mediators and calcium channel blockers are an important group of pharmaceuticals [93].

Starch is another important biological macromolecule whose functional properties depend on the degree and type of processing that is carried out. The commercial isolation and purification of starch makes use of a number of physicochemical and biological (enzymic) processes [94]. Starch can be commercially produced from a range of plant sources (over 50 different plants are commonly used), The major use of starch is for consumption and as an energy source however there is a move towards non-petrochemicals (including starch) as starting materials for a number of non-food applications [95]. Apart from its biologically based origin, starch polymers are also biodegradable and are a renewable source [96]. As a polymer, starch consists of glucose units. Linear starch molecules are termed amylose while starch molecules that have branched units are termed amylopectin (Fig. 8.3).

Starches with a high amount of (or exclusively consisting of) amylopectin are termed 'waxy' starches. Starch, in its natural form, is semi-crystalline and processing can affect this structure as well as degrading the polymer itself. One of the major results of cooking is to convert the semi-crystalline starch granules (which are hard to digest) into a more digestible form. This latter form is more easily hydrolysed by enzymes in vivo and this leads to a rapid release of glucose. The consumption

Fig. 8.3 Structure of amylose

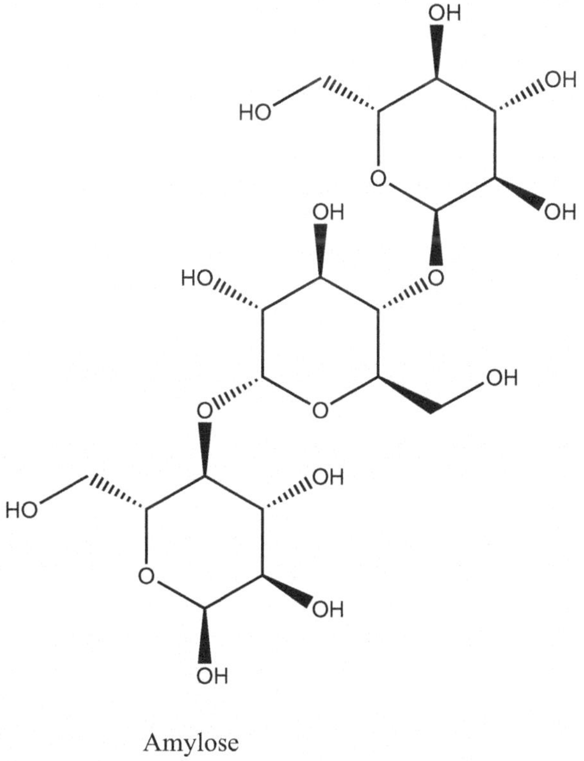

Amylose

of rapidly metabolised starch may be responsible for the increase in obesity rates. Another aspect of the heating and subsequent cooling of starch is that it may partially recrystallise. Such recrystallised starch is not as easily digested and may have a prebiotic function in the intestine (see below). Since the components of starch are relatively simple there is not a significant amount of chemistry associated with starch itself. It can, however, participate in Maillard-type reactions. In addition, if starch is heated until it pyrolyses, it forms caramels which are commonly used for colouring foods and for imparting flavour. One other use of starch is in the encapsulation of secondary natural products including, for example, ferulic acid [97] and with phenolic compounds more generally [98]. This approach can be extended to other food ingredients with functional and nutritional attributes [99]. This is not a simple enhancement of bioavailability and improved physiological transport of secondary natural products since some phenolic compounds (for example, isoflavones) can also inhibit the degradation of starch by forming complexes [100]. As noted previously, heating and then cooling starch can lead to recrystallization of solubilised starch to give resistant starch which is not digested in the small intestine but travels to the large intestine where it is digested by (and encourages the growth of) certain beneficial bacteria leading to the formation of short chain fatty acids (SCFAs). This effect will be considered in more detail later however it should not be confused with the concept of probiotics. The idea of probiotics (the literal opposite of antibiotics) was first specifically expressed by Lilley and Stillwell in 1965 and were initially defined as substances secreted by some bacteria which prolonged the logarithmic phase of growth of other species [101]. Probiotics were later defined as living species that were non-pathogenic and could help to restore gut microbial health [102]. Metchnikoff had previously, in 1910, suggested that factors were produced that could prolong life [103] however the link to intestinal microbial health (and the health of the host organism) was not clearly enunciated until Fuller in 1989 who wrote about probiotics in the context of "the natural balance of the gut microflora can be restored and the animal returned to its normal nutrition, growth and health status" [104]. Interestingly Fuller wrote about probiotics as a way of reestablishing the natural, protective microbial flora of the gastrointestinal tract which has been disrupted. While there is a great deal of evidence regarding the effects of probiotics, much of it is conflicting [105] and there is confusion about whether effects are due to gut colonisation of 'healthy bacteria' or some other component. It has been noted that supplementation "failed to support the ability of probiotics to cause persistent changes in gut microbiota," in healthy individuals [106]. A more precise and functional concept was later summarised. The concept of prebiotics was first suggested by Gibson and Roberfroid [107] and they introduced the following concept "prebiotics are nondigestible food ingredients that beneficially affect the host by selectively stimulating the growth and/or activity of one or a limited number of bacterial species already resident in the colon". Importantly, prebiotics were not a somewhat nebulous bacterially derived growth factor but were defined as food ingredients. The definition of prebiotics most commonly adopted is "a substrate that is selectively utilized by host microorganisms conferring a health benefit". This definition does not specify a chemical structure but focuses on the functional properties of the substance (or group of substances) [108]. Resistant starch is not hydrolysed in the small intestine but enters the large intestine where it is used as a carbon source by certain bacteria [109]. The role of SCFAs has been studied. It would appear that, like quorum sensing molecules, they have cross-species signalling capability [110]. They are produced by the action of gut bacteria but will then bind to mammalian intestinal receptors. There is evidence of a link between quorum sensing and SCFA production [111] however further studies are required both within the intestine and at the gut/neural interface [112].

8.4 Processing Contaminants—Acrylamide and Heterocyclic Amines

The formation of acrylamide as a consequence of the Maillard reaction has already been mentioned. While extensive work has been carried out on the mechanism of formation of acrylamide [113], its importance to human health remains a subject of debate [114]. The effects of chronic (lifetime) exposure to acrylamide remains unknown and, as with many chronic toxic entities, the development of long-term, predictive testing systems is a challenge [115]. Acrylamide can be seen as an example of a toxic entity where mitigation strategies can take place throughout the food system. The choice of which strategies to adopt should balance the risk to health with the desirable qualities of the Maillard products. The first of these is the reduction of asparagine and reducing sugar levels in uncooked materials. This can be carried out by washing or soaking the material. This may, however, have a deleterious effect on processing [116]. One possibility to enhance the removal of asparagine specifically is the use of the bidentate compound, citric acid. While this has been employed as part of a successful strategy, citric acid itself lowers the pH and imparts a sour taste therefore its use is necessarily limited due to these organoleptic constraints. A second strategy is the use of the enzyme asparaginase to convert asparagine to citric acid [117]. A third possibility is the prevention of the biosynthesis of asparagine in plants whose processing involves a Maillard reaction. In this case selection of varieties with a low level of expression of a specific biosynthetic gene (for example, *TaASN-B2)* or with a gene deletion using CRISPR or a similar technology may be effective (to varying degrees) [118]. The removal of asparagine can be combined with processing strategies to reduce acrylamide levels without impairing the organoleptic attributes of the product [119]. It is unlikely that any single strategy will be effective and therefore a 'toolkit' approach using multiple and complementary approaches is likely to be adopted.

The production of heterocyclic aromatic amines (HAAs) is another processing side effect that gives rise to concern. They occur (together with polycyclic aromatic hydrocarbons—PAHs) when meat is cooked to a high temperature [120]. They are both mutagenic and carcinogenic [121] and measures to reduce their formation and occurrence have been adopted. As with acrylamide, the formation of HAAs is a consequence of the Maillard reaction [122]. Their formation and metabolism have been studied [123] with a view towards lowering the levels without adversely affecting the product. Not surprisingly, the mitigation strategies adopted are similar to those employed in the reduction of acrylamide levels and involve modulation of heat [124]. The possibility of using a biological method (probiotics) to remove HAAs from the large intestine has been suggested [125] and a similar approach has been suggested for acrylamide removal [126]. The possibility of transforming acrylamide or HAAs into some other, less toxic (or less electrophilic) compound [127] has not been suggested to be a valid approach.

8.5 Processing to Remove Natural Toxins

Some natural toxic compounds are heat labile and can, therefore, be eliminated by processing involving heat. It should be noted, however, that there can be differences in the stability of some toxins when subjected to dry or wet heat systems. A range of other processing methods are used to remove toxic secondary natural products from foods and some of these have been reviewed [128]. While some natural products are acutely toxic, many of them, as noted previously, are employed for their broadly aversive properties rather than acute toxicity [129]. In addition to the toxicity of specific entities, biotic or abiotic stress can increase the levels of secondary natural products and the production of cocktails of toxicants with the potential for synergistic effects should be noted (See Sect. 8.6). In addition to perceived and actual risks associated with foods, people also consider the benefits however while benefits are considered based upon heuristics and experience, risks are evaluated on the basis of cognitive information processing. This means that, while benefits and risks are, to an extent, considered to be at opposite ends of the spectrum, the perception is not a continuum flowing from one end to the other [130]. In addition, while natural toxins occur far more widely than synthetic chemicals and have a wider range of potentially toxic effects, this is not necessarily reflected in consumer perception [131]. Mitigation of the effects of natural toxins should take into account the physicochemical nature of the toxin including solubility, the opportunity for biological exacerbation of the toxic effect if the tissue is damaged (for example as occurs with cyanogenic glycosides and glucosinolates) and the removal of parts of the plant that contain large stores of the toxin (for example, peel and seeds). In some cases, secondary natural products are less prevalent, post germination and this may also decrease the risk [132]. In addition to traditional methods, high pressure processing, microwave digestion, supercritical fluid extraction and other more recent processing techniques have been assessed for their effects on individual toxins.

In the case of microbial spoilage, the storing of crops under poor conditions can lead to the formation of a range of mycotoxins. The most effective means of preventing the formation of microbial toxins is prevention of microbial growth rather than post-storage removal. In addition, microbial spoilage is often heterogeneous and therefore obtaining a representative sample for chemical analysis is challenging [133].

8.6 Mixtures of Bioactive Principles in Foods

The formation of natural products is regulated but, nonetheless, a single organism may contain a range of products from a single, or multiple biosynthetic pathways. The manipulation of pathways with common starting materials or intermediates is a key aspect of bioengineering to optimise production of specific products [134]. Both specific and non-specific enzymes are involved in biosynthesis [135] and this may increase the numbers of distinct entities formed. Genome mining has greatly increased knowledge about the occurrence of biosynthetic pathways—even in cases where the secondary natural products themselves are not detected [136]. Combinatorial biosynthesis is used to create novel products through the manipulation of biosynthetic pathway enzymes [137]. The employment of terpene synthases to produce a large number of different terpenes [138] is a demonstration of what has been described as biosynthetic enzyme plasticity. An understanding of biosynthetic pathways and enzyme plasticity is an important aspect of being able to predict the formation of different secondary natural products.

One stage beyond the reporting of the occurrence of individual compounds is the development of in silico methods to predict uptake and toxicity [139]. The use of such methods can be either compound or biological function (for example, disease) specific [140]. The occurrence of multiple products (and/or multiple families of products) makes biological function predictions of mixtures present in extracts or in foods much more difficult to carry out. Overlaid upon multiple biological activities are the multiplicity of dosages experienced. There are several different approaches to determining the potential interactions between natural products (both individually and collectively) and biological systems. These include the modulation of enzyme activity, up and down regulation of expression and interactions with biological targets. Some examples will illustrate the principles involved.

When plants that contain glucosinolates are subjected to cellular damage, a thioglucosidase enzyme (myrosinase) comes into contact with its glucosinolate substrate. The glucose moiety is cleaved and the residual aglucone can undergo a number of reactions forming isothiocyanates, nitriles and other products. This reaction does not have a specific requirement for an activator or cofactor however the rate of the reaction is considerably enhanced by the presence of ascorbic acid and this is the result of an allosteric effect mediated by ascorbic acid [141]. A large number of natural products have also been found to act as allosteric modulators through interaction with a range of biological targets [142] and this may suggest other approaches to the use of natural products as pharmacologically active molecules by considering combinations of allosteric modifiers and bioactive agents whose effect may be modulated by allostery. Natural products have also been found to be allosteric modulators of specific biological receptors [143]. Since some natural products are also agonists or antagonists of certain receptors then positive and negative synergistic effects can be predicted to occur. Another area of interest for the interaction of natural products with biological systems is the modulation of metabolism. One of the most frequently cited examples is the inhibition of the xenobiotic metabolising enzyme CYP3A4 by furanocoumarins found in grapefruit [144]. This increases the physiological concentration of the xenobiotic by reducing its rate of conjugation and elimination from the body. A range of other natural products have been found to have an effect on the Cytochrome P450 enzyme system [145]. The activation of Phase II detoxification enzymes has also been noted and studied [146] and such effects may also play a role in the interactions of natural products with living systems. The development of 'interaction metabolomics' has been suggested to be a novel approach to the discovery of synergy between the effects of natural products [147] using a combination of bioactivity and molecular recognition using mass spectrometry. It remains, however, essentially a reductionist approach and cannot be used to investigate unknown synergistic activities.

8.7 Fermentation for Processing and Taste/Flavour. Yeast and Lactic Fermentations

Fermentation, as noted previously, is an important technique for processing. It is used to enhance product quality and for preservation. Although fermentation is a biological process, the final product is more important than the process that is used. It is, therefore, more closely related to natural product chemistry. Fermentation is defined, however, as a biological process in which carbohydrates are broken down in an anaerobic process to give smaller oxidised units with the production of energy. Compared to aerobic breakdown of carbohydrates (respiration), fermentation gives less energy and the final products are organic molecules (ethanol or organic acids) that are not completely decomposed and hence still are a potential energy store. Two of the main routes employed by microorganisms in the fermentative process are yeast fermentations to give ethanol and lactic acid bacteria fermentations whose end products are organic acids. Detailed descriptions of both can be found in textbooks and review papers [148]. One important aspect of fermentation is its role in food preservation. The drop in pH which accompanies lactic acid fermentations prevents the growth of many pathogenic and spoilage bacteria. In the context of natural products, the use of fermentation to develop new chemical 'white space' has been suggested and mixed fermentations may also increase the production of certain natural products [149]. This may be a consequence of the growth of competing organisms being perceived as a biotic stress. Unsurprisingly, quorum sensing has been implicated in some fermentation processes [150] and in quality aspects of food fermentation [151]. There may be an opportunity to manipulated mixed fermentations by addition of specific quorum sensing molecules or inhibitors. The potential for quorum sensing has also been found in *Saccharomyces cerevisiae* fermentations where tryptophol and 2-phenyl ethanol were identified as quorum sensing molecules [152]. The organisms responsible for fermentation have been found to have the capacity to carry out biotransformations on secondary natural products [153] and may also reduce the levels of anti-nutritional compounds [154]. In addition to natural fermentation and the selection of strains that produce natural products with desirable properties, the use of omics technology may further optimise the characteristics of the products of fermentation [64]. These include the beneficial biological effects for the consumer and desirable organoleptic properties. As noted in the recipe given in Sect. 8.1, the use of wine can change the characteristics including, in particular, the extent (and the rate) to which different natural products are able to be extracted. Clearly lactic acid based fermentation which results in a lowering of the pH will be likely to have a positive effect on the extractability of ionisable, nitrogenous natural products such as alkaloids.

8.8 Effect of Processing on Enzymes and Protein Structure

The effect of processing on biological macromolecules has been noted previously (Sect. 8.3). In the case of proteins, the application of heat will, to varying degrees, result in denaturation. This can have a positive effect in the removal of discontinuous epitopes responsible for some allergic reactions however many enzymes associated with the biotransformation of natural products (for example, myrosinase and alliinase) are heat labile and this can reduce the organoleptic properties of products. In some cases, if organisms are capable of growing at higher temperatures, the enzymes (and other proteins) they produce are more thermostable [155]. The production of heat shock proteins can also be a response to abiotic stress. In addition to heat, other processing techniques including high pressure [156] and extrusion [157] can also have an effect.

Processing can have a number of significant effects on natural products both directly (for example the formation of small molecule processing contaminants) and through their interaction with other components such as macromolecules including starch, proteins and enzymes. Processing is, in many cases, a continuous process which proceeds through a series of intermediate stages to local, unstable equilibria. In this way, the study of processing may provide a paradigm for the investigation of complex living systems. The chemistry driven approach to processing is often to try to examine the different procedures that are used and their effect on specific characteristics. Process engineering suggests that a more multivariate approach may be required and this may suggest approaches to the study of the transitions in complex biological systems.

References

1. Figiel A, et al. Composition of oregano essential oil (Origanum vulgare) as affected by drying method. J Food Eng. 2010;98: 240247.
2. Gil V, Macleod AJ. The effects of pH on glucosinolate degradation by a thioglucoside glucohydrolase preparation. Phytochemistry. 1980;19:2547–51.

3. Abe K, Hori Y, Myoda T. Volatile compounds of fresh and processed garlic. Exp Ther Med. 2020;19(2):1585–93. https://doi.org/10.3892/etm.2019.8394. Epub 2019 Dec 27. PMID: 32010343; PMCID: PMC6966211.
4. Huang WW, Sallah-Ud-Din R, Dlamini WN, Berekute AK, Getnet ME, Yu KP. Effectiveness of a covered oil-free cooking process on the abatement of air pollutants from cooking meats. Heliyon. 2023;9(9): e19531. https://doi.org/10.1016/j.heliyon.2023.e19531. PMID: 37809458; PMCID: PMC10558720.
5. Peleg M, Normand MD, Corradini MG. The Arrhenius equation revisited. Crit Rev Food Sci Nutr. 2012;52(9):830–51. https://doi.org/10.1080/10408398.2012.667460. PMID: 22698273.
6. Kilic S, Oz E, Turhan S, Amarowickz R, Toldra F, Brennan C, Oz F. The effect of sealing process on the textural and sensory properties and protein profile of cooked meat. Int J Food Sci Technol. 2023;58:4800–11. https://doi.org/10.1111/ijfs.16590.
7. Lund MN, Ray CA. Control of Maillard reactions in foods: strategies and chemical mechanisms. J Agric Food Chem. 2017;65(23):45374552. https://doi.org/10.1021/acs.jafc.7b00882. Epub 2017 May 30. PMID: 28535048.
8. Simon SA. Interactions between salt and acid stimuli: a lesson in gustation from simultaneous epithelial and neural recordings. J Gen Physiol. 2002;120(6):787–91. https://doi.org/10.1085/jgp.20028735. PMID: 12451049; PMCID: PMC2229563.
9. Katsumata T, Nakakuki H, Tokunaga C, Fujii N, Egi M, Phan TH, Mummalaneni S, DeSimone JA, Lyall V. Effect of Maillard reacted peptides on human salt taste and the amiloride-insensitive salt taste receptor (TRPV1t). Chem Senses. 2008;33(7):665–80. https://doi.org/10.1093/chemse/bjn033. Epub 2008 Jul 5. PMID: 18603652; PMCID: PMC2533421.
10. Waterhouse AL, Sacks GL, Jeffery DW. Aldehydes, ketones, and related compounds. In: Waterhouse AL, Sacks GL, Jeffery DW, editors. Understanding wine chemistry. https://doi.org/10.1002/9781118730720.ch9.
11. Henry CJK, Heppell N. Nutritional losses and gains during processing: future problems and issues†. Proc Nutr Society. 2002;61(1):145–8. https://doi.org/10.1079/PNS2001142.
12. Monda A, de Stefano MI, Villano I, Allocca S, Casillo M, Messina A, Monda V, Moscatelli F, Dipace A, Limone P, Di Maio G, La Marra M, Di Padova M, Chieffi S, Messina G, Monda M, Polito R. Ultra-processed food intake and increased risk of obesity: a narrative review. Foods. 2024;13(16):2627. https://doi.org/10.3390/foods13162627. PMID: 39200554; PMCID: PMC11353718.
13. Robinson E, Jones A. Causality or confounding? Applying E values to examine associations between ultra-processed food consumption and risk of weight gain. Int J Obes (Lond). 2024;48(9):1342–6. https://doi.org/10.1038/s41366-024-01566-8. Epub 2024 Jun 15. PMID: 38879728; PMCID: PMC11347361.
14. Melse-Boonstra A. Bioavailability of micronutrients from nutrient-dense whole foods: zooming in on dairy, vegetables, and fruits. Front Nutr. 2020;24(7):101. https://doi.org/10.3389/fnut.2020.00101. PMID: 32793622; PMCID: PMC7393990.
15. Van Loo-Bouwman CA, Naber TH, Schaafsma G. A review of vitamin A equivalency of β-carotene in various food matrices for human consumption. Br J Nutr. 2014;111(12):2153–66. https://doi.org/10.1017/S0007114514000166. Epub 2014 Feb 11. PMID: 24513222.
16. Riedl J, Linseisen J, Hoffmann J, Wolfram G. Some dietary fibers reduce the absorption of carotenoids in women. J Nutr. 1999;129(12):21706. https://doi.org/10.1093/jn/129.12.2170. PMID: 10573545.
17. Weaver CM, Dwyer J, Fulgoni VL, King JC, Leveille GA, MacDonald RS, Ordovas J, Schnakenberg D. Processed foods: contributions to nutrition. Am J Clin Nutr. 2014;99(6):1525–42. https://doi.org/10.3945/ajcn.114.089284. Epub 2014 Apr 23. PMID: 24760975; PMCID: PMC6410904.
18. Sharan M, Vijay D, Dhaka P, Bedi JS, Gill JPS. Biofilms as a microbial hazard in the food industry: a scoping review. J Appl Microbiol. 2022;133(4):2210–34. https://doi.org/10.1111/jam.15766. Epub 2022 Aug 30. PMID: 35945912.
19. Albuquerque TG, Bragotto APA, Costa HS. Processed food: nutrition, safety, and public health. Int J Environ Res Public Health. 2022;19(24):16410. https://doi.org/10.3390/ijerph192416410. PMID: 36554295; PMCID: PMC9778909.

20. Drewnowski A, Detzel P, Klassen-Wigger P. Perspective: achieving sustainable healthy diets through formulation and processing of foods. Curr Dev Nutr. 2022;6(6):nzac089. https://doi.org/10.1093/cdn/nzac089. PMID: 35711570; PMCID: PMC9197573.
21. Zahir A, Khan IA, Nasim M, Azizi MN, Azi F. Food process contaminants: formation, occurrence, risk assessment and mitigation strategies—a review. Food Addit Contam Part A Chem Anal Control Expo Risk Assess. 2024;41(10):1242–74. https://doi.org/10.1080/19440049.2024.2381210. Epub 2024 Jul 22. PMID: 39038046.
22. Zahir A, Ge Z, Khan IA. Public health risks associated with food process contaminants—a review. J Food Prot. 2025;88(2):100426. https://doi.org/10.1016/j.jfp.2024.100426. Epub 2024 Dec 4. PMID: 39643160.
23. Gaston TE, Mendrick DL, Paine MF, Roe AL, Yeung CK. "Natural" is not synonymous with "Safe": toxicity of natural products alone and in combination with pharmaceutical agents. Regul Toxicol Pharmacol. 2020;113:104642. https://doi.org/10.1016/j.yrtph.2020.104642. Epub 2020 Mar 18. PMID: 32197968; PMCID: PMC7211136.
24. Fletcher MT, Netzel G. Food safety and natural toxins. Toxins (Basel). 2020;12(4):236. https://doi.org/10.3390/toxins12040236. PMID: 32276351; PMCID: PMC7232292.
25. Wooding S, Gunn H, Ramos P, Thalmann S, Xing C, Meyerhof W. Genetics and bitter taste responses to goitrin, a plant toxin found in vegetables. Chem Senses. 2010;35(8):685–92. https://doi.org/10.1093/chemse/bjq061. Epub 2010 Jun 14. PMID: 20551074.
26. Tuzim K, Korolczuk A. An update on extra-oral bitter taste receptors. J Transl Med. 2021;19(1):440. https://doi.org/10.1186/s12967-021-03067-y. Erratum in: J Transl Med. 2021 Nov 26;19(1):478. https://doi.org/10.1186/s12967-021-03137-1. PMID: 34674725; PMCID: PMC8529754.
27. Jeon TI, Seo YK, Osborne TF. Gut bitter taste receptor signalling induces ABCB1 through a mechanism involving CCK. Biochem J. 2011;438(1):33–7. https://doi.org/10.1042/BJ20110009. PMID: 21592089; PMCID: PMC3319131.
28. Liszt KI, Wang Q, Farhadipour M, Segers A, Thijs T, Nys L, Deleus E, Van der Schueren B, Gerner C, Neuditschko B, Ceulemans LJ, Lannoo M, Tack J, Depoortere I. Human intestinal bitter taste receptors regulate innate immune responses and metabolic regulators in obesity. J Clin Invest. 2022;132(3): e144828. https://doi.org/10.1172/JCI144828. PMID: 34784295; PMCID: PMC8803326.
29. Dolan LC, Matulka RA, Burdock GA. Naturally occurring food toxins. Toxins (Basel). 2010;2(9):2289–332. https://doi.org/10.3390/toxins2092289. Epub 2010 Sep 20. PMID: 22069686; PMCID: PMC3153292.
30. Bender AE, Reaidi GB. Toxicity of kidney beans (Phaseolus Vulgaris) with particular reference to lectins. J Plant Foods. 1982;4(1):15–22. https://doi.org/10.1080/0142968X.1982.11904243.
31. Omotayo OP, Omotayo AO, Mwanza M, Babalola OO. Prevalence of mycotoxins and their consequences on human health. Toxicol Res. 2019;35(1):1–7. https://doi.org/10.5487/TR.2019.35.1.001. Epub 2018 Jan 15. PMID: 30766652; PMCID: PMC6354945.
32. Raters M, Matissek R. Thermal stability of aflatoxin B1 and ochratoxin A. Mycotoxin Res. 2008;24(3):130–4. https://doi.org/10.1007/BF03032339. PMID: 23604747.
33. Eskola M, Kos G, Elliott CT, Hajšlová J, Mayar S, Krska R. Worldwide contamination of food-crops with mycotoxins: Validity of the widely cited 'FAO estimate' of 25. Crit Rev Food Sci Nutr. 2020;60(16):2773–89. https://doi.org/10.1080/10408398.2019.1658570. Epub 2019 Sep 3. PMID: 31478403.
34. Krska, R., Bueschl, C., Eskola, M., & Elliott, C. (2024). Chronical health risks from natural toxins in our food—from a European perspective (Chinese and English versions). Science and Technology of Cereals, Oils and Foods, 32(4), 1–15. https://doi.org/10.16210/j.cnki.10077561.2024.04.001.
35. Pekar J, Ret D, Untersmayr E. Stability of allergens. Mol Immunol. 2018;100:14–20. https://doi.org/10.1016/j.molimm.2018.03.017. Epub 2018 Mar 30. PMID: 29606336; PMCID: PMC6020993.

36. Deller MC, Kong L, Rupp B. Protein stability: a crystallographer's perspective. Acta Crystallogr F Struct Biol Commun. 2016;72(Pt 2):72–95. https://doi.org/10.1107/S2053230X15024619. Epub 2016 Jan 26. PMID: 26841758; PMCID: PMC4741188.
37. Verhoeckx KCM, Vissers YM, Baumert JL, Faludi R, Feys M, Flanagan S, Herouet-Guicheney C, Holzhauser T, Shimojo R, van der Bolt N, Wichers H, Kimber I. Food processing and allergenicity. Food Chem Toxicol. 2015;80:223–40. https://doi.org/10.1016/j.fct.2015.03.005. Epub 2015 Mar 14. PMID: 25778347.
38. Teodorowicz M, van Neerven J, Savelkoul H. Food processing: the influence of the maillard reaction on immunogenicity and allergenicity of food proteins. Nutrients. 2017;9(8):835. https://doi.org/10.3390/nu9080835. PMID: 28777346; PMCID: PMC5579628.
39. Matsumura M, Signor G, Matthews BW. Substantial increase of protein stability by multiple disulphide bonds. Nature. 1989;342(6247):291–3. https://doi.org/10.1038/342291a0. PMID: 2812028.
40. Francis OL, Wang KY, Kim EH, Moran TP. Common food allergens and cross-reactivity. J Food Allergy. 2020;2(1):17–21. https://doi.org/10.2500/jfa.2020.2.200020. PMID: 39022146; PMCID: PMC11250430.
41. Velísek J, Davídek J, Kubelka V, Janícek G, Svobodová Z, Simicová Z. New chlorine-containing organic compounds in protein hydrolysates. J Agric Food Chem. 1980;28(6):1142–4. https://doi.org/10.1021/jf60232a022. PMID: 7451739.
42. Lee BQ, Khor SM. 3-Chloropropane-1,2-diol (3-MCPD) in soy sauce: a review on the formation, reduction, and detection of this potential carcinogen. Compr Rev Food Sci Food Saf. 2015;14(1):48–66. https://doi.org/10.1111/1541-4337.12120. PMID: 33401813.
43. Xing H, Chen S, Wang X, Li J, Ren F. 3-Monochloropropane-1,2-diol causes spermatogenesis failure in male rats via Sertoli cell dysfunction but not testosterone reduction. Toxicol Lett. 2022;1(360):1–10. https://doi.org/10.1016/j.toxlet.2022.01.006. Epub 2022 Jan 19. PMID: 35063618.
44. Knutsen HK, Alexander J, Barregård L, Bignami M, Brüschweiler B, Ceccatelli S, Cottrill B, Dinovi M, Edler L, Grasl-Kraupp B, Hoogenboom LR, Nebbia CS, Oswald IP, Petersen A, Rose M, Roudot AC, Schwerdtle T, Vleminckx C, Vollmer G, Wallace H, Lampen A, Morris I, Piersma A, Schrenk D, Binaglia M, Levorato S, Hogstrand C. Update of the risk assessment on 3monochloropropane diol and its fatty acid esters. EFSA Panel on Contaminants in the Food Chain (CONTAM) EFSA J. 2018;16(1):e05083. https://doi.org/10.2903/j.efsa.2018.5083. PMID: 32625654; PMCID: PMC7009698.
45. IARC. IARC monographs on the evaluation of carcinogenic risk of chemicals to humans, vol. 101. Lyon, France: International Agency for Research on Cancer; 2016. pp. 349–74.
46. Maillard LC. Action des acides amines sur les sucres; formation de melanoidines par voie méthodique. Comptes Rendus. 1912;154:66–8.
47. Mottram DS, Wedzicha BL, Dodson AT. Acrylamide is formed in the Maillard reaction. Nature. 2002;419(6906):448–9. https://doi.org/10.1038/419448a. PMID: 12368844.
48. Davies CG, Wedzicha BL. Kinetics of the inhibition of ascorbic acid browning by sulphite. Food Addit Contam. 1992;9(5):471–7. https://doi.org/10.1080/02652039209374099. PMID: 1298651.
49. Kaur N, Brock N, Raffan S, Halford NG. Low asparagine wheat: Europe's first field trial of genome edited wheat amid rapidly changing regulations on acrylamide in food and genome editing of crops. Breed Sci. 2024;74(1):37–46. https://doi.org/10.1270/jsbbs.23058. Epub 2024 Mar 20. PMID: 39246437; PMCID: PMC11375425.
50. Jia R, Wan X, Geng X, Xue D, Xie Z, Chen C. Microbial L-asparaginase for application in acrylamide mitigation from food: current research status and future perspectives. Microorganisms. 2021;9(8):1659. https://doi.org/10.3390/microorganisms9081659. PMID: 34442737; PMCID: PMC8400838.
51. Hanley AB, Offen C, Clarke M, Ing B, Roberts M, Burch R. Acrylamide reduction in processed foods. Adv Exp Med Biol. 2005;561:387–92.
52. van Egmond HP. Natural toxins: risks, regulations and the analytical situation in Europe. Anal Bioanal Chem. 2004;378(5):1152–60. https://doi.org/10.1007/s00216-003-2373-4. Epub 2003 Dec 13. PMID: 14673562.

53. Einstein A, Szilárd L. Refrigeration (Appl: 16 December 1927; Priority: Germany, 16 December 1926) U.S. patent 1,781,541, 11 November 1930.
54. Zhao Y, Takhar PS. Freezing of foods: mathematical and experimental aspects. Food Eng Rev. 2017;9:1–12. https://doi.org/10.1007/s12393016-9157-z.
55. Giannakourou M, Taoukis P. Changes during food freezing and frozen storage. Foods. 2021;10(11):2525. https://doi.org/10.3390/foods10112525. PMID: 34828806; PMCID: PMC8618057.
56. Nwankwo CS, Okpomor EO, Dibagar N, Wodecki M, Zwierz W, Figiel A. Recent developments in the hybridization of the freeze-drying technique in food dehydration: a review on chemical and sensory qualities. Foods. 2023;12(18):3437. https://doi.org/10.3390/foods12183437. PMID: 37761146; PMCID: PMC10528370.
57. Zhang SJ, Duzdevich D, Ding D, Szostak JW. Freeze-thaw cycles enable a prebiotically plausible and continuous pathway from nucleotide activation to nonenzymatic RNA copying. Proc Natl Acad Sci U S A. 2022;119(17):e2116429119. https://doi.org/10.1073/pnas.2116429119. Epub 2022 Apr 21. PMID: 35446612; PMCID: PMC9169909.
58. Cao E, Chen Y, Cui Z, Foster PR. Effect of freezing and thawing rates on denaturation of proteins in aqueous solutions. Biotechnol Bioeng. 2003;82(6):684–90. https://doi.org/10.1002/bit.10612. PMID: 12673768.
59. Liu M, Li J, Ma H, Qin G, Niu M, Zhang X, Zhang J, Wei Y, Han J, Liang Y, Zhang S, Yin L, Zhu H, Huang Y, Li L, Zheng X, Liu C. Structural and physicochemical characteristics of wheat starch as influenced by freezethawed cycles and antifreeze protein from Sabina chinensis (Linn.) Ant. cv. Kaizuca leaves. Food Chem X. 2023;20:100927. https://doi.org/10.1016/j.fochx.2023.100927. PMID: 38144810; PMCID: PMC10740099.
60. Zhang B, Wang K, Hasjim J, Li E, Flanagan BM, Gidley MJ, Dhital S. Freezedrying changes the structure and digestibility of B-polymorphic starches. J Agric Food Chem. 2014;62(7):1482–91. https://doi.org/10.1021/jf405196m. Epub 2014 Feb 6. PMID: 24471496.
61. Maicas S. The role of yeasts in fermentation processes. Microorganisms. 2020;8(8):1142. https://doi.org/10.3390/microorganisms8081142. PMID: 32731589; PMCID: PMC7466055.
62. Galimberti A, Bruno A, Agostinetto G, Casiraghi M, Guzzetti L, Labra M. Fermented food products in the era of globalization: tradition meets biotechnology innovations. Curr Opin Biotechnol. 2021;70:36–41. https://doi.org/10.1016/j.copbio.2020.10.006. Epub 2020 Nov 21. PMID: 33232845.
63. Hilgendorf K, Wang Y, Miller MJ, Jin YS. Precision fermentation for improving the quality, flavor, safety, and sustainability of foods. Curr Opin Biotechnol. 2024;86: 103084. https://doi.org/10.1016/j.copbio.2024.103084. Epub 2024 Feb 23. PMID: 38394936.
64. Mannaa M, Han G, Seo YS, Park I. Evolution of food fermentation processes and the use of multi-omics in deciphering the roles of the microbiota. Foods. 2021;10(11):2861. https://doi.org/10.3390/foods10112861. PMID: 34829140; PMCID: PMC8618017.
65. Huangfu W, Cao S, Li S, Zhang S, Liu M, Liu B, Zhu X, Cui Y, Wang Z, Zhao J, Shi Y. In vitro and in vivo fermentation models to study the function of dietary fiber in pig nutrition. Appl Microbiol Biotechnol. 2024;108(1):314. https://doi.org/10.1007/s00253-024-13148-9. PMID: 38683435; PMCID: PMC11058960.
66. Hu B, Liu C, Jiang W, Zhu H, Zhang H, Qian H, Zhang W. Chronic in vitro fermentation and in vivo metabolism: extracellular polysaccharides from Sporidiobolus pararoseus regulate the intestinal microbiome of humans and mice. Int J Biol Macromol. 2021;1(192):398–406. https://doi.org/10.1016/j.ijbiomac.2021.09.127. Epub 2021 Sep 25. PMID: 34571128.
67. Macfarlane GT, Macfarlane S. Fermentation in the human large intestine: its physiologic consequences and the potential contribution of prebiotics. J Clin Gastroenterol. 2011;45(Suppl):S120–7. https://doi.org/10.1097/MCG.0b013e31822fecfe. PMID: 21992950.
68. Campbell A, Gdanetz K, Schmidt AW, et al. H2 generated by fermentation in the human gut microbiome influences metabolism and competitive fitness of gut butyrate producers. Microbiome. 2023;11:133. https://doi.org/10.1186/s40168-023-01565-3.

69. Teigen L, Biruete A, Khoruts A. Impact of diet on hydrogen sulfide production: implications for gut health. Curr Opin Clin Nutr Metab Care. 2023;26(1):55–8. https://doi.org/10.1097/MCO.0000000000000881. Epub 2022 Sep 27. PMID: 36542535; PMCID: PMC10413438.
70. Krautkramer KA, Fan J, Bäckhed F. Gut microbial metabolites as multi-kingdom intermediates. Nat Rev Microbiol. 2021;19:77–94. https://doi.org/10.1038/s41579-020-0438-4.
71. Kuziel GA, et al. Functional diversification of dietary plant small molecules by the gut microbiome. Cell. 188(7):1967–1983.e22.
72. Nowak D, Jakubczyk E. The freeze-drying of foods-the characteristic of the process course and the effect of its parameters on the physical properties of food materials. Foods. 2020;9(10):1488. https://doi.org/10.3390/foods9101488. PMID: 33080983; PMCID: PMC7603155.
73. Owusu-Apenten R, Vieira E. Food drying. In: Elementary food science. Food science text series. Cham: Springer; 2023. https://doi.org/10.1007/978-3-030-65433-7_14.
74. Thamkaew G, Sjöholm I, Galindo FG. A review of drying methods for improving the quality of dried herbs. Crit Rev Food Sci Nutr. 2020;61(11):1763–86. https://doi.org/10.1080/10408398.2020.1765309.
75. Luyben KCAM, Liou JK, Bruin S. Enzyme degradation during drying. Biotechnol Bioeng. 1982;24:533–52. https://doi.org/10.1002/bit.2602403030.
76. Vázquez-Duhalt R, et al. Role of enzyme hydrophobicity in biocatalysis in organic solvents. Enzyme Microb Technol. 1992;14(1992):837–41.
77. ElNaker NA, Daou M, Ochsenkühn MA, Amin SA, Yousef AF, Yousef LF. A metabolomics approach to evaluate the effect of lyophilization versus oven drying on the chemical composition of plant extracts. Sci Rep. 2021;11(1):22679. https://doi.org/10.1038/s41598-021-02158-6. PMID: 34811431; PMCID: PMC8608909.
78. Ellgaard L, McCaul N, Chatsisvili A, Braakman I. Co- and post-translational protein folding in the ER. Traffic. 2016;17(6):615–38. https://doi.org/10.1111/tra.12392. Epub 2016 Apr 22. PMID: 26947578.
79. Ramazi S, Zahiri J. Posttranslational modifications in proteins: resources, tools and prediction methods. Database (Oxford). 2021;2021:baab012. https://doi.org/10.1093/database/baab012. PMID: 33826699; PMCID: PMC8040245.
80. Sundquist WI, Kräusslich HG. HIV-1 assembly, budding, and maturation. Cold Spring Harb Perspect Med. 2012;2(7):a006924. https://doi.org/10.1101/cshperspect.a006924. Erratum in: Cold Spring Harb Perspect Med. 2012 Aug;2(8). https://doi.org/10.1101/cshperspect.a015420. PMID: 22762019; PMCID: PMC3385941.
81. Konvalinka J, Kräusslich HG, Müller B. Retroviral proteases and their roles in virion maturation. Virology. 2015;479–480:403–17. https://doi.org/10.1016/j.virol.2015.03.021. Epub 2015 Mar 26. PMID: 25816761.
82. Lv Z, Chu Y, Wang Y. HIV protease inhibitors: a review of molecular selectivity and toxicity. HIV AIDS (Auckl). 2015;8(7):95–104. https://doi.org/10.2147/HIV.S79956. PMID: 25897264; PMCID: PMC4396582.
83. Li W, Li F, Zhang X, Lin HK, Xu C. Insights into the post-translational modification and its emerging role in shaping the tumor microenvironment. Signal Transduct Target Ther. 2021;6(1):422. https://doi.org/10.1038/s41392-021-00825-8. Erratum in: Signal Transduct Target Ther. 2022 Jan 31;7(1):31. https://doi.org/10.1038/s41392-022-00901-7. PMID: 34924561; PMCID: PMC8685280.
84. Li Y, Zhang R, Hei H. Advances in post-translational modifications of proteins and cancer immunotherapy. Front Immunol. 2023;22(14):1229397. https://doi.org/10.3389/fimmu.2023.1229397. PMID: 37675097; PMCID: PMC10477431.
85. Kilara A, Sharkasi TY. Effects of temperature on food proteins and its implications on functional properties. Crit Rev Food Sci Nutr. 1986;23(4):323–95. https://doi.org/10.1080/10408398609527429. PMID: 3519084.
86. Kim TD, Ryu HJ, Cho HI, Yang CH, Kim J. Thermal behavior of proteins: heat-resistant proteins and their heat-induced secondary structural changes. Biochemistry. 2000;39(48):14839–46. https://doi.org/10.1021/bi001441y. PMID: 11101300.

87. Ishino S, Ishino Y. DNA polymerases as useful reagents for biotechnology—the history of developmental research in the field. Front Microbiol. 2014;29(5):465. https://doi.org/10.3389/fmicb.2014.00465. PMID: 25221550; PMCID: PMC4148896.
88. Bhatnagar BS, Bogner RH, Pikal MJ. Protein stability during freezing: separation of stresses and mechanisms of protein stabilization. Pharm Dev Technol. 2007;12(5):505–23. https://doi.org/10.1080/10837450701481157. PMID: 17963151.
89. Talley K, Alexov E. On the pH-optimum of activity and stability of proteins. Proteins. 2010;78(12):2699–706. https://doi.org/10.1002/prot.22786. PMID: 20589630; PMCID: PMC2911520.
90. Di Russo NV, Estrin DA, Martí MA, Roitberg AE. pH-Dependent conformational changes in proteins and their effect on experimental pK(a)s: the case of Nitrophorin 4. PLoS Comput Biol. 2012;8(11):e1002761. https://doi.org/10.1371/journal.pcbi.1002761. Epub 2012 Nov 1. PMID: 23133364; PMCID: PMC3486867.
91. Möller J, Schroer MA, Erlkamp M, Grobelny S, Paulus M, Tiemeyer S, Wirkert FJ, Tolan M, Winter R. The effect of ionic strength, temperature, and pressure on the interaction potential of dense protein solutions: from nonlinear pressure response to protein crystallization. Biophys J. 2012;102(11):2641–8. https://doi.org/10.1016/j.bpj.2012.04.043. Epub 2012 Jun 5. PMID: 22713580; PMCID: PMC3368150.
92. Nørby JG, Esmann M. The effect of ionic strength and specific anions on substrate binding and hydrolytic activities of Na, K-ATPase. J Gen Physiol. 1997;109(5):555–70. https://doi.org/10.1085/jgp.109.5.555. PMID: 9154904; PMCID: PMC2217059.
93. Jones KE, Hayden SL, Meyer HR, Sandoz JL, Arata WH, Dufrene K, Ballaera C, Lopez Torres Y, Griffin P, Kaye AM, Shekoohi S, Kaye AD. The evolving role of calcium channel blockers in hypertension management: pharmacological and clinical considerations. Curr Issues Mol Biol. 2024;46(7):6315–27. https://doi.org/10.3390/cimb46070377. PMID: 39057019; PMCID: PMC11275245.
94. Das R, Kayastha AM. An overview on starch processing and key enzymes. In: Wu J, Xia W, editors. Industrial starch debranching enzymes. Singapore: Springer; 2023. https://doi.org/10.1007/978-981-19-7026-9_1.
95. Adewale P, Yancheshmeh MS, Lam E. Starch modification for non-food, industrial applications: Market intelligence and critical review. Carbohydr Polym. 2022;291:119590. https://doi.org/10.1016/j.carbpol.2022.119590. Epub 2022 May 10. PMID: 35698403.
96. Zarski A, Bajer K, Kapuśniak J. Review of the most important methods of improving the processing properties of starch toward non-food applications. Polymers (Basel). 2021;13(5):832. https://doi.org/10.3390/polym13050832. PMID: 33803238; PMCID: PMC7967182.
97. Karunaratne R, Zhu F. Physicochemical interactions of maize starch with ferulic acid. Food Chem. 2016;15(199):372–9. https://doi.org/10.1016/j.foodchem.2015.12.033. Epub 2015 Dec 8. PMID: 26775984.
98. Zhu F. Interactions between starch and phenolic compound. Trends Food Sci Technol. 2015;43:129–43.
99. Zhu F. Encapsulation and delivery of food ingredients using starch based systems. Food Chem. 2017;15(229):542–52. https://doi.org/10.1016/j.foodchem.2017.02.101. Epub 2017 Feb 22. PMID: 28372213.
100. Xu S, Huang J, Lian S, Chen Y, Chi C, Chen W, Chen X, Yang Y, Weng L. Effect of isoflavone structures on the formation of starch-isoflavone complexes: experimental and molecular dynamics analysis. Int J Biol Macromol. 2025;294: 139439. https://doi.org/10.1016/j.ijbiomac.2024.139439. Epub 2025 Jan 3. PMID: 39756732.
101. Lilly DM, Stilwell RH. Probiotics: growth-promoting factors produced by microorganisms. Science. 1965;147(3659):747–8. https://doi.org/10.1126/science.147.3659.747. PMID: 14242024.
102. Williams NT. Probiotics. Am J Health Syst Pharm. 2010;67(6):449–58. https://doi.org/10.2146/ajhp090168. PMID: 20208051.

103. Podolsky SH. Metchnikoff and the microbiome. Lancet. 2012;380(9856):1810–1. https://doi.org/10.1016/s0140-6736(12)62018-2. PMID: 23189332.
104. Fuller R. Probiotics in man and animals. J Appl Bacteriol. 1989;66(5):365–78 PMID: 2666378.
105. Suez J, Zmora N, Segal E, et al. The pros, cons, and many unknowns of probiotics. Nat Med. 2019;25:716–29. https://doi.org/10.1038/s41591-019-0439-x.
106. Khalesi S, Bellissimo N, Vandelanotte C, Williams S, Stanley D, Irwin C. A review of probiotic supplementation in healthy adults: helpful or hype? Eur J Clin Nutr. 2019;73(1):24–37. https://doi.org/10.1038/s41430-0180135-9. Epub 2018 Mar 26. PMID: 29581563.
107. Gibson GR, Roberfroid MB. Dietary modulation of the human colonic microbiota: introducing the concept of prebiotics. J Nutr. 1995;125(6):1401–12. https://doi.org/10.1093/jn/125.6.1401. PMID: 7782892.
108. Gibson G, Hutkins R, Sanders M, et al. Expert consensus document: the international scientific association for probiotics and prebiotics (ISAPP) consensus statement on the definition and scope of prebiotics. Nat Rev Gastroenterol Hepatol. 2017;14:491–502. https://doi.org/10.1038/nrgastro.2017.75.
109. Bojarczuk A, et al. Health benefits of resistant starch: a review of the literature. J Funct Foods. 2022;93(3): 105094.
110. Morrison DJ, Preston T. Formation of short chain fatty acids by the gut microbiota and their impact on human metabolism. Gut Microbes. 2016;7(3):189–200. https://doi.org/10.1080/19490976.2015.1134082. Epub 2016 Mar 10. PMID: 26963409; PMCID: PMC4939913.
111. Yang M, Meng F, Gu W, Fu L, Zhang F, Li F, Tao Y, Zhang Z, Wang X, Yang X, Li J, Yu J. Influence of polysaccharides from Polygonatum kingianum on short-chain fatty acid production and quorum sensing in Lactobacillus faecis. Front Microbiol. 2021;17(12): 758870. https://doi.org/10.3389/fmicb.2021.758870. PMID: 34867887; PMCID: PMC8635744.
112. Uhlig F, Hyland NP. Making sense of quorum sensing at the intestinal mucosal interface. Cells. 2022;11:1734. https://doi.org/10.3390/cells11111734.
113. Lingnert H, Grivas S, Jägerstad M, Skog K, Törnqvist M, Åman P. Acrylamide in food: mechanisms of formation and influencing factors during heating of foods. Scand J Nutr. 2002;46(4):159–72. https://doi.org/10.1080/110264802762225273.
114. Zhao M, Zhang B, Deng L. The mechanism of acrylamide-induced neurotoxicity: current status and future perspectives. Front Nutr. 2022;25(9): 859189. https://doi.org/10.3389/fnut.2022.859189. PMID: 35399689; PMCID: PMC8993146.
115. Govindaraju I, Sana M, Chakraborty I, Rahman MH, Biswas R, Mazumder N. Dietary acrylamide: a detailed review on formation, detection, mitigation, and its health impacts. Foods. 2024;13(4):556. https://doi.org/10.3390/foods13040556. PMID: 38397533; PMCID: PMC10887767.
116. Batuwita BKHH, Jayasinghe JMJK, Marapana RAUJ, et al. Reduction of asparagine and reducing sugar content, and utilization of alternative food processing strategies in mitigating acrylamide formation—a review. Food Bioprocess Technol. 2025;18:2101–44. https://doi.org/10.1007/s11947-024-03565-z.
117. Alam S, Pranaw K, Tiwari R, Khare SK. Recent development in the uses of asparaginase as food enzyme. In: Parameswaran B, Varjani S, Raveendran S, editors. Green bio-processes. Energy, environment, and sustainability. Singapore: Springer; 2019. https://doi.org/10.1007/978-981-13-3263-0_5.
118. Raffan S, Sparks C, Huttly A, Hyde L, Martignago D, Mead A, Hanley SJ, Wilkinson PA, Barker G, Edwards KJ, Curtis TY, Usher S, Kosik O, Halford NG. Wheat with greatly reduced accumulation of free asparagine in the grain, produced by CRISPR/Cas9 editing of asparagine synthetase gene TaASN2. Plant Biotechnol J. 2021. https://doi.org/10.1111/pbi.13573.
119. Liyanage DWK, et al. Processing strategies to decrease acrylamide formation, reducing sugars and free asparagine content in potato chips from three commercial cultivars. Food Control, 2021;119(2021):107452.
120. Robbana-Barnat S, Rabache M, Rialland E, Fradin J. Heterocyclic amines: occurrence and prevention in cooked food. Environ Health Perspect. 1996;104(3):280–8. https://doi.org/10.1289/ehp.96104280. PMID: 8919766; PMCID: PMC1469314.

121. Layton DW, Bogen KT, Knize MG, Hatch FT, Johnson VM, Felton JS. Cancer risk of heterocyclic amines in cooked foods: an analysis and implications for research. Carcinogenesis. 1995;16(1):39–52. https://doi.org/10.1093/carcin/16.1.39. PMID: 7834804.
122. Skog KI, Johansson MA, Jägerstad MI. Carcinogenic heterocyclic amines in model systems and cooked foods: a review on formation, occurrence and intake. Food Chem Toxicol. 1998;36(9–10):87996. https://doi.org/10.1016/s0278-6915(98)00061-1. PMID: 9737435.
123. Bellamri M, Walmsley SJ, Turesky RJ. Metabolism and biomarkers of heterocyclic aromatic amines in humans. Genes Environ. 2021;43(1):29. https://doi.org/10.1186/s41021-021-00200-7. PMID: 34271992; PMCID: PMC8284014.
124. Olalekan Adeyeye SA, Ashaolu TJ. Heterocyclic amine formation and mitigation in processed meat and meat products: a mini-review. J Food Prot. 2021;84(11):1868–77. https://doi.org/10.4315/JFP-20-471. PMID: 33956955.
125. Khosravi-Darani K, Barzegar F, Baghdadi M. Detoxification of heterocyclic aromatic amines by probiotic to inhibit medical hazards. Mini Rev Med Chem. 2019;19(15):1196–203. https://doi.org/10.2174/1389557519666190318102201. PMID: 30887924.
126. Maher A, Miśkiewicz K, Rosicka-Kaczmarek J, Nowak A. Detoxification of acrylamide by potentially probiotic strains of lactic acid bacteria and yeast. Molecules. 2024;29(20):4922. https://doi.org/10.3390/molecules29204922. PMID: 39459290; PMCID: PMC11510321.
127. Gates KS. An overview of chemical processes that damage cellular DNA: spontaneous hydrolysis, alkylation, and reactions with radicals. Chem Res Toxicol. 2009;22(11):1747–60. https://doi.org/10.1021/tx900242k. PMID: 19757819; PMCID: PMC2806061.
128. Urugo MM, Tringo TT. Naturally occurring plant food toxicants and the role of food processing methods in their detoxification. Int J Food Sci. 2023;27(2023):9947841. https://doi.org/10.1155/2023/9947841. PMID: 37153649; PMCID: PMC10159748.
129. Madariaga-Mazón A, Hernández-Alvarado RB, Noriega-Colima KO, Osnaya-Hernández A, Martinez-Mayorga K. Toxicity of secondary metabolites. Phys Sci Rev. 2019;4(12). https://doi.org/10.1515/psr2018-0116.
130. Ueland Ø, Gunnlaugsdottir H, Holm F, Kalogeras N, Leino O, Luteijn JM, Magnússon SH, Odekerken G, Pohjola MV, Tijhuis MJ, Tuomisto JT, White BC, Verhagen H. State of the art in benefit-risk analysis: consumer perception. Food Chem Toxicol. 2012;50(1):67–76. https://doi.org/10.1016/j.fct.2011.06.006. Epub 2011 Jun 12. PMID: 21683114.
131. Kher SV, De Jonge J, Wentholt MTA, Deliza R, de Andrade JC, Cnossen HJ, Luijckx NBL, Frewer LJ. Consumer perceptions of risks of chemical and microbiological contaminants associated with food chains: a cross-national study. Int J Consum Stud. 2013;37:73–83. https://doi.org/10.1111/j.14706431.2011.01054.x.
132. Li X, Li J, Dong S, et al. Effects of germination on tocopherol, secoisolarlciresinol diglucoside, cyanogenic glycosides and antioxidant activities in flaxseed (Linum usitatissimum L.). Int J Food Sci Technol. 2019;54(7):2346–54. https://doi.org/10.1111/ijfs.14098.
133. Shephard GS. Current status of mycotoxin analysis: a critical review. J AOAC Int. 2016;99(4):842–8. https://doi.org/10.5740/jaoacint.16011 1. PMID: 27455927.
134. Kuska J, O'Reilly E. Engineered biosynthetic pathways and biocatalytic cascades for sustainable synthesis. Curr Opin Chem Biol. 2020;58:146–54. https://doi.org/10.1016/j.cbpa.2020.08.006. Epub 2020 Nov 2. PMID: 33152607.
135. Chlubnova I, Legentil L, Dureau R, Pennec A, Almendros M, Daniellou R, Nugier-Chauvin C, Ferrières V. Specific and non-specific enzymes for furanosyl-containing conjugates: biosynthesis, metabolism, and chemo-enzymatic synthesis. Carbohydr Res. 2012;15(356):44–61. https://doi.org/10.1016/j.carres.2012.04.002. Epub 2012 Apr 11. PMID: 22554502.
136. Sélem-Mojica N, Aguilar C, Gutiérrez-García K, Martínez-Guerrero CE, Barona-Gómez F. EvoMining reveals the origin and fate of natural product biosynthetic enzymes. Microb Genom. 2019;5(12):e000260. https://doi.org/10.1099/mgen.0.000260. Epub 2019 Apr 4. PMID: 30946645; PMCID: PMC6939163.
137. Skellam E, Rajendran S, Li L. Combinatorial biosynthesis for the engineering of novel fungal natural products. Commun Chem. 2024;7:89. https://doi.org/10.1038/s42004-024-01172-9.

138. Xing B, Lei Z, Bai Z, Zang G, Wang Y, Zhang C, Chen M, Zhou Y, Ding J, Yang D, Ma M. Structural biology of terpene synthases. Methods Enzymol. 2024;699:59–87. https://doi.org/10.1016/bs.mie.2024.03.012. Epub 2024 Mar 31. PMID: 38942516.
139. Durán-Iturbide NA, Díaz-Eufracio BI, Medina-Franco JL. In silico ADME/Tox profiling of natural products: a focus on BIOFACQUIM. ACS Omega. 2020;5(26):16076–84. https://doi.org/10.1021/acsomega.0c01581. PMID: 32656429; PMCID: PMC7346235.
140. Gheidari D, Mehrdad M, Karimelahi Z. Virtual screening, ADMET prediction, molecular docking, and dynamic simulation studies of natural products as BACE1 inhibitors for the management of Alzheimer's disease. Sci Rep. 2024;14:26431. https://doi.org/10.1038/s41598-02475292-6.
141. Ohtsuru M, Hata T. Studies on the activation mechanism of the Myrosinase by L-ascorbic acid. Agric Biol Chem. 1973;37(8):1971–2. https://doi.org/10.1080/00021369.1973.10860935.
142. Bruder M, Polo G, Trivella DBB. Natural allosteric modulators and their biological targets: molecular signatures and mechanisms. Nat Prod Rep. 2020;37(4):488–514. https://doi.org/10.1039/c9np00064j. Epub 2020 Feb 12. PMID: 32048675.
143. Alexeev M, Grosenbaugh DK, Mott DD, Fisher JL. The natural products magnolol and honokiol are positive allosteric modulators of both synaptic and extra-synaptic GABA(A) receptors. Neuropharmacology. 2012;62(8):2507–14. https://doi.org/10.1016/j.neuropharm.2012.03.002. Epub 2012 Mar 12. PMID: 22445602; PMCID: PMC3652012.
144. Bailey DG, Dresser G, Arnold JM. Grapefruit-medication interactions: forbidden fruit or avoidable consequences? CMAJ. 2013;185(4):309–16. https://doi.org/10.1503/cmaj.120951. Epub 2012 Nov 26. PMID: 23184849; PMCID: PMC3589309.
145. Zuo HL, Huang HY, Lin YC, Cai XX, Kong XJ, Luo DL, Zhou YH, Huang HD. Enzyme activity of natural products on Cytochrome P450. Molecules. 2022;27(2):515. https://doi.org/10.3390/molecules27020515. PMID: 35056827; PMCID: PMC8779343.
146. Keum YS. Regulation of Nrf2-mediated phase II detoxification and anti-oxidant genes. Biomol Ther (Seoul). 2012;20(2):144–51. https://doi.org/10.4062/biomolther.2012.20.2.144. PMID: 24116287; PMCID: PMC3792210.
147. Vidar WS, Baumeister TUH, Caesar LK, Kellogg JJ, Todd DA, Linington RG, M Kvalheim O, Cech NB. Interaction metabolomics to discover synergists in natural product mixtures. J Nat Prod. 2023;86(4):655–71. https://doi.org/10.1021/acs.jnatprod.2c00518. Epub 2023 Apr 13. PMID: 37052585; PMCID: PMC10152448.
148. Voidarou C, Antoniadou M, Rozos G, Tzora A, Skoufos I, Varzakas T, Lagiou A, Bezirtzoglou E. Fermentative foods: microbiology, biochemistry, potential human health benefits and public health issues. Foods. 2020;10(1):69. https://doi.org/10.3390/foods10010069. PMID: 33396397; PMCID: PMC7823516.
149. Pettit RK. Mixed fermentation for natural product drug discovery. Appl Microbiol Biotechnol. 2009;83(1):19–25. https://doi.org/10.1007/s00253009-1916-9. Epub 2009 Mar 21. PMID: 19305992.
150. Britton SJ, Neven H, Maskell DL. Microbial smalltalk: does quorum sensing play a role in beer fermentation? J Am Soc Brew Chem. 2020;79(3):231–9. https://doi.org/10.1080/03610470.2020.1843928.
151. Johansen P, Jespersen L. Impact of quorum sensing on the quality of fermented foods. https://doi.org/10.1016/j.cofs.2017.01.001.
152. Avbelj M, Zupan J, Raspor P. Quorum-sensing in yeast and its potential in wine making. Appl Microbiol Biotechnol. 2016;100(18):7841–52. https://doi.org/10.1007/s00253-016-7758-3. Epub 2016 Aug 9. PMID: 27507587.
153. Filannino P, Bai Y, Di Cagno R, Gobbetti M, Gänzle MG, Di Cagno R. Metabolism of phenolic compounds by Lactobacillus spp. during fermentation of cherry juice and broccoli puree. Food Microbiol. 2015;46:272–9. https://doi.org/10.1016/j.fm.2014.08.018.
154. Abu-Salem FM, Mohamed R, Gibriel A, Rasmy NMH. Levels of some antinutritional factors in tempeh produced from some legumes and jojobas seeds. Int Sch Sci Res Innov. 2014;8:296–301.

155. Ward OP, Moo-Young M. Thermostable enzymes. Biotechnol Adv. 1988;6(1):39–69. https://doi.org/10.1016/0734-9750(88)90573-3. PMID: 14543441.
156. Julien R, Royer CA. Lessons from pressure denaturation of proteins. J R Soc Interface. 2018;15(2018):0244.
157. Wittek P, Zeiler N, Karbstein HP, Emin MA. High moisture extrusion of soy protein: investigations on the formation of anisotropic product structure. Foods. 2021;10(1):102. https://doi.org/10.3390/foods10010102. PMID: 33418980; PMCID: PMC7825338.

Chapter 9
Secondary Natural Products and Microbes

Contents

9.1 Recipe—Paella ... 256
9.2 Probiotics and Their Metabolites ... 259
9.3 The Physiological Microbiome ... 262
9.4 Quorum Sensing in Microbes ... 268
9.5 Manipulation of Quorum Sensing by Secondary Natural Products ... 272
9.6 Antibiotics in Foods (Lantibiotics and Other Bacteriocins) ... 274
9.7 Bioactive Compounds from the Genus Allium ... 276
References ... 279

Abstract As scientific thought develops then definitions can change. One example of this is the term 'probiotic'. It was originally defined as a substance produced by one microorganism, that stimulate the growth of another. Later the term was used to describe the microorganism itself and so a strictly functional definition morphed into a function/identity definition. One of the most important attributes of probiotics is in the stimulation of production of short chain fatty acids (SCFAs) which bind to Free Fatty Acid Receptors (FFARs) and this has a number of physiological consequences including, potentially an effect on Alzheimer's disease and other neurological disorders. The importance of physiological microbiomes in health has been noted. Secondary natural products may interact with microbiomes in different ways including influencing virulence, biofilm formation and the induction of host inflammation. In addition to the effects of exogenous natural products, the physiological microbiome is an important source of endogenous secondary natural products. There are instances of cross talk between quorum sensing systems from different species and kingdoms and this may enhance the establishment of equilibrium states in complex biological systems. One of the potentially most impactful aspects of microbiome manipulation is the induction by natural products of the expression of an enzyme (lactonase) that can break down the potent and widespread bacterial quorum sensing molecule, homoserine lactone. More generally quorum sensing inhibition (QSI) by natural products is an active area of investigation and may be important in finding solutions to anti-microbial resistance (AMR).

Keywords Probiotics · Quorum sensing · SCFAs · FFARs · QSI · AMR

B. Hanley, *Secondary Natural Products in Foods and Food Systems*,
https://doi.org/10.1007/978-3-032-00292-1_9

9.1 Recipe—Paella

The effect of individual natural products on living systems can be amplified both by the primary interaction and its consequences on other parts of the system but also by compounds with multiple effects on different biological targets. For example saffron has both a flavour and a taste profile and these contribute separately to its biological impact. While interaction with specific receptors can activate a signal transduction pathway leading to a range of downstream effects in a target species, this can be amplified by the reaction of a particular secondary natural product with one living system which then precipitates effects in another living system. An example of this, which will be examined in more detail later, is the effect of molecules on quorum sensing properties of organisms (particularly microbial species) or on their growth and proliferation in other ways (for example, prebiotics which, as noted previously, can include resistant starch and may also be formed in the recipe outlined in this chapter—paella). This type of effect underpins the concept of control of one organism by a community of others (the microbiome) and this will be examined in more detail. Studies of natural products includes analysis and synthesis. The analysis may reveal a core structure which is then modified and the synthesis to prove a particular structure is accomplished by first of all breaking the molecule (conceptually if not actually) into fragments that can be combined to give the original structure (retrosynthesis). Recipes can be considered to be similar to these processes of modification of a core structure and retrosynthesis to reveal the constituent parts (ingredients). Once a basic recipe is established, modification of it by the addition of other components can be carried out. This can be considered to have an analogy with the creation of a core chemical structure (for example, a specific alkaloid) and its subsequent modification to give substances with the same core structure but with different properties. There are, however, important differences. Unlike natural products, recipes do not arise from the random selection of natural sources in a Darwinian fashion. They require planning (intelligent design?) in order to be created in the first place. To create a recipe using principles of random selection would be analogous to designing a molecule de novo by putting fragments together with just a vague outline of the target molecule and its properties. While it might be argued that this is the essence of combinatorial chemistry, it is an approach that relies on a very effective selection process and such a provision was unavailable to chefs (or to natural product chemists). Semi-random combination was the approach taken by Perkin (and others) as they developed synthetic schemes before there was an appreciation of functional groups in organic and natural product chemistry—or the analytical tools to probe sub-structures. As noted previously, chemists adopted the proportionality that underlies the 'affinity' approach to inorganic chemistry and combined this with the calculation of empirical formulas. Using these simple (but very precise) tools, they sought to combine molecules in proportions that added up to the known empirical formula of the desired product. This was the basis of Perkin's attempt to synthesise quinine (see below). In that way natural product and organic chemistry resembled both cooking and alchemy. There were knowledge gaps in the understanding of chemistry, cooking and alchemy and these hampered progress. One difference was that the properties of ingredients (and the consequences of their use) was better understood than the properties of secondary natural products or how constituents might be combined to best effect. Trial and error were reasonable approaches to take when devising either recipes or synthetic schemes. The macro properties (texture, colour) of the results of a recipe were observable and the micro properties (taste, odour) were able to be tested empirically (albeit subjectively). The results of a recipe may be interesting (and palatable) within a range of properties. In the case of the synthesis of natural products, the only option was to be successful and to create the precise molecule of interest with identical properties to its naturally occurring doppelganger. When Perkin failed to synthesise quinine and instead made the first, commercially viable aniline dye, none of the techniques he had available would have been able to tell him how close he was to the initially desired product. In the case of recipes, however, even a simple understanding of basic tastes (salt, sweet, bitter and sour), the creation of sauces, the development of fermentation, different cooking techniques (boiling, roasting etc.) all contributed to the capability for the planned manipulation of food to give an acceptable end product whose core attributes could then be refined.

As noted previously, some culinary ingredients have more than a single purpose. In the case of saffron, it imparts both tase and colour. For the reductionist chemist, one important question is, are the two activities of saffron imparted by the same secondary natural product or are there two (or more) substances involved. The main compound responsible for the colour that addition of saffron imparts to food is a carotenoid, crocin. It is a diester of the disaccharide gentiobiose and the dicarboxylic acid crocetin (Fig. 9.1).

The flavour (and, in particular, the aroma) of saffron is a consequence of the presence of safranal [1]. Safranal occurs naturally as the corresponding glycoside (which is bitter). In this case, drying the naturally occurring compound gives safranal (Fig. 9.2).

This is an example of the impact of processing upon a secondary natural product (see Chap. 8).

Paella—Colour, Taste.....Action!

Paella is a dish that can have a number of ingredients and, depending on the mixture used, can vary in taste, colour and aroma. In the case of the colour, it can either be intrinsic to some of the ingredients (red, yellow or green peppers) or imparted by the addition of a component (saffron). The development of flavour is dependent both on the herbs and spices used and the cooking method and time.

Take a skillet pan and add 2 tablespoons of olive oil. Heat and add a cubed, white (Spanish) onion and fry gently for about 10–15 min until the onion is softened and translucent (this also affects the release of phenolic compounds from onion and may be significant in the health promoting properties of the vegetable [2]. ***While the onion is cooking, prepare the vegetables. Take a small red, yellow and green pepper and chop each into strips. Crush two garlic cloves. Add the peppers***

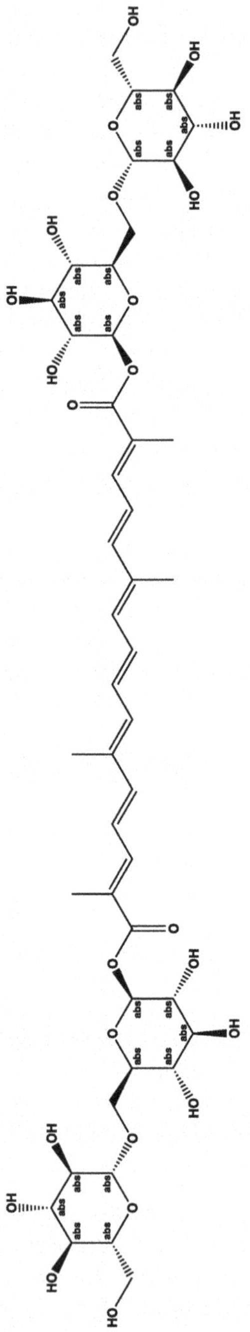

Fig. 9.1 The structure of crocin

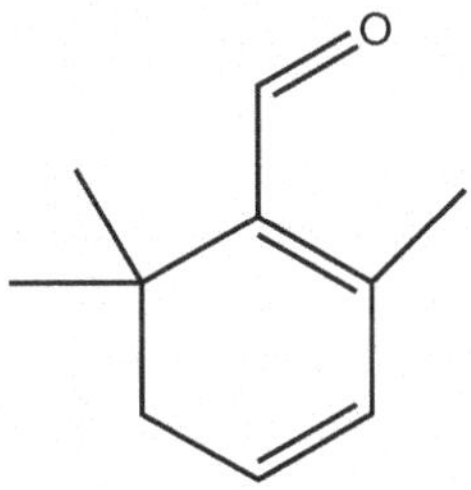

Fig. 9.2 The structure of safranal

and garlic cloves to the pan and fry for about 5 min while agitating gently with a wooden spatula. If you want, add some (half a breast or so) of cubed chopped raw chicken—you can also opt for a vegetarian choice by omitting the chicken. Fry, stirring for 5 min. Add smoked paprika (1/4tsp) cayenne pepper (1/4 tsp) and turmeric (1/2 tsp). Add paella rice (150 g) and stir for 2 min. Add saffron (1 tsp) and a good glass of dry white wine and some vegetable or chicken stock (450 mls) then cover and simmer gently for about 30 min (the time taken to prepare the broad beans). Take shelled broad beans (200 g) and pour over boiling water and simmer for 2 min. Drain and squeeze each bean to remove the skin. Add the beans to the skillet. Discard the skins. Cover with a tight fitting lid and turn off the heat. Add 12 halved mini plum tomatoes and 12 halved, pitted, Kalamata olives. Leave for 15 min. Stir and serve. Put salt and pepper on the table. It is also possible to add cooked prawns at the same time as the broad beans but this is optional.

9.2 Probiotics and Their Metabolites

While microorganisms, like other species, can be classified according to their relationships and by both genetics and phylogeny, they can also be described according to their function. This is the case with probiotics. The modern definition is that probiotics are bacterial species. For an earlier definition based upon the biological effects of specific entities, "to describe substances produced by one microorganism, that stimulate the growth of another" [3] see Chap. 8). According to the current definition, probiotics are mostly Gram-positive bacteria that include species belonging to the *Lactobacillus* and *Bifidobacterium* genera. Other microorganisms with similar properties have also been described including other bacterial species and yeasts. A functional definition based on their positive influence on the surrounding microflora or, ultimately, on the health of the consuming species leads, in essence, to classification by secondary metabolite profile. The beneficial effect on the consuming species appears to be a function of the prevention of growth of pathogens, restoration (or maintenance) of a 'healthy' microflora and the production of Short Chain Fatty Acids (SCFAs). Pathogen growth in vivo is initiated by the binding of the microbial pathogen to a mucosal surface. Inhibition of this adhesion is one of the ways that probiotics can inhibit pathogen growth thereby preventing the manifestation of pathogen-stimulated disease [4]—in particular, inflammation. Probiotics are postulated to also have a positive effect on the host immune system. Modulation of the immune system is a crucial aspect of the maintenance of a healthy intestinal system and resistance to pathogens [5]. One of the challenges in confirming effective intestinal immune modulation is the difficulty in carrying out any interventions and measuring the effects of these. While model systems (particularly cocultures) have been developed [6], there remains a lack of definitive data regarding the influence of administered probiotics in vivo [7]. One of the primary problems is measuring the extent to which probiotics are able to reach the large intestine and successfully colonise the space. Many studies have relied upon the measurement of bacterial species in stool samples, which only provides a partial picture, and the use of more invasive methods (for example, colonoscopy) is relatively rare. Trying to maximise the possible impact of positive changes in the intestine with a concomitant beneficial effect on gut and systemic health has led to the suggested use of a synergistic mixture of pro and prebiotics together with plants that have been found to contain secondary natural products with anti-inflammatory properties [8]. Many of the effects associated with probiotics have been found through animal (particularly rodent) experiments [9] and it has been suggested that the effects on human intestinal microflora are relatively short lived [10]. This may suggest that longer term colonisation of the large intestine by significant amounts of administered probiotics does not take place easily and the exogenous probiotic population is unstable and transient once administered. The properties of probiotics are strain-specific and several model systems have been used to simulate transit through the gastrointestinal system [11]. Such model systems have demonstrated that many products on the market which claim to contain microorganisms that are able to deliver probiotic effects to the large intestine are, in fact, unable to do so [12]. Encapsulation has been suggested to be the most effective way of ensuring delivery of probiotic bacteria and mixed strains of bacteria exhibit the greatest level of activity against relevant pathogenic bacteria [13]. In addition to the production of SCFAs (which will be discussed in more detail later), probiotics are also implicated in the production of a range of micronutrients and secondary natural products. Secondary natural products from probiotics (referred to as postbiotics) have been studied and antioxidative properties identified [14]. In addition, postbiotics also include antibiotics and bacteriostatic compounds which serve to limit the growth of pathogens. One of the most important classes of compounds produced are bacteriocins [15]. Bacteriocins will be discussed in more detail later (Sect. 9.6) however they generally have a narrow specificity and therefore have potential for use against colonisation by specific pathogens rather than the destruction of the whole gut microbiome and therefore they can be more easily assessed by in vivo model systems [16]. One of the reasons for the interest in the interaction of secondary natural products with probiotics is the biotransformative capability of microbes generally [17]. Although few studies have been carried out, the impact of probiotics on drug uptake and efficacy has been noted [18] and further studies on combining specific prodrugs with probiotics with defined biotransformative capabilities is a likely growth area. Furthermore, the potential of probiotics to protect against the consequences of ingestion of toxic natural products such as aflatoxin B1 has been noted [19].

Possibly the greatest amount of interest in the biological activity of probiotics arises from their capability to produce SCFAs. The impact of probiotics on the production of SFCAs has been extensively studied and reviewed [20]. While the addition of probiotic bacteria has been proposed as a way of influencing SCFA production, it is a process that meets with variable levels of success—probably because of the immunological properties of the gut rather than any intrinsic property of the bacteria. The physiological properties of the intestine (and the GI tract generally) are affected by the functional properties of the microbiome and the interactions that occur within the microbiome and between the microbiome and the host [21]. These will be examined in more detail in Sect. 9.3.In this section the physiological consequences of SCFA production will be examined in more detail. The production of SCFAs is one of the most biologically relevant aspects of the interaction between microbes and the host. The major SCFAs produced in the large intestine are acetate (C2), propionate (C3) and butyrate (C4). Others are also produced in smaller amounts. The major physiological effects of SCFAs occurs as a result of their interaction with Free Fatty Acid Receptor 2 (FFAR2). Other members of the class (FFAR1,3 and 4) also occur. Of the 4 subtypes; FFAR2 and FFAR3 are activated by short-chain fatty acids, while FFAR1 and FFAR4 are activated by long-chain fatty acids. FFARs have also been implicated in colorectal cancer [22]. All of the members of the class of FFARs are GPCRs and this has important consequences for their physiological activity. They are important metabolic mediators and respond to fatty acids as agonists. The interaction of SCFAs with FFARs is responsible for many of the effects observed [23]. While a detailed description of the various physiological effects associated with SCFAs is outside the scope of this chapter, some of these effects may have a broader significance. For example, butyrate is implicated in the strengthening of the mucus layer of the gut [24]. The mechanism is considered to involve upregulation of mucin production. The discovery that mucins may be involved in protecting neural tissue from inflammation-related damage [25] and in the prevention of age-related cognitive decline further suggests that a wider role for SCFAs might lie in the prevention of cognitive impairment such as that caused by Alzheimer's disease. In addition, it is known that an increase in certain intestinal bacterial species with age results in a decrease in butyrate production [26]. Other mechanisms for the interaction of SCFAs with neural cells have been noted including the downregulation of histone deacetylase [27] and the induction of the proliferation of neural progenitor cells [28]. The impact of SCFAs is, then, significant in extra-intestinal physiology. The presence of FFARs in the brain has been noted and a direct effect of a secondary natural product, fenchol, a monoterpene, (Fig. 9.3) has been suggested to involve activation of FFAR2 [29].

The importance (and ubiquity) of FFARs is adding to the understanding of their potential role in a number of disease states [30]. A range of FFAR agonists have been developed and these form part of a substantial drug development programme [31]. In addition to the effects noted previously (and, in particular, the effects on metabolic processes), FFARs have been implicated in immune modulation and, for example,

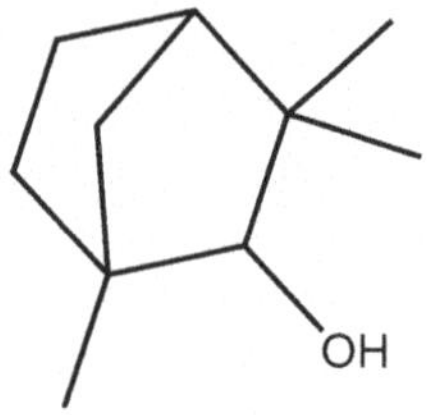

Fig. 9.3 The structure of fenchol

as a potential target for the treatment of irritable bowel disease [32]. While the physiological activity of another class of GPCRs—bitter taste receptors- is dependent in part on where they are found in the body, this may also be a functional attribute of FFARs. FFARs (in particular FFAR4), when expressed in the oral cavity, appear to be responsive to the taste of fats [33]. Further studies might suggest that different GPCRs may be used interchangeably and may have allosteric as well as active binding sites and their biological function may vary depending on the binding of specific ligands (agonists or antagonists) and their location.

9.3 The Physiological Microbiome

The physiological microbiome can be considered to be a function of the intramicrobiome interactions and the environment and the relationship to the host. Secondary natural products can act as modulators in all cases. The interaction of individual natural products with microbiomes illustrates the nature of the complex relationships that exist. The microorganisms that comprise a major part of the microbiome are responsive to chemical signals—generated externally from the environment, by other members of the microbial community and by the host. Determining the consequences and significance of each of the potential inputs individually can be challenging. The complexity of biological systems can mean that isolation and determination of causative factors is often a complex process with overlapping causes and effects. One approach is to make use of array technology and this will be illustrated in the following examples. This combines attributes of chemistry and biology—the sensitivity of one with the selectivity of the other. Biological systems can be precisely probed using appropriate biological tools which can then be quantified using a sensitive chemical marker system. There is, however, an important caveat. The use of a biological/chemical approach may not enable the system to be deconvoluted or simplified. In such cases, direct use of a chemical system may not enable the differentiation between different aspects of a biological system. Even when biological specificity is introduced, the system may not be able to be interpreted using chemical methods. For example, the nuclear magnetic resonance (NMR) spectrum of a biological macromolecule such as a receptor protein is very complex. By binding an agonist or an antagonist it might be hoped that the significant aspects of the receptor-ligand complex may be able to be differentiated and correlated with the biological activity. Biological specificity (the binding of a ligand) might help to delineate the biologically significant parts of the structure. However data from even simplified biological systems are often difficult to interpret and may require stable isotope labelling [34], the use of very specific techniques such as fluorine spectroscopy [35] or the use of other highly specialised techniques [36] which are not commonly available. One alternative is to sacrifice some levels of granularity and to focus on the information that is required rather than that which it is possible to generate. This is the approach taken in the use of the biological specificity of antibodies to develop high throughput screening [37]. This involves the identification of a structural characteristic by a biological molecule (an antibody). In another example, fluorescence in situ hybridisation (FISH) can be used to rapidly identify bacterial species [38] and to correlate this with clinical endpoints. The biological specificity (hybridisation to a specific DNA sequence) is combined with a fluorescent reporting system. In some ways, FISH can be considered to be the modern microscopical, molecular equivalent of Gram staining. It is a molecular probe that differentiates between species but with a much higher level of precision than bacterial staining. Interestingly, however, Gram staining provides information about the properties of the bacterial cell while FISH simply determines identity. The investigation of the interactions of the host/microbiome/secondary natural product triad can be considered in the following terms.

- Which natural products provoke a response in the microbial part of the microbiome either directly or as a consequence of an invoked host response?
- Which secondary natural products which originate from the microbes can cause a physiological response in other microbes or in the host and what are the consequences of that response?

One aspect of the interaction of secondary natural products with microbes is quorum sensing and this will be described in detail in Sect. 9.4. Other secondary natural products have been found which have an effect on microbes by different mechanisms. These effects may arise from the natural products acting directly on the microbes themselves or as a consequence of an induced effect from the host (for example, modulation of innate immunity or enhanced induction of phagocytosis). Two of the most significant effects of natural products on microbes are microbiostatic and microbiocidal effects. The use of natural products to inhibit growth or to kill microbes is part of the arsenal that is available and is part of a strategy to control deleterious microbes which has been referred to biocontrol [39] however biocontrol is a much broader area than the discovery of microbiocidal entities. Natural products for biocontrol and their use as bactericidal agents have been described and reviewed [40]. Much of the interest in secondary natural products as agents to kill microbes is driven by the need to find new therapeutic agents and to combat antimicrobial resistance (AMR) by finding new chemical white space in which to operate [41]. While the discovery of novel microbiocidal compounds with novel mechanisms of action has been the subject of much effort, there is also significant interest in the discovery and use of microbiostatic substances which inhibit growth but are not lethal. In some cases, they can induce dormancy. Agar diffusion assays are commonly used to assess the capability of substances to arrest microbial growth [42] however these are most often used to determine the effects on individual target pathogens rather than mixed populations of microbes. In some cases, however, activity against mixtures of pathogens (bacterial and fungal) are tested and efficacy has been observed [43]. The interplay between the microbiota, the host and natural products has been reviewed [44]. The host response to natural products can also affect the properties of microbial components of the microbiome. One of the main purposes of the immune system is to protect the host from the effects of pathogenic microbial agents. Some natural products have been found to increase the cytotoxic activity of natural killer cells, macrophages, and neutrophils while others promote an anti-inflammatory response to prevent hyperreactivity in the host [45]. Once again, much of the interest derives from a need to develop effective measures to combat specific pathogens rather than trying to maintain a microbiological balance.

While plants and other species are able to produce highly biologically active (and often subtle) defence compounds, microbes are often in direct competition with each other for resources and space and the production of anti-microbial compounds by competing microbes is common. In terms of the development of microbial protective agents, the possibility of using the human intestinal microbiome as a source of bioactive natural products has been reviewed [46]. Nonetheless, the mechanisms that underlie the host: microbiome interactions remain largely obscure—even if some of the biologically active entities have been identified. The activities associated with quorum sensing will be investigated in Sect. 9.4 however one of the key interactions that are involved in both microbe: microbe interactions and microbial community: host interactions are those associated with ecosystems. While considering these dualities, the importance of SCFAs has already been mentioned and will not be extensively further reviewed. In the context of microbe: microbe competition mentioned previously, the discovery of antibiotics such as penicillin and cephalosporins are among the most significant advances in the practical use of secondary natural products. The 'Age of Antibiotics' began with the discovery of penicillin by Fleming [47]. Prior to this, the most significant antibiotics included the original 'magic bullet', Salvarsan (first used clinically in 1909 [48]) and the sulphonamides (Fig. 9.4).

The latter were used to treat respiratory infections such as the pneumonia that Churchill contracted during the second world war in 1943 [49]. Perhaps the most significant aspect of both Salvarsan and sulphonamides is that they are synthetic molecules and naturally-occurring analogues were discovered later (see below). The original discovery of the antibiotic properties of sulphonamides was a consequence of the synthesis of azo derivatives to be used as dyestuffs by IG Farbenindustrie in 1909. One of these was found to have antiseptic properties however it was not in widespread use until after a compound, Prontosil, was patented in 1932 and used to treat Staphylococcal infections [50]. In 1935, researchers at the Pasteur Institute in Paris found that the active agent of Prontisil was, in fact, a sulfanilamide metabolite (Fig. 9.5) [51].

This led to the synthesis and testing of a large number of sulphonamides including, from the British pharmaceutical company May and Baker, the compound known as M&B 693 which was used to treat bacterial pneumonia. Sulphonamides operate by inhibiting the bacterial dihydropteroate synthase (DHPS) and, although they are

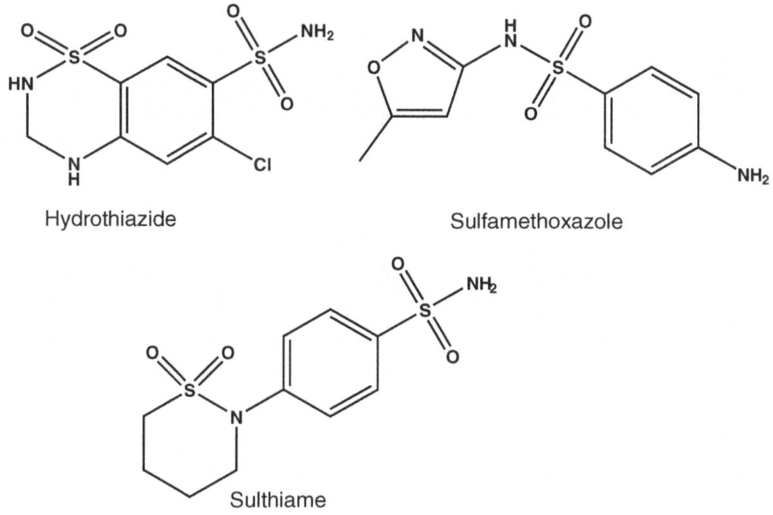

Fig. 9.4 Structure of representative sulphonamides

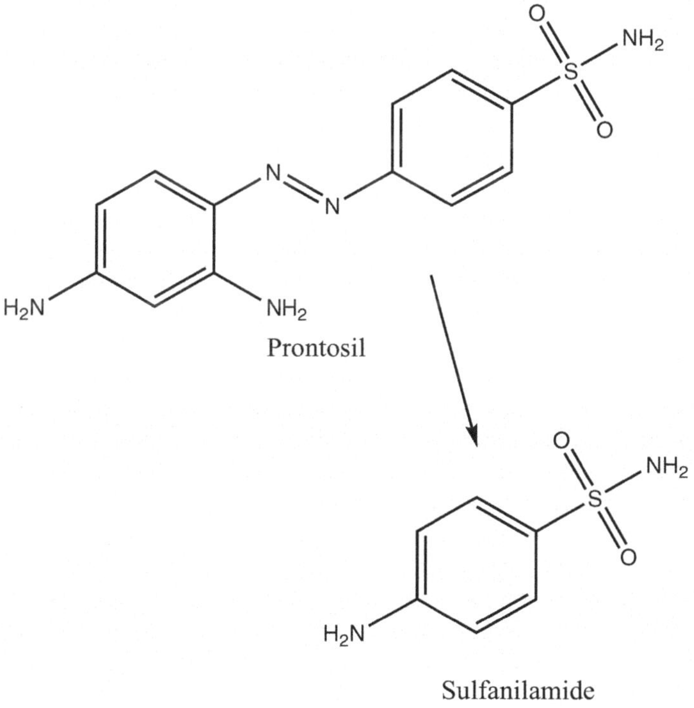

Fig. 9.5 The active agent of prontisil

largely synthetic in origin, resistance has been found [52] In an interesting reversal of the usual direction of research into natural products (isolation from a natural source and characterisation followed by synthesis of the parent compound and analogues in an attempt to develop more pharmacologically potent derivatives), some natural sulfonamides have more recently been discovered [53]. Comparatively few natural products containing a sulphonamide moiety have been found [54] however they have been shown to contain some interesting biosynthetic characteristics [55].

As noted previously, secondary natural products which originate from the microbes can cause a physiological response in other microbes or in the host. The antibiotic properties noted previously (for example, penicillins and cephalosporins) are one example of that effect. One of the most substantial sources of microbially-derived secondary natural products is the human intestinal microbiome. One method used metagenomics to probe biosynthetic gene clusters (BGCs). This led to the discovery of lactocillin, an analogue of the antimicrobial thiopeptide, thiocillin (Fig. 9.6) being found to be able to be produced by a human microbiome organism.

This approach of using metagenomics to identify BGCs and to identify similar compounds to known products from unusual sources has continued [56]. An alternative approach, which has been termed 'functional genomics' involves taking DNA from a given ecological or environmental niche and creating a library in a suitable host then screening the products for interesting biological activities [57]. While the identification of molecules that help to prevent pathogen outgrowth and facilitate the survival of a 'healthy' microbiome is a primary target, other aspects of functionality are able to be investigated. This includes, in particular, the rhizosphere and its interaction with plants and natural products both from the microbial and the vegetative parts of the system [58]. In addition, other ecological microbiomes have the potential to develop new, small molecule chemistry [59]. While the direct effect of microbes on mammalian hosts can be investigated by using specific biologically-based assay systems (for example, binding to receptors or enzyme inhibition or activation), the broader effects of the range of molecules (effectors) is more challenging to ascertain. In addition, the development of biologics has accelerated the pace of discovery of substances for pharmacological and other uses. The development of transcription activators and the use of heterologous gene expression in suitable hosts has increased the amounts of scarce secondary natural products that can be obtained and enhanced their commercial exploitation. This approach, combined with genome mining [60] has added substantially to the numbers of commercially viable natural products that are available. While antibiotics are an obvious target [61, 62] modulation of the biological activity of the host [63], effects on the immune system [64] may be significant. Effects on population dynamics of the microbial part of the microbiome and therefore on the host are perhaps of greatest potential. One aspect of that interplay is the manipulation and modulation of quorum sensing and this will be examined in more detail in Sect. 9.4).

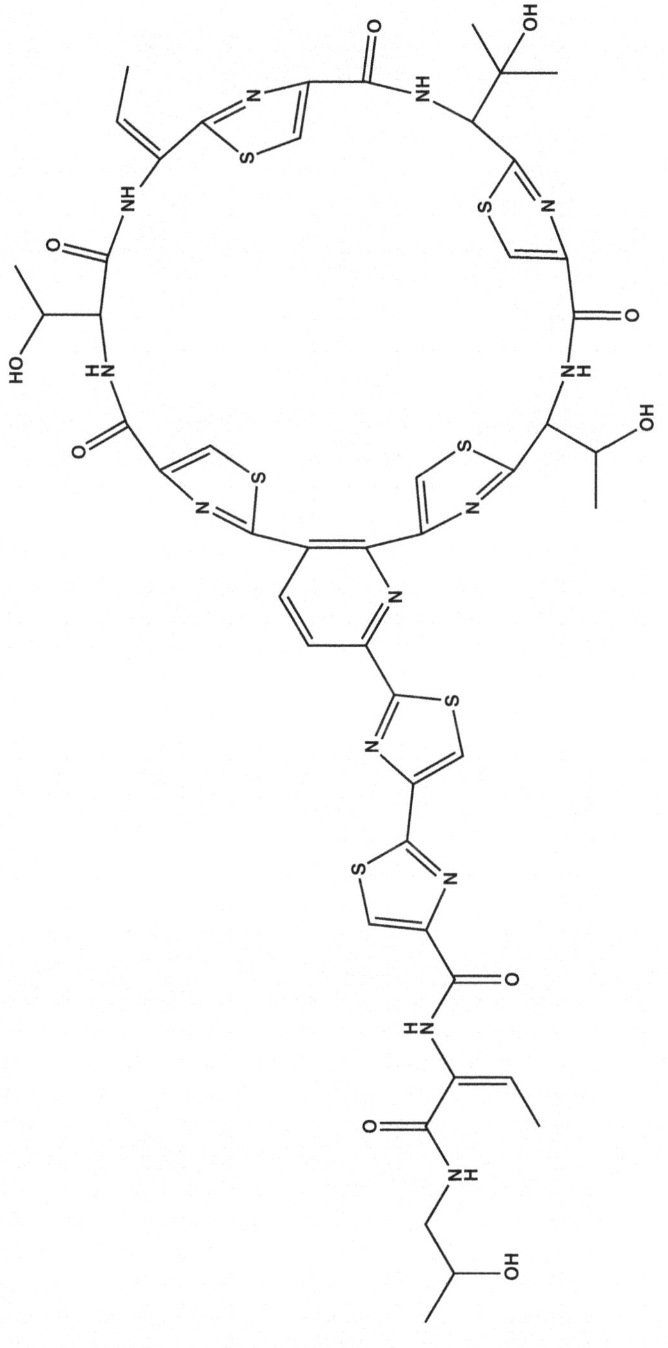

Fig. 9.6 Structures of thiocillin

9.4 Quorum Sensing in Microbes

Quorum sensing has been extensively studied from the standpoint of microbial (particularly bacterial) communication and associated properties such as virulence and swarming. It also is a key regulatory mechanism to prevent dysbiosis and the maintenance of healthy mammalian biological epithelial integrity [65]. More recently, it has become apparent that two key factors are important modulators (and consequences) of quorum sensing.

- Cross-species and cross-kingdom communication (referred to in passing in the discussion above about the rhizosphere).
- The role of small molecules in the modulation of quorum sensing effects (for example, the induction of expression of lactonases by secondary natural products). This will be discussed in more detail in Sect. 9.5.

Quorum sensing has been defined as a system that allows bacteria to communicate information about cell density using small 'autoinducer' (AI) molecules (see below) and to make use of that information to control specific aspects related to growth such as virulence and gene expression [66]. The fact that the production of AI molecules seems to affect microbial growth suggests that these (and their production) might be a potential target to control the growth and proliferation of specific microbes (for example, pathogens) in a microbial population. Just as penicillins and cephalosporins are able to interfere with a specific aspect of bacterial cell growth (the creation of the bacterial cell wall), by analogy, inhibitors of quorum sensing mechanisms may modulate the growth and virulence of certain microbes. For this to be an effective strategy, the production and use of quorum sensing must be sufficiently widespread but also sufficiently speciesspecific for the Ehrlich principle of a specific 'magic bullet' (Zauberkugel) to apply [67]. The involvement of quorum sensing in several processes including bioluminescence, sporulation, competence, antibiotic production, biofilm formation, and virulence factor secretion suggests that development of quorum sensing inhibitors could be part of a strategy to inhibit pathogens which targets a range of biological events influenced by quorum sensing. The process of quorum sensing in bacteria has been extensively studied. During their cycle of reproduction, bacteria produce small signalling molecules (called autoinducers). In the case of Gram-negative bacteria, the molecules are acyl homoserine lactones (Fig. 9.7) and these are able to diffuse through the bacterial cell wall.

In the case of Gram-positive bacteria, the most common autoinducer molecules are peptides that must be actively transported through the cell wall in order to be effective. As the numbers of cells increases, the number of the autoinducers also increases until the extracellular concentration of the autoinducers exceeds that of the intracellular concentration and the export of autoinducer molecules ceases. At this point, the autoinducer molecules begin to find targets within the cell and this initiates signalling cascades that affect gene transcription. It has been suggested that increasing the extracellular concentration of autoinducers may affect the growth phase of bacteria such that some aspects of the proliferative process may be disrupted

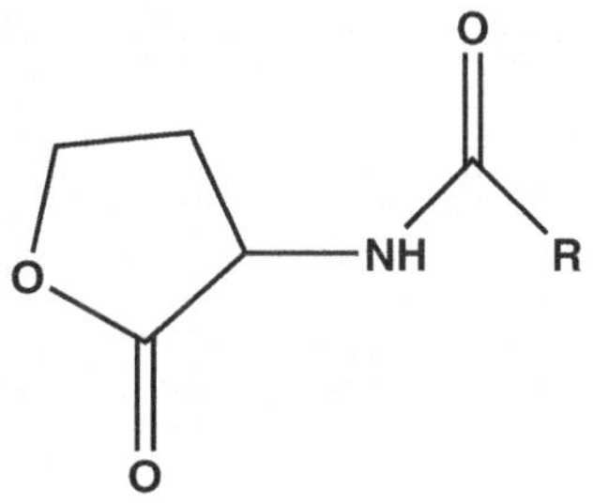

Fig. 9.7 Structure of acyl homoserine lactones

[68]. A number of other strategies involving the modulation of bacterial cell–cell communication can be envisioned.

While quorum sensing can be a sensitive tool with different (and complementary) levels of control requiring both the synthesis of the autoinducer molecules and receptors that can bind them to induce a downstream effect, some bacteria can operate and take advantage of quorum sensing while only having part of the machinery. For example, some bacteria can make use of the quorum sensing system but do not produce quorum sensing autoinducers themselves. This type of cross-species communication is an important aspect of the functioning of the microbial part of the microbiome. An example of this is the co-existence of *Staphylococcus aureus* (a common food poisoning microbe) and *Bacillus* species in the human intestine. The latter can disrupt the *Staphylococcus* quorum sensing system (preventing colonisation) [69] and also produce a compound, fengycin (Fig. 9.8) that competes at autoinducer binding sites.

The crosstalk between bacteria is a significant aspect of the mammalian gut microbiome [70]. As the extent of the interrelationships between quorum sensing systems and the importance of communication within microbial communities has become apparent, the importance of the production of multiple quorum sensing molecules and systems and the cross-talk between them has been revealed [71]. This can help to begin to explain the evolutionary development of the range of quorum sensing systems and how and why they are deployed [72]. The lack of specificity at one level of quorum sensing may appear, on the basis of an individual species, to be somewhat counter-intuitive but, on a microbial community level, it creates a series of robust and balanced community profiles without either the development of a monoculture or a dependence upon certain 'keystone species' as the sustaining force behind quorum sensing [73]. The possibility of cross-talk between autoinducers is not only a modern concept and was described quite early in the investigation of quorum sensing systems [74]. While cross-species communication within bacterial systems has been extensively studied [75], the possibility of interkingdom communication has also been noted [76]. Both cross-species and cross-genera quorum sensing has been observed [77] with a range of biological systems being affected and it was found that a seaweed species (*Ulva*) could make use of bacterial quorum sensing molecules and used homoserine lactones to modulate calcium signalling [78]. Despite this early demonstration of the potential inter-kingdom influence of quorum sensing, many

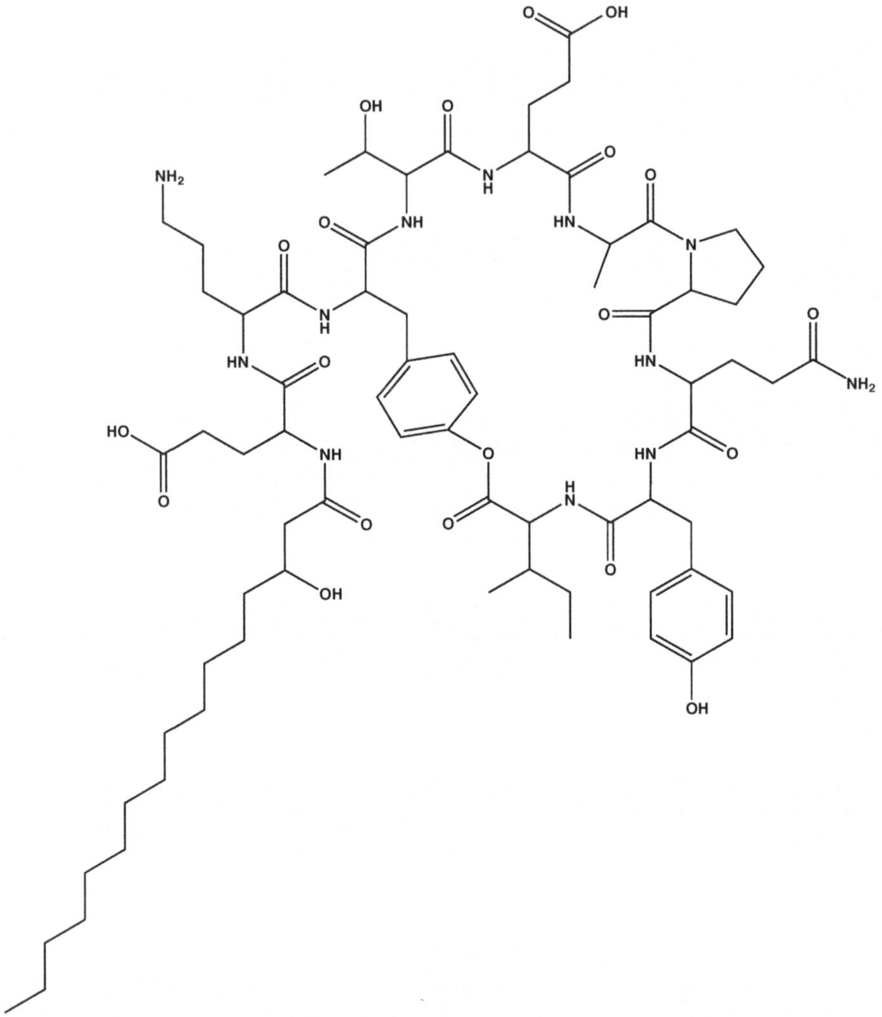

Fig. 9.8 Structure of fengycin

researchers have concentrated on the 'low hanging fruit' of their own specialism in terms of biological systems [79]. The mycorrhizal interactions between bacteria and plants have been known about and appreciated for a considerable time and quorum sensing plays a role in this symbiotic relationship [80]. The most frequently quoted example of bacterial/plant interaction is the nitrogen fixing bacteria associated with nodules formed by legumes [81]. The detailed molecular interactions that underpin plant/microbe interactions has been less well studied however it is known that one particular group of quorum sensing molecules—acyl homoserine lactones—have a particular role and are both taken up and transported within certain plants [82]. Plants can also produce quorum sensing mimics and inhibitors in order to encourage symbiotic relationships with bacteria [83] or resist pathogenicity [84]. The evolution of the interaction between plants and bacteria via quorum sensing molecules has been suggested by examining different species of plants which diverge from land plants at different points of evolutionary development [85] and different acyl homoserine lactones have different levels of activity in these plants in terms of spore generation. In multicellular organisms (unlike unicellular, undifferentiated bacteria), cells are able to sense and collectively control the cell density. Synthetic quorum sensing systems have been introduced into mammalian cells and these cells have been shown to be responsive to a specific chemical signal [86]. In addition, mammalian cells appear to have the capacity to inactivate certain quorum sensing molecules [87] thereby inhibiting pathogenicity. This will be discussed in more detail in Sect. 9.5 when lactonase production will be examined. One aspect of the interaction between the quorum sensing system in mammals and bacterial species is illustrated by the interaction of the bitter taste receptor T2R38 with a number of bacterial quorum sensing molecules [88]. The bitter receptor T2R38 has been suggested to be the primary target for acyl homoserine lactone signalling molecules and provide a mechanism whereby the bacterial signalling molecule stimulates phagocytosis [89]. The use of quorum sensing as a mechanism for immune response modulation has been suggested [90] and this may also suggest strategies for immune response regulation based on small secondary natural products. While the discussion of quorum sensing has moved from bacteria to other species of increasing complexity (yeasts, fungi, plants, mammals), it has become clear that less complex species such as viruses also make use of quorum sensing mechanisms [91]. Organisms that rely on symbiotic species to survive (epibionts) also utilise quorum sensing [92] and these signalling molecules may affect other species in the ecosystem. Indeed, this may be found to be a crucial factor in the role of epibionts in microbial systems.

One of the analogies that have been adopted for quorum sensing and signalling diversity is that of the Tower of Babel. The concept of the Tower of Babel was that, by legend, humans sought to build a tower so that they may storm the citadel of heaven. God then created different languages and thereby destroyed their capacity to communicate and so the Tower could not be completed [93]. In a similar way, multiple signalling pathways suggest that living systems are separated by 'language' and it is only an external deity that has the required oversight. It is perhaps interesting to note that the scientific study of the convergent evolution of language is opposed to the biblical interpretation of how different languages came about [94]. The biblical version can either be a literal interpretation of a biblical text or can be considered as a variation on the theme of "intelligent design" [95]—a crucial intervention by some external entity. From a scientific perspective, the development of different signalling pathways suggests the evolution of these solutions had multiple start points and therefore evolved in different ways to arrive at similar solutions (small molecule/receptor interactions reflecting the viability of populations). In short, whether it is termed "Creationism" or "Intelligent Design", theological opposition to evolution in the context of quorum sensing and signalling is unscientific and defies rational thought. It requires a discontinuity to exist and be explained as something that cannot be rationally interpreted and so an external agency must be invoked. Science does not require absolute convergence nor absolute linearity since non-equivalent changes may offer advantages in different ways. Indeed, the idea that different solutions to the same set of challenges or that different purposes can be found for the same structures (as seems to be the case with sweet and bitter receptors) is an important scientific, evolutionary-guided principle. It is, therefore, consistent with the theme of coevolution of species (and the same species under different environmental pressures) that organisms should adapt the biological investment in secondary natural products and any associated biological targets and systems and evolve these for a range of uses and environments. It is also important that any evolutionary advantage can be gained both from the co-evolution of molecules and targets and the use to which they are put. In this way, investment in completely new systems (which will consume more resources) can be minimised while retaining an ecological and survival advantage. Signalling diversity is an important evolutionary concept [72] and, rather than creating sets of mutually indecipherable cues, it allows for communication between and within species by different routes.

9.5 Manipulation of Quorum Sensing by Secondary Natural Products

While the manipulation of quorum sensing by natural products has been the subject of several studies, with varying degrees of success (see below), the reverse—the manipulation of natural product formation by quorum sensing mechanisms has also been considered [96]. The importance of quorum sensing and quorum sensing mechanisms to various aspects of microbial growth has been noted. A co-evolutionary argument would suggest that other species which have developed in the same ecological niche as those that use quorum sensing, will have developed methods for the manipulation of this mode of communication. This is likely to be a factor in the survival of organisms subjected to attack by pathogenic microbes with quorum sensing capability and in the success in any competition for valuable resources. Since many species rely on the flexibility in structure and function afforded by secondary natural products for defensive purposes, some of these are likely to be used to counter the quorum sensing mechanisms of invasive microbes (particularly pathogens). Such an activity is termed 'quorum quenching'. Since a large number of Gram negative bacteria make use of acyl homoserine lactones [97], these and the systems they modulate, are likely to be primary targets for secondary natural products from other species. Gram positive bacteria make use of a number of signalling pathways (particularly autoinducing peptides) [98] and quenching activity that influences these is likely to be more specific. Finally, organisms other than bacteria can (as noted previously) make use of quorum sensing and signalling pathways even if, in some cases, they do not have the biosynthetic machinery to be able to produce quorum sensing molecules.

Quorum quenching is the disruption of quorum sensing signalling. This can occur at a number of levels—effects on gene expression, disruption of the quorum sensing molecules themselves and mechanistic quorum sensing inhibitors (QSIs) (antagonists) of the different consequences of the quorum sensing process can all be employed [99]. Although quorum sensing is primarily a prokaryotic activity, as noted previously, some eukaryotes, including yeasts, algae, plants and mammals can also detect quorum sensing molecules. The possibility of quorum quenching (and related effects) being a manipulable function of microbiomes has been noted [100]. In the case of homoserine lactones, enzymic degradation of the signalling molecule is one of the most commonly observed mechanisms of quorum quenching. This can be the result of the action of lactonases, amidases, reductases and oxidases. A human lactonase (paraoxonase) has been found to confer some protection against Gram negative pathogens by cleaving the homoserine lactone ring [101]. Two other members of the class (PON2 and PON3) have been found with differing (but overlapping) specificities [102]. In addition to their role as lactonases, PON enzymes are also implicated in the innate immune system and in the control of inflammation. Certain secondary natural products (for example catechins) have been found to enhance PON enzyme activity [103] and the assay of other small molecules for such enhancing activity would be of value. It is also important to determine if the activity of the PON enzymes is due to microbially-relevant lactonase activity or if some other biological property is responsible for any observed clinical effects. This clarification will be facilitated by multi-omics studies [104]. It has been noted that lactonase activity is associated with a number of *Bacillus* species—even though, as Gram positive bacteria, they do not generally make use of homoserine lactones for signalling. It is possible that secretion of a lactonase will provide the species with an inhibitory capability against the growth of Gram negative bacteria [105]. While lactonases have been the subject of the most significant amount of study, the further characterisation of the other enzymes involved in quorum quenching is likely to provide links to a range of classes of secondary natural products. In addition, lactonases are complex enzymes with a number of structural features and this may increase the possibility of allosteric effects being found with some natural products [106]. A second possibility is the use of QSIs and rapid throughput assays have been developed [107]. This type of rapid screening is necessary since the numbers of active molecules has been shown to be fairly low. Screening of a library of 200,000 molecules gave just two acyl homoserine lactone positive QSIs [108]. The application of drug design principles in the screening of secondary natural products has enabled a number of likely molecules to be identified in silico [109]. Mechanistically-based QSI strategies are likely to be a more effective way of discovering novel bioactive molecules than mass screening [110]. Enough information is already available for structure/activity relationships to be probed for some QQ approaches with various modes of action, however a disparate number of natural structures have been found to be active as QSIs and this is illustrated by the assays that have been carried out. QSIs have been found in extracts from a number of food sources including star anise, grape seed extract, some aromatic herbs, grapefruit extracts and ginger and turmeric extracts. Active substances include benzoates, phenylpropanoids, stilbenes, flavonoids, gallotannins, proanthocyanidins, coumarins, alkaloids and terpenes. It has also been suggested that the mechanism of action of some antibiotics may be attributable to QSI activity

[111]. If this is a viable hypothesis then comparison of the mechanism of action of different antibiotics in terms of QSI activity may be a useful start point for investigations. Down regulation of the biosynthesis of quorum sensing molecules is part of the natural modulation of the quorum sensing system [112] therefore downregulation of the biosynthetic pathway [113] itself or inhibition of the constituent biosynthetic enzymes by small molecules is a valid approach to control of quorum sensing and attributes of microbial systems such as virulence or biofilm formation. It may also be of significance that, as noted previously, quorum sensing molecules can bind to certain bitter receptors in extra-oral organs and may, therefore be responsible for nonsensory outputs. Inhibition of the formation or translocation of quorum sensing molecules or direct competition for binding sites by QSIs may affect these activities also. In addition, some bitter compounds, in addition to binding to bitter receptors, may have agonist or antagonist activity with regard to the quorum sensing system.

9.6 Antibiotics in Foods (Lantibiotics and Other Bacteriocins)

The use of lactic acid bacteria for food preservation led to the discovery of a class of antimicrobial compounds (bacteriocins). The most significant and widely used of these are the lantibiotics such as nisin which was discovered in 1928 [114, 115]. Structurally, lantibiotics are polyethers that contain the amino acids lanthionine or methyl lanthionine (Fig. 9.9).

They are classified according to structure and mode of action. Type A lantibiotics are elongated molecules and operate by disrupting membrane structures. Type B lantibiotics are globular and disrupt enzyme function in target bacteria. They are particularly valuable for food use since they are active against many food spoilage organisms however exhibit low toxicity to humans. Bacteriocins generally are classified into four types depending on whether not they are post-translationally modified, their structure and their molecular weight and genetic origin [116]. Bacteriocins have potential uses as antimicrobials [117] however they typically have a narrow range of activity. They are commonly found in lactic acid bacteria [118] and, while they have been used in food preservation over a considerable period of time, there are few reports of the development of resistance by target bacteria (although see below for a mechanistic-based analysis and a cautionary observation). The use of

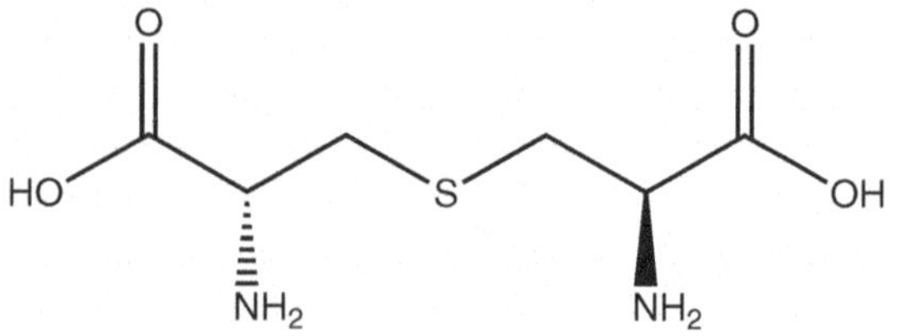

Fig. 9.9 Structure of lanthionine

bacteriocins has been suggested as an alternative to the use of cocktails of drugs to mitigate the effects of the build up of anti-microbial resistance [119]. They have a range of structures and mechanisms of action [120] unlike the example of using mixtures of compounds that affect common features of bacterial growth (for example the mevalonate pathway combined with penicillin combining to inhibit bacterial cell wall synthesis—[121], an approach of using bacteriocins is, in conceptual terms, similar to the traditional use of antibiotic mixtures in an attempt to evade resistance mechanisms. The other particular advantage that is attributed to bacteriocins is that they appear to have limited (or no) biological activity against commensal bacteria and are, therefore, mostly regarded as safe [122]. Some lantibiotics appear to have a dual mode of action and this may both increase their effectiveness and inhibit the development of specific resistance mechanisms [123]. They are also able to be engineered to give molecules with enhanced biological activity. While bacteriocins have fewer issues with resistance than other antibiotics, the possibility of resistance evolving may become important if they are increasingly used in a clinical setting. Resistance can be classified as a consequence of three mechanistic developments:

- Inactivation by irreversible enzymatic cleavage/modification;
- Target modification at the site of antibiotic binding; or
- Reduced drug accumulation due to low permeability or to increased drug efflux

As noted previously, xenobiotic efflux may be a consequence of the intake of some secondary natural products. There has been some suggestion of a resistance mechanism for bacteriocins involving target modification [124] however enzymically oriented resistance has also been reported [125]. Although they have been the subject of considerable interest, the absence of any clinically used bacteriocins suggests that commercialisation for use in a medical setting may not be a viable option [126]. The problem may be that bacteriocins, for all their unique properties, do not represent a paradigm shift in how to deal with pathogenic, infectious agents. They are, in essence, antimicrobials with a limited spectrum of activity and a high level of specificity. The most well known (and widely used) lantibiotic, nisin, is inhibited by substances produced by many human pathogens [127] and there is every reason to suppose that increased use of it (and other bacteriocins) will result in the spread of resistance. It seems likely that bacteriocins will be another dead end in the ongoing struggle to overcome clinical antimicrobial resistance. While the production of antimicrobial compounds by microbes is part of a direct competition between similar organisms for resources, plant anti-microbial compounds occupy a different chemical space [128] and operate from a mechanistically different premise including synergy with other antimicrobial compounds [129] and decreasing bacterial resistance rather than bactericidal activity per se [130]. This is illustrated by the bioactive principles produced by plants of the genus *Allium* whose biological activity includes directly plant protective effects and other biological functions.

9.7 Bioactive Compounds from the Genus Allium

The development of anti-microbial resistance (AMR) in plants is a challenge [131] and increases the susceptibility of economically important species to attack by pathogens. The development of AMR has been an issue from the very beginnings of chemically-based antimicrobial development. Indeed, even before the widespread use of beta lactam antibiotics in a clinical environment, enzymes that could inactivate the penicillin antibiotics had been found [132]. Because of the prevalence of AMR among bacterial species, perhaps exacerbated by the direct competition between microbes for space and resources, there has been increased interest in non-microbially derived anti-microbial substances [133]. While many of these lack the specificity of the magic bullet proposed by Ehrlich [134], their broader spectrum of anti-microbial activities decreases the likelihood of resistance being acquired and offers scope for synergistic activities to be found. One example of this are the bioactive S-containing (and other) compounds found in plants of the genus *Allium.* These have been extensively studied and therefore provide a useful, illustrative example. Other possible mechanisms that have been cited includes the use of plant bioactive agents to enhance the activity of traditional antibiotics and to thereby increase their effectiveness—particularly against microbes that demonstrate resistance [135]. The list of multi-drug-resistant microbes continues to increase so the importance of synergistic effects between bioactive agents from different sources with different mechanisms of action is likely to be an increasingly important factor. In the case of phytochemicals, a range of mechanisms have been proposed to explain their broad spectrum antimicrobial activity [136]. This has encouraged researchers to consider the application of some plant bioactives such as allicin (from garlic) in combination with other antibiotics [137]. One of the problems is that, in many cases, synergistic activity is found between extracts that contain a large number of components and specific antibiotics. It can be difficult to determine the mechanism of action and, in addition, the existence of synergy between the different phytochemicals must also be considered. In the case of plants of the genus *Allium*, most interest has been generated by the Scontaining compounds created after the plant tissue is disrupted. This combination of a relatively biologically inactive 'prodrug' and an effector enzyme is also observed in the case of the 'mustard oil bomb' in the genus *Brassica* [138]. The reaction in the case of *Allium* species involves the cleavage of alk(en)yl-L-cysteine sulphoxides by the compartmentalised enzyme, alliinase (a C-S lyase), to give the corresponding thiosulphinates (Fig. 9.10) which can then combine in various ways to give a range of S-compounds [139].

The first member of the class of alk(en)yl-cysteine sulphoxides to be identified was allylcysteine sulphoxide from garlic, which was given the trivial name alliin. Soon three additional related compounds, S-methylcysteine sulfoxide (methiin), S-1-propenylcysteine sulfoxide (isoalliin), and S-n-propylcysteine sulfoxide (propiin) were isolated from onion. These are the four major members of the class and, between them, account for much of the biological activity (and sensory properties) of plants of the genus. The biological activity of these S-containing compounds can be considered

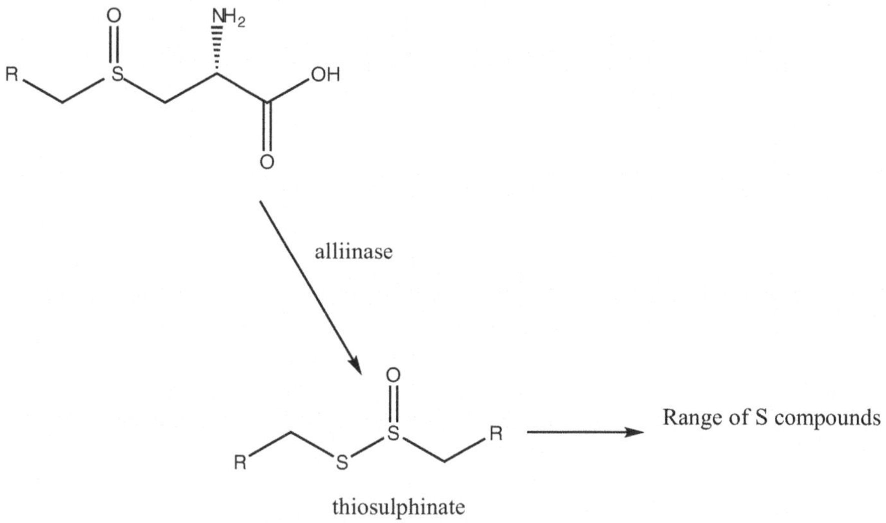

Fig. 9.10 Formation of thiosulphinates

to be of two distinct types. There is the activity associated with being of benefit to the host (producer) plant and then there is the benefit that accrues from the consumption of the plants by other species. These will be considered briefly as two distinct categories.

Extracts from plants of the genus *Allium* have a long history of use in preventing microbial growth. Indeed, extracts of *Allium fistulosum* (Welsh onion) have been approved for use as a soil bactericidal agent [140]. Despite this regulated use of *Allium fistulosum*, garlic (*Allium sativum*) is considered to be the most potent antimicrobial plant of the genus [141]. Omics studies have been used to try to elucidate more details of the nature of the plant protection effects of alliums [142]. These studies suggest that the plants of the genus may contain multiple forms of the alliinase enzyme both in a vacuole (which is released upon tissue damage) and in the cytosol (which does not require tissue damage to be liberated) and production of the cleaved S-containing compounds may therefore occur both in the presence and absence of plant tissue damage. In addition to the production of allicin, there is another important S-containing compound that is produced specifically in onion. This compound (propanethiol-S-oxide) is responsible for the lachrymatory properties of onion [143].

Regarding the mechanism of action of antimicrobial S compounds produced by *Allium* species, one of the activities attributed to allicin is as a specific quorum sensing (and virulence) inhibitor [144]. In addition, another S-containing compound from garlic, ajoene, has been implicated in a quorum sensing inhibition effect through the inhibition of regulatory sRNA [145]. An in silico analysis has also suggested a quorum sensing disruptive mode of action in the case of *Escherichia coli* [146]. Other non-S-containing metabolites from garlic have also been implicated in quorum sensing inhibition [147] and this multiple target effect may enhance the biological relevance of secondary natural products as phytoprotectants. Garlic extract has also

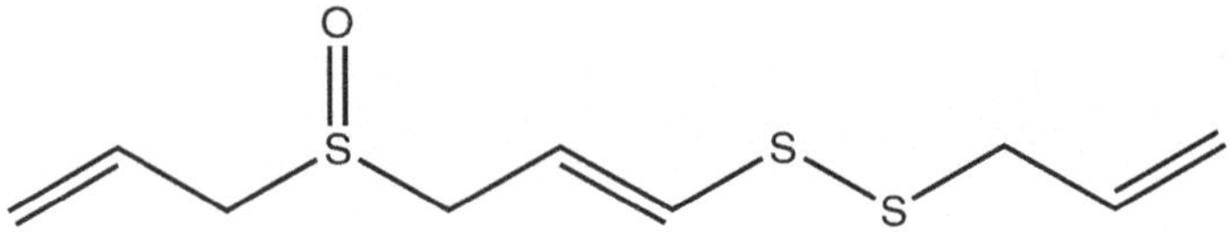

Fig. 9.11 Structure of ajoene

been implicated in an anti-fungal strategy, again based upon quorum sensing inhibition [148]. While the presence of plant protective substances in a range of *Allium* species has been demonstrated this could, in some cases, be a defence against insect or herbivore predation rather than antimicrobial activity [149].

The use of plants of the genus *Allium* as medicinal entities for consuming species has been extensively reported. Unlike the antimicrobial effects, the beneficial physiological effects are not directly linked to any benefit for the producer species. Many of the effects are a consequence of the presence of a range of organosulphur compounds however phenolic compounds such as quercetin have also been found. A number of effects have been noted however in many cases these are observational rather than being mechanistically proven [150]. Flavonols (rather than sulphur compounds) have been proposed to be responsible for some of the observed effects [151] and this may be linked to immunomodulatory effects. In many cases, however, the application of pharmacological methods has provided results that are equivocal and are observational rather than mechanistic. In addition, although some candidates have been suggested, it is unclear which component of the plants are responsible for the observed activity. [152]. The reductionist approach applied to the components produced, in particular, by components of garlic, has provided some of the impetus for studies of the bioactive principle ajoene. Ajoene is formed in crushed garlic as a mixture of E- and Z-4,5,9trithiadodeca-1,6,11-triene 9-oxide (Fig. 9.11).

In addition to the quorum sensing inhibition mentioned previously, it appears to inhibit platelet aggregation caused by all known platelet aggregation agonists [153] and has been proposed to have anti-proliferative properties. It has been shown to inhibit HMGCoA reductase thereby inhibiting cholesterol biosynthesis and disrupting the mevalonate pathway. It also appears to interfere with protein prenylation [154] and this may explain its effect on cell proliferation.

The interaction of secondary natural products with microbes and the microbiome is of particular interest. The role of quorum sensing would appear to offer significant possibilities—particularly if used alongside more conventional antimicrobial therapies.

Certainly the concept of modulation of microbial communities rather than their absolute elimination would appear to be a more sustainable strategy in the light of the prevalence of AMR (except for the case of some lethal pathogens). Food-related microbial modulators tend to have a more broad spectrum effect than those derived from other microbial species and, while they may lack the specificity envisioned in the Ehrlich 'magic bullet' approach (as exemplified by lantibiotics), they are less likely to be subjected to the development of resistance by microbial species and have significant possibilities to be used as synergistic microbial modulators.

Some of the other effects of bioactive compounds from foods and food systems have been mentioned however these are complicated both by the complexity of consuming species and the effects of confounders. Such investigations are also hampered by the lack of suitable models and the interpretation of small, marginal effects over considerable periods of time.

References

1. Esmaealzadeh D, Moodi Ghalibaf A, Shariati Rad M, Rezaee R, Razavi BM, Hosseinzadeh H. Pharmacological effects of Safranal: an updated review. Iran J Basic Med Sci. 2023;26(10):1131–43. https://doi.org/10.22038/IJBMS.2023.69824.15197. PMID: 37736506; PMCID: PMC10510479.
2. Cattivelli A, Conte A, Martini S, Tagliazucchi D. Influence of cooking methods on onion phenolic compounds bioaccessibility. Foods. 2021;10(5):1023. https://doi.org/10.3390/foods10051023. PMID: 34066759; PMCID: PMC8151956.
3. Lilly DM, Stillwell RH. Probiotics: growth-promoting factors produced by microorganisms. Science. 1965;147(3659):747–8. https://doi.org/10.1126/science.147.3659.747. PMID: 14242024.
4. Bermudez-Brito M, Plaza-Díaz J, Muñoz-Quezada S, Gómez-Llorente C, Gil A. Probiotic mechanisms of action. Ann Nutr Metab. 2012;61(2):160–74. https://doi.org/10.1159/000342079. Epub 2012 Oct 2 PMID: 23037511.
5. Mazziotta C, Tognon M, Martini F, Torreggiani E, Rotondo JC. Probiotics mechanism of action on immune cells and beneficial effects on human health. Cells. 2023;12(1):184. https://doi.org/10.3390/cells12010184. PMID: 36611977; PMCID: PMC9818925.
6. Beterams A, De Paepe K, Maes L, Wise IJ, De Keersmaecker H, Rajkovic A, Laukens D, Van de Wiele T, Calatayud AM. Versatile human in vitro triple coculture model coincubated with adhered gut microbes reproducibly mimics pro-inflammatory host-microbe interactions in the colon. FASEB J. 2021;35: e21992. https://doi.org/10.1096/fj.202101135R.
7. Zmora N, Zilberman-Schapira G, Suez J, et al. Personalized gut mucosal colonization resistance to empiric probiotics is associated with unique host and microbiome features. Cell. 2018;174(6):1388–405.
8. Eliaz I. The failure of probiotics-and the strategy of microbiome synergy. Integr Med (Encinitas). 2020;19(3):8–10. PMID: 33132772; PMCID: PMC7572142.
9. Wang X, Zhang P, Zhang X. Probiotics regulate gut microbiota: an effective method to improve immunity. Molecules. 2021;26(19):6076. https://doi.org/10.3390/molecules26196076. PMID: 34641619; PMCID: PMC8512487.
10. Han S, Lu Y, Xie J, Fei Y, Zheng G, Wang Z, Liu J, Lv L, Ling Z, Berglund B, Yao M, Li L. Probiotic gastrointestinal transit and colonization after oral administration: a long journey. Front Cell Infect Microbiol. 2021;10(11): 609722. https://doi.org/10.3389/fcimb.2021.609722. PMID: 33791234; PMCID: PMC8006270.
11. Papadimitriou K, Zoumpopoulou G, Foligné B, Alexandraki V, Kazou M, Pot B, Tsakalidou E. Discovering probiotic microorganisms: in vitro, in vivo, genetic and omics approaches. Front Microbiol. 2015;17(6):58. https://doi.org/10.3389/fmicb.2015.00058. PMID: 25741323; PMCID: PMC4330916.
12. Naissinger da Silva M, Tagliapietra BL, Flores VDA, Pereira Dos Santos Richards NS. In vitro test to evaluate survival in the gastrointestinal tract of commercial probiotics. Curr Res Food Sci. 2021;4:320–325. https://doi.org/10.1016/j.crfs.2021.04.006. PMID: 34095855; PMCID: PMC8165489.

13. Bernatek M, Żukiewicz-Sobczak W, Lachowicz-Wiśniewska S, Piątek J. Factors determining effective probiotic activity: evaluation of survival and antibacterial activity of selected probiotic products using an "in vitro" study. Nutrients. 2022;14(16):3323. https://doi.org/10.3390/nu14163323. PMID: 36014829; PMCID: PMC9413312.
14. Chávarri Hueda M, Diez-Gutiérrez L. Evaluation of antioxidant activity of postbiotics in cell cultures. In: Dharumadurai D, editors. Postbiotics. Methods and protocols in food science. New York, NY: Humana;2024. https://doi.org/10.1007/9781-0716-3421-9_32.
15. Fantinato V, Camargo HR, Sousa ALOP. Probiotics study with Streptococcus salivarius and its ability to produce bacteriocins and adherence to KB cells. Rev Odontol UNESP. 2019;48:1–9. https://doi.org/10.1590/1807-2577.02919.
16. Benítez-Chao DF, León-Buitimea A, Lerma-Escalera JA, Morones-Ramírez JR. Bacteriocins: an overview of antimicrobial, toxicity, and biosafety assessment by in vivo models. Front Microbiol. 2021;15(12): 630695. https://doi.org/10.3389/fmicb.2021.630695. PMID: 33935991; PMCID: PMC8083986.
17. Tembeni B, Idowu OE, Benrkia R, et al. Biotransformation of selected secondary metabolites by Alternaria species and the pharmaceutical, food and agricultural application of biotransformation products. Nat Prod Bioprospect. 2024;14:46. https://doi.org/10.1007/s13659-024-00469-5.
18. Purdel C, Ungurianu A, Adam-Dima I, Margină D. Exploring the potential impact of probiotic use on drug metabolism and efficacy. Biomed Pharmacother. 2023;161: 114468. https://doi.org/10.1016/j.biopha.2023.114468. Epub 2023 Mar 1 PMID: 36868015.
19. Ariti KD, Tadesse BR. Harnessing probiotics to combat aflatoxin: a natural approach to food safety: a review. Int J Biomed Eng Clin Sci. 2024;10(1):917. https://doi.org/10.11648/j.ijbecs.20241001.12.
20. Markowiak-Kopeć P, Śliżewska K. The effect of probiotics on the production of short-chain fatty acids by human intestinal microbiome. Nutrients. 2020;12(4):1107. https://doi.org/10.3390/nu12041107. PMID: 32316181; PMCID: PMC7230973.
21. Lin D, Medeiros DM. The microbiome as a major function of the gastrointestinal tract and its implication in micronutrient metabolism and chronic diseases. Nutr Res. 2023;112:30–45. https://doi.org/10.1016/j.nutres.2023.02.007.
22. Binienda A, Owczarek K, Sałaga M, Fichna J. Synthetic free fatty acid receptor (FFAR) 2 agonist 4-CMTB and FFAR4 agonist GSK13764 inhibit colon cancer cell growth and migration and regulate FFARs expression in in vitro and in vivo models of colorectal cancer. Pharmacol Rep. 2024;76(6):1403–1414. https://doi.org/10.1007/s43440-024-00667-5. Epub 2024 Oct 21. PMID: 39432182; PMCID: PMC11582145.
23. Kimura I, Ichimura A, Ohue-Kitano R, Igarashi M. Free fatty acid receptors in health and disease. Physiol Rev. 2020;100:171–210. https://doi.org/10.1152/physrev.00041.2018.
24. Liang L, Liu L, Zhou W, Yang C, Mai G, Li H, Chen Y. Gut microbiota-derived butyrate regulates gut mucus barrier repair by activating the macrophage/WNT/ERK signaling pathway. Clin Sci. 2022;136:291–307. https://doi.org/10.1042/CS20210778.
25. Shi SM, Suh RJ, Shon DJ, et al. Glycocalyx dysregulation impairs blood–brain barrier in ageing and disease. Nature. 2025. https://doi.org/10.1038/s41586-02508589-9.
26. Odamaki T, Kato K, Sugahara H, Hashikura N, Takahashi S, Xiao J-Z, Abe F, Osawa R. Age-related changes in gut microbiota composition from newborn to centenarian: a cross-sectional study. BMC Microbiol. 2016;16:90. https://doi.org/10.1186/s12866-016-0708-5.
27. Mirzaei R, Bouzari B, Hosseini-Fard SR, Mazaheri M, Ahmadyousefi Y, Abdi M, Jalalifar S, Karimitabar Z, Teimoori A, Keyvani H, et al. Role of microbiotaderived short-chain fatty acids in nervous system disorders. Biomed Pharmacother Biomed Pharmacother. 2021;139: 111661. https://doi.org/10.1016/j.biopha.2021.111661.
28. Yang LL, Millischer V, Rodin S, MacFabe DF, Villaescusa JC, Lavebratt C. Enteric short-chain fatty acids promote proliferation of human neural progenitor cells. J Neurochem. 2020;154:635–46. https://doi.org/10.1111/jnc.14928.
29. Razazan A, Karunakar P, Mishra SP, Sharma S, Yadav H. Fenchol ameliorates Alzheimer's disease like phenotypes by modulating microbiome/proteolysis/senescence axis. Alzheimer's Dement. 2020;16: e044718. https://doi.org/10.1002/alz.044718.

30. FalomirLockhart LJ, Cavazzutti GF, Giménez E, Toscani AM. Fatty acid signaling mechanisms in neural cells: fatty acid receptors. Front Cellul Neurosci. 13:162.
31. Governa P, Caroleo MC, Carullo G, Aiello F, Cione E, Manetti F. FFAR1/GPR40: one target, different binding sites, many agonists, no drugs, but a continuous and unprofitable tug-of-war between ligand lipophilicity, activity, and toxicity. Bioorg Med Chem Lett. 2021;41:127969. https://doi.org/10.1016/j.bmcl.2021.127969. Epub 2021 Mar 23. PMID: 33771587.
32. Bartoszek A, Moo EV, Binienda A, Fabisiak A, Krajewska JB, Mosińska P, Niewinna K, Tarasiuk A, Martemyanov K, Salaga M, Fichna J. Free fatty acid receptors as new potential therapeutic target in inflammatory bowel diseases. Pharmacol Res. 2020;152: 104604. https://doi.org/10.1016/j.phrs.2019.104604. Epub 2019 Dec 14 PMID: 31846762.
33. Costanzo A, Liu D, Nowson C, et al. A low-fat diet up-regulates expression of fatty acid taste receptor gene FFAR4 in fungiform papillae in humans: a co-twin randomised controlled trial. Br J Nutr. 2019;122(11):1212–20. https://doi.org/10.1017/S0007114519002368.
34. Nygaard R, Zou Y, Dror RO, Mildorf TJ, Arlow DH, Manglik A, Pan AC, Liu CW, Fung JJ, Bokoch MP, Thian FS, Kobilka TS, Shaw DE, Mueller L, Prosser RS, Kobilka BK. The dynamic process of β(2)-adrenergic receptor activation. Cell. 2013;152(3):532–42. https://doi.org/10.1016/j.cell.2013.01.008. PMID: 23374348; PMCID: PMC3586676.
35. Horst R, Liu JJ, Stevens RC, Wüthrich K. β_2-adrenergic receptor activation by agonists studied with 1^9F NMR spectroscopy. Angew Chem Int Ed Engl. 2013;52(41):10762–5. https://doi.org/10.1002/anie.201305286. Epub 2013 Aug 16. PMID: 23956158; PMCID: PMC3925757.
36. Bostock MJ, et al. The role of NMR spectroscopy in mapping the conformational landscape of GPCRs. Curr Opin Struct Biol. 2019;57:145–56.
37. Ren H, Li J, Zhang N, et al. Function-based high-throughput screening for antibody antagonists and agonists against G protein-coupled receptors. Commun Biol. 2020;3:146. https://doi.org/10.1038/s42003-020-0867-7.
38. Shah JS, Ramasamy R. Fluorescence in situ hybridization (FISH) tests for identifying protozoan and bacterial pathogens in infectious diseases. Diagnostics (Basel). 2022;12(5):1286. https://doi.org/10.3390/diagnostics12051286. PMID: 35626441; PMCID: PMC9141552.
39. Ayaz M, Li CH, Ali Q, Zhao W, Chi YK, Shafiq M, Ali F, Yu XY, Yu Q, Zhao JT, Yu JW, Qi RD, Huang WK. Bacterial and fungal biocontrol agents for plant disease protection: journey from lab to field, current status, challenges, and global perspectives. Molecules. 2023;28(18):6735. https://doi.org/10.3390/molecules28186735. PMID:37764510;PMCID:PMC10537577.
40. Ramesh S, Radhakrishnan P. Natural products as eco-friendly bactericides for plant growth and development. In: Singh J, Yadav A, editors. Natural bioactive products in sustainable agriculture. Singapore: Springer;2020. https://doi.org/10.1007/978-981-15-3024-1_10.
41. Liu Y, Zhu J, Liu Z, Zhi Y, Mei C, Wang H. Flavonoids as promising natural compounds for combating bacterial infections. Int J Mol Sci. 2025;26:2455. https://doi.org/10.3390/ijms26062455.
42. Philip N, Leishman S, Bandara H, Walsh L. Growth inhibitory effects of antimicrobial natural products against cariogenic and health-associated oral bacterial species. Oral Health Prev Dent. 2020;18(3):537–42. https://doi.org/10.3290/j.ohpd.a44307. PMID: 32515425; PMCID: PMC11654489.
43. Hafidh RR, Abdulamir AS, Vern LS, Abu Bakar F, Abas F, Jahanshiri F, Sekawi Z. Inhibition of growth of highly resistant bacterial and fungal pathogens by a natural product. Open Microbiol J. 2011;5:96–106. https://doi.org/10.2174/1874285801105010096. Epub 2011 Aug 16. PMID: 21915230; PMCID: PMC3171003.
44. Qadri H, Shah AH, Almilaibary A, Mir MA. Microbiota, natural products, and human health: exploring interactions for therapeutic insights. Front Cell Infect Microbiol. 2024;5(14):1371312. https://doi.org/10.3389/fcimb.2024.1371312. PMID: 39035357; PMCID: PMC11257994.
45. Gasmi A, Shanaida M, Oleshchuk O, Semenova Y, Mujawdiya PK, Ivankiv Y, Pokryshko O, Noor S, Piscopo S, Adamiv S, Bjørklund G. Natural ingredients to improve immunity. Pharmaceuticals (Basel). 2023;16(4):528. https://doi.org/10.3390/ph16040528. PMID: 37111285; PMCID: PMC10143734.

46. Milshteyn A, et al. Accessing Bioactive Natural Products from the Human Microbiome. Cell Host & Microbe 23(6):725–736.
47. Tan SY, Tatsumura Y. Alexander fleming (1881–1955): discoverer of penicillin. Singapore Med J. 2015;56(7):366–7. https://doi.org/10.11622/smedj.2015105. PMID: 26243971; PMCID: PMC4520913.
48. Jones HW. Report of a series of cases of syphilis terated by Ehrlich's arsenobenzol at the Walter Reed General Hospital Distict of Columbia. Boston Med Surg J. 1911;164:381–3. https://doi.org/10.1056/NEJM191103161641103.
49. Christensen SB. Drugs that changed society: history and current status of the early antibiotics: salvarsan, sulfonamides, and β-lactams. Molecules. 2021;26(19):6057. https://doi.org/10.3390/molecules26196057. PMID: 34641601; PMCID:PMC8512414.
50. Otten H. Domagk and the development of the sulphonamides. J Antimicrob Chemother. 1986;17:689–90. https://doi.org/10.1093/jac/17.6.689.
51. Tréfouël J, Tréfouël J, Nitti F, Bovet D. Activité du p-aminophénylsulfamide sur les infections streptococciques expérimentales de la souris et du lapin. C R Soc Biol (Paris). 1935;120:756–8.
52. Venkatesan M, Fruci M, Verellen LA, Skarina T, Mesa N, Flick R, Pham C, Mahadevan R, Stogios PJ, Savchenko A. Molecular mechanism of plasmid-borne resistance to sulfonamide antibiotics. Nat Commun. 2023;14(1):4031. https://doi.org/10.1038/s41467-023-39778-7. PMID: 37419898; PMCID: PMC10328974.
53. Xu Y, Liang X, Li Y, et al. Arylsulfonamides from the roots and rhizomes of tupistra chinensis baker. Nat Prod Commun. 2020;15(4). https://doi.org/10.1177/1934578X20921665.
54. Mujumdar P, Teruya K, Tonissen KF, Vullo D, Supuran CT, Peat TS, Poulsen SA. An unusual natural product primary sulfonamide: synthesis, carbonic anhydrase inhibition, and protein X-ray structures of psammaplin C. J Med Chem. 2016;59(11):5462–70. https://doi.org/10.1021/acs.jmedchem.6b00443. Epub 2016 May 23 PMID: 27172398.
55. Awakawa T, Barra L, Abe I. Biosynthesis of sulfonamide and sulfamate antibiotics in actinomycete. J Ind Microbiol Biotechnol. 2021;48(3–4):kuab001. https://doi.org/10.1093/jimb/kuab001. PMID: 33928358; PMCID: PMC9113183.
56. Milshteyn A, Schneider JS, Brady SF. Mining the metabiome: identifying novel natural products from microbial communities. Chem Biol. 2014;21:1211–23.
57. Brady SF, Simmons L, Kim JH, Schmidt EW. Metagenomic approaches to natural products from free-living and symbiotic organisms. Nat Prod Rep. 2009;26:1488–503.
58. Pantigoso HA, Newberger D, Vivanco JM. The rhizosphere microbiome: plantmicrobial interactions for resource acquisition. J Appl Microbiol. 2022;133(5):2864–76. https://doi.org/10.1111/jam.15686. PMID: 36648151; PMCID: PMC9796772.
59. Zhang JW, Wang R, Liang X, Han P, Zheng YL, Li XF, Gao DZ, Liu M, Hou LJ, Dong HP. Novel gene clusters for natural product synthesis are abundant in the mangrove swamp microbiome. Appl Environ Microbiol. 2023;89(6):e0010223. https://doi.org/10.1128/aem.00102-23. Epub 2023 May 16. PMID: 37191511; PMCID: PMC10304795.
60. Zerikly M, Challis GL. Strategies for the discovery of new natural products by genome mining. ChemBioChem. 2009;10:625–33. https://doi.org/10.1002/cbic.200800389.
61. Sengupta S, Pabbaraja S, Mehta G. Natural products from the human microbiome: an emergent frontier in organic synthesis and drug discovery. Org Biomol Chem. 2024;22(20):4006–30. https://doi.org/10.1039/d4ob00236a. PMID: 38669195.
62. Jiang K, Pang X, Li W, Xu X, Yang Y, Shang C, Gao X. Interbacterial warfare in the human gut: insights from Bacteroidales' perspective. Gut Microbes. (2025);17(1). https://doi.org/10.1080/19490976.2025.2473522.
63. Yuhao W, et al., The gut microbiota reprograms intestinal lipid metabolism through long noncoding RNA Snhg9.Science 2023;381:851–857. https://doi.org/10.1126/science.ade0522.
64. Wang X, Yuan W, Yang C, Wang Z, Zhang J, Xu D, Sun X, Sun W. Emerging role of gut microbiota in autoimmune diseases. Front Immunol. 2024;3(15):1365554. https://doi.org/10.3389/fimmu.2024.1365554. PMID:38765017;PMCID:PMC11099291.

65. Hou K, Wu ZX, Chen XY, et al. Microbiota in health and diseases. Sig Transduct Target Ther. 2022;7:135. https://doi.org/10.1038/s41392-022-00974-4.
66. Rutherford ST, Bassler BL. Bacterial quorum sensing: its role in virulence and possibilities for its control. Cold Spring Harb Perspect Med. 2012;2(11): a012427. https://doi.org/10.1101/cshperspect.a012427. PMID:23125205;PMCID:PMC3543102.
67. Strebhardt K, Ullrich A. Paul Ehrlich's magic bullet concept: 100 years of progress. Nat Rev Cancer. 2008;8:473–80. https://doi.org/10.1038/nrc2394.
68. Lu HD, Spiegel AC, Hurley A, Perez LJ, Maisel K, Ensign LM, Hanes J, Bassler BL, Semmelhack MF, Prud'homme RK. Modulating Vibrio cholerae quorum-sensingcontrolled communication using autoinducer-loaded nanoparticles. Nano Lett. 2015;15(4):2235–41. https://doi.org/10.1021/acs.nanolett.5b00151. Epub 2015 Mar 3. PMID: 25651002; PMCID: PMC4390484.
69. Le KY, Otto M. Quorum-sensing regulation in staphylococci-an overview. Front Microbiol. 2015;6:1174. https://doi.org/10.3389/fmicb.2015.01174.
70. Bivar Xavier K. Bacterial interspecies quorum sensing in the mammalian gut microbiota. C R Biol. 2018;341(5):297–9. https://doi.org/10.1016/j.crvi.2018.03.006. Epub 2018 Apr 7. Erratum in.
71. Sanders JG, Akl H, Hagen SJ, et al. Crosstalk enables mutual activation of coupled quorum sensing pathways through "jump-start" and "push-start" mechanisms. Sci Rep. 2023;13:19230. https://doi.org/10.1038/s41598-02346399-z.
72. Aframian N, Eldar A. A bacterial tower of babel: quorum-sensing signaling diversity and its evolution. Annu Rev Microbiol. 2020;74:587–606. https://doi.org/10.1146/annurev-micro-012220-063740. Epub 2020 Jul 17. PMID: 32680450; PMCID: PMC7611908.
73. Wellington S, Greenberg EP. Quorum sensing signal selectivity and the potential for interspecies cross talk. mBio 2019;10:1–14. https://doi.org/10.1128/mBio.00146-19.
74. Bassler BL, Greenberg EP, Stevens AM. Cross-species induction of luminescence in the quorum-sensing bacterium Vibrio harveyi. J Bacteriol. 1997;179(12):4043–5. https://doi.org/10.1128/jb.179.12.4043-4045.1997. PMID:9190823;PMCID:PMC179216.
75. Majumdar S, Pal S. Cross- species communication in bacterial world. J Cell Commun Signal. 2017;11(2):187–90. https://doi.org/10.1007/s12079-017-0383-9. Epub 2017 Feb 20. PMID: 28220296; PMCID: PMC5440349.
76. Gilbert SF. Inter-kingdom communication and the sympoietic way of life. Front Cell Dev Biol. 2024;12(12):1427798. https://doi.org/10.3389/fcell.2024.1427798. PMID:39071805;PMCID:PMC11275584.
77. Dusane DH, Matkar P, Venugopalan VP, et al. Cross-Species induction of antimicrobial compounds, biosurfactants and quorum-sensing inhibitors in tropical marine epibiotic bacteria by pathogens and biofouling microorganisms. Curr Microbiol. 2011;62:974–80. https://doi.org/10.1007/s00284-010-9812-1.
78. Joint I, Tait K, Wheeler G. Cross-kingdom signalling: exploitation of bacterial quorum sensing molecules by the green seaweed Ulva. Philos Trans R Soc Lond B Biol Sci. 2007;362(1483):1223–33. https://doi.org/10.1098/rstb.2007.2047. PMID:17360272;PMCID:PMC2435585.
79. Coolahan M, Whalen KE. A review of quorum-sensing and its role in mediating interkingdom interactions in the ocean. Commun Biol. 2025;8:179. https://doi.org/10.1038/s42003-025-07608-9.
80. Oldroyd GED. Speak, friend, and enter: signalling systems that promote beneficial symbiotic associations in plants. Nat Rev Microbiol. 2013;11:252–63.
81. Buhian WP, Bensmihen S. Mini-review: nod factor regulation of phytohormone signaling and homeostasis during rhizobia-legume symbiosis. Front. Plant Sci. 2018;9.
82. Sieper T, et al. N-acyl-homoserine lactone uptake and systemic transport in barley rest upon active parts of the plant. New Phytol. 2014;201:545–55.
83. Corral-Lugo A, Daddaoua A, Ortega A, Espinosa-Urgel M, Krell T. Rosmarinic acid is a homoserine lactone mimic produced by plants that activates a bacterial quorum-sensing regulator. Sci. Signal. 2016;9.

84. Koh CL, et al. Plant-derived natural products as sources of anti-quorum sensing compounds. Sensors (Switzerland). 2013;13:6217–28.
85. Vesty EF, Whitbread AL, Needs S, et al. Cross-kingdom signalling regulates spore germination in the moss Physcomitrella patens. Sci Rep. 2020;10:2614. https://doi.org/10.1038/s41598-020-59467-5.
86. Ma Y, Budde MW, Mayalu MN, Zhu J, Lu AC, Murray RM, Elowitz MB. Synthetic mammalian signaling circuits for robust cell population control. Cell. 2022;185(6):967–79.e12. https://doi.org/10.1016/j.cell.2022.01.026. Epub 2022 Mar 1. PMID: 35235768; PMCID: PMC8995209.
87. Chun CK, Ozer EA, Welsh MJ, Zabner J, Greenberg EP. Inactivation of a Pseudomonas aeruginosa quorum-sensing signal by human airway epithelia. Proc Natl Acad Sci USA. 2004;101(10):3587–90. https://doi.org/10.1073/pnas.0308750101. Epub 2004 Feb 17. PMID: 14970327; PMCID: PMC373506.
88. Verbeurgt C, Veithen A, Carlot S, Tarabichi M, Dumont JE, Hassid S, Chatelain P. The human bitter taste receptor T2R38 is broadly tuned for bacterial compounds. PLoS ONE. 2017;12(9): e0181302. https://doi.org/10.1371/journal.pone.0181302. PMID: 28902853; PMCID: PMC5597121.
89. Maurer S, Wabnitz GH, Kahle NA, Stegmaier S, Prior B, Giese T, Gaida MM, Samstag Y, Hänsch GM. Tasting pseudomonas aeruginosa biofilms: human neutrophils express the bitter receptor T2R38 as sensor for the quorum sensing molecule N-(3-Oxododecanoyl)-l-homoserine lactone. Front Immunol. 2015;24(6):369. https://doi.org/10.3389/fimmu.2015.00369. PMID: 26257736; PMCID: PMC4513437.
90. Blandizzi AL, Pacher C, Guilliams P, Haskó MG. Rethinking communication in the immune system: the quorum sensing concept. Trends Immunol. 2019;40(2):88–97. https://doi.org/10.1016/j.it.2018.12.002. Epub 2019 Jan 2. PMID: 30611647; PMCID: PMC6619492.
91. Hernandez DE, Sintim HO. Quorum sensing autoinducer-3 finally yields to structural elucidation. ACS Cent Sci. 2020;6(2):93–96. https://doi.org/10.1021/acscentsci.0c00033. Epub 2020 Feb 12. PMID: 32123727; PMCID: PMC7047277.
92. Saha S, Bulzu P A, Urajová P, Mareš J, Konert G, Câmara Manoel J, Macho M, Ewe D, Hrouzek P, Masojídek J, Ghai R, Saurav K. Quorum-Sensing signals from epibiont mediate the induction of novel microviridins in the mat-forming cyanobacterial genus nostoc. mSphere. 20210;6(4):e0056221. https://doi.org/10.1128/mSphere.00562-21. Epub 2021 Jul 14. PMID: 34259556; PMCID: PMC8386392.
93. Harding TS. Science at the tower of babel. Philosophy of Science. 1938;5(3): 338353. https://doi.org/10.1086/286514.
94. Pleyer M, Hartmann S. Constructing a consensus on language evolution? convergences and differences between biolinguistic and usage-based approaches. Front Psychol. 2019;14(10):2537. https://doi.org/10.3389/fpsyg.2019.02537. PMID: 31803099; PMCID: PMC6868443.
95. Pennock RT. Tower of babel: the evidence against the new creationism ISBN 0262-16180-X, ISBN 0-262-66165-9.
96. Gonzales M, Jacquet P, Gaucher F, Chabrière É, Plener L, Daudé D. AHL-based quorum sensing regulates the biosynthesis of a variety of bioactive molecules in bacteria. J Nat Prod. 2024;87(4):1268–84. https://doi.org/10.1021/acs.jnatprod.3c00672. Epub 2024 Feb 23 PMID: 38390739.
97. Fuqua C, Greenberg E. Listening in on bacteria: acyl-homoserine lactone signalling. Nat Rev Mol Cell Biol. 2002;3:685–95. https://doi.org/10.1038/nrm907.
98. Bhatt VS. Quorum sensing mechanisms in gram positive bacteria. In: Bramhachari PV, editor, Implication of quorum sensing system in biofilm formation and virulence. Singapore: Springer;2018. https://doi.org/10.1007/978-981-13-2429-1_20.
99. Grandclément C, Tannières M, Moréra S, Dessaux Y, Faure D. Quorum quenching: role in nature and applied developments. FEMS Microbiol Rev. 2016;40(1):86–116. https://doi.org/10.1093/femsre/fuv038.

100. Falà AK, Álvarez-Ordóñez A, Filloux A, Gahan CGM, Cotter PD. Quorum sensing in human gut and food microbiomes: significance and potential for therapeutic targeting. Front Microbiol. 2022;25(13):1002185. https://doi.org/10.3389/fmicb.2022.1002185. PMID: 36504831; PMCID: PMC9733432.
101. Mackness M, Mackness B. Human paraoxonase-1 (PON1): gene structure and expression, promiscuous activities and multiple physiological roles. Gene. 2015;567(1):12–21. https://doi.org/10.1016/j.gene.2015.04.088. Epub 2015 May 9. PMID: 25965560; PMCID: PMC4458450.
102. Draganov DI, Teiber JF, Speelman A, Osawa Y, Sunahara R, La Du BN. Human paraoxonases (PON1, PON2, and PON3) are lactonases with overlapping and distinct substrate specificities. J Lipid Res. 2005;46(6):1239–47. https://doi.org/10.1194/jlr.M400511-JLR200. Epub 2005 Mar 16 PMID: 15772423.
103. Lou-Bonafonte JM, Gabás-Rivera C, Navarro MA, Osada J. The search for dietary supplements to elevate or activate circulating paraoxonases. Int J Mol Sci. 2017;18(2):416. https://doi.org/10.3390/ijms18020416. PMID: 28212288; PMCID: PMC5343950.
104. Petrič B, Kunej T, Bavec A. A multi-omics analysis of PON1 lactonase activity in relation to human health and disease. OMICS. 2021;25(1):38–51. https://doi.org/10.1089/omi.2020.0160. Epub 2020 Dec 11. PMID: 33306925; PMCID: PMC8045895.
105. Dong YH, Gusti AR, Zhang Q, et al. Identification of quorum-quenching Nacyl homoserine lactonases from Bacillus species. Appl Environ Microb. 2002;68:17549.
106. Hiblot J, Bzdrenga J, Champion C, et al. Crystal structure of VmoLac, a tentative quorum quenching lactonase from the extremophilic crenarchaeon Vulcanisaeta moutnovskia. Sci Rep. 2015;5:8372. https://doi.org/10.1038/srep08372.
107. Rasmussen TB, Skindersoe ME, Bjarnsholt T, et al. Identity and effects of quorum-sensing inhibitors produced by Penicillium species. Microbiology-SGM. 2005;151: 132540.
108. Muh U, Schuster M, Heim R, et al. Novel Pseudomonas aeruginosa quorumsensing inhibitors identified in an ultra-high-throughput screen. Antimicrob Agents Ch. 2006;50:3674–9.
109. Naga NG, Shaaban MI. Quorum sensing and quorum sensing inhibitors of natural origin. In: Cruz JN. editors, Drug discovery and design using natural products. Cham: Springer;2023. https://doi.org/10.1007/978-3-031-352058_13.
110. Escobar-Muciño E, Arenas-Hernández MMP, Luna-Guevara ML. Mechanisms of inhibition of quorum sensing as an alternative for the control of E. coli and Salmonella. Microorganisms. 2022;10(5):884. https://doi.org/10.3390/microorganisms10050884. PMID: 35630329; PMCID: PMC9143355.
111. Skindersoe ME, Alhede M, Phipps R, Yang L, Jensen PO, Rasmussen TB, Bjarnsholt T, Tolker-Nielsen T, Høiby N, Givskov M. Effects of antibiotics on quorum sensing in Pseudomonas aeruginosa. Antimicrob Agents Chemother. 2008;52(10):3648–63. https://doi.org/10.1128/AAC.01230-07. Epub 2008 Jul 21. PMID: 18644954; PMCID: PMC2565867.
112. Maillé É, Ruffin M, Adam D, Messaoud H, Lafayette SL, McKay G, Nguyen D, Brochiero E. Quorum sensing down-regulation counteracts the negative impact of pseudomonas aeruginosa on CFTR channel expression, function and rescue in human airway epithelial cells. Front Cell Infect Microbiol. 2017;10(7):470. https://doi.org/10.3389/fcimb.2017.00470. PMID: 29177135; PMCID: PMC5686086.
113. Vadakkan K, Choudhury AA, Gunasekaran R, Hemapriya J, Vijayanand S. Quorum sensing intervened bacterial signaling: pursuit of its cognizance and repression. J Genet Eng Biotechnol. 2018;16(2):239–52. https://doi.org/10.1016/j.jgeb.2018.07.001. Epub 2018 Jul 7. PMID: 30733731; PMCID: PMC6353778.
114. Rogers LA. The inhibiting effect of streptococcus lactis on Lactobacillus bulgaricus. J Bacteriol. 1928;16:321–5.
115. McAuliffe O, Ross RP, Hill C. Lantibiotics: structure, biosynthesis and mode of action. FEMS Microbiol Rev. 2001;25(3):285–308. https://doi.org/10.1111/j.1574-6976.2001.tb00579.x.
116. Mokoena MP. Lactic acid bacteria and their bacteriocins: classification, biosynthesis and applications against uropathogens: a mini-review. Molecules (Basel, Switzerland). 2017;22(8):1255. https://doi.org/10.3390/molecules22081255.

117. Darbandi A, Asadi A, Mahdizade Ari M, Ohadi E, Talebi M, Halaj Zadeh M, Darb Emamie A, Ghanavati R, Kakanj M. Bacteriocins: properties and potential use as antimicrobials. J Clin Lab Anal. 2022;36(1):e24093. https://doi.org/10.1002/jcla.24093. Epub 2021 Dec 1. PMID: 34851542; PMCID: PMC8761470.
118. Perez RH, Zendo T, Sonomoto K. Novel bacteriocins from lactic acid bacteria (LAB): various structures and applications. Microb Cell Fact. 2014;13(Suppl 1):S3.
119. Koulenti D, Song A, Ellingboe A, Abdul-Aziz MH, Harris P, Gavey E, et al. Infections by multidrug-resistant Gram-negative Bacteria: what's new in our arsenal and what's in the pipeline? Int J Antimicrob Agents. 2019;53:211–24. https://doi.org/10.1016/j.ijantimicag.2018.10.011.
120. Antoshina DV, Balandin SV, Ovchinnikova TV. Structural features, mechanisms of action, and prospects for practical application of class II Bacteriocins. Biochemistry (Mosc). 2022;87(11):1387–403. https://doi.org/10.1134/S0006297922110165. PMID: 36509729.
121. Dewachter L, Dénéréaz J, Liu X, de Bakker V, Costa C, Baldry M, Sirard JC, Veening JW. Amoxicillin-resistant Streptococcus pneumoniae can be resensitized by targeting the mevalonate pathway as indicated by sCRilecs-seq. Elife. 2022;24(11): e75607. https://doi.org/10.7554/eLife.75607. PMID:35748540;PMCID:PMC9363119.
122. Benítez-Chao DF, León-Buitimea A, Lerma-Escalera JA, Morones-Ramírez JR. Bacteriocins: an overview of antimicrobial, toxicity, and biosafety assessment *by in vivo* models. Front Microbiol. 2021;15(12): 630695. https://doi.org/10.3389/fmicb.2021.630695. PMID:33935991;PMCID:PMC8083986.
123. Bierbaum G, Sahl HG. Lantibiotics: mode of action, biosynthesis and bioengineering. Curr Pharm Biotechnol. 2009;10(1):2–18. https://doi.org/10.2174/138920109787048616. PMID: 19149587.
124. Inglis RF, Scanlan P, Buckling A. Iron availability shapes the evolution of bacteriocin resistance in pseudomonas aeruginosa. ISME J. 2016;10:2060–6. https://doi.org/10.1038/ismej.2016.15.
125. Jarvis B. Resistance to nisin and production of nisin-inactivating enzymes by several Bacillus species. J Gen Microbiol. 1967;47:33–48. https://doi.org/10.1099/00221287-47-1-33.
126. Sugrue I, Ross RP, Hill C. Bacteriocin diversity, function, discovery and application as antimicrobials. Nat Rev Microbiol. 2024;22:556–71. https://doi.org/10.1038/s41579-024-010 45-x.
127. Khosa S, Lagedroste M, Smits SH. Protein defense systems against the lantibiotic nisin: function of the immunity protein nisi and the resistance protein NSR. Front Microbiol. 2016;12(7):504. https://doi.org/10.3389/fmicb.2016.00504. PMID: 27148193; PMCID: PMC4828448.
128. Chassagne F, Samarakoon T, Porras G, Lyles JT, Dettweiler M, Marquez L, Salam AM, Shabih S, Farrokhi DR, Quave CL. A systematic review of plants with antibacterial activities: a taxonomic and phylogenetic perspective. Front Pharmacol. 2021;8(11): 586548. https://doi.org/10.3389/fphar.2020.586548. PMID: 33488385; PMCID: PMC7821031.
129. Khameneh B, Iranshahy M, Soheili V, et al. Review on plant antimicrobials: a mechanistic viewpoint. Antimicrob Resist Infect Control. 2019;8:118. https://doi.org/10.1186/s13756-019-0559-6.
130. Álvarez-Martínez FJ, Barrajón-Catalán E, Herranz-López M, Micol V. Antibacterial plant compounds, extracts and essential oils: an updated review on their effects and putative mechanisms of action. Phytomedicine. 2021;90: 153626. https://doi.org/10.1016/j.phymed.2021.153626. Epub 2021 Jul 9 PMID: 34301463.
131. Sundin GW, Wang N. Antibiotic resistance in plant-pathogenic bacteria. Annu Rev Phytopathol. 2018;25(56):161–80. https://doi.org/10.1146/annurev-phyto080417-045946. Epub 2018 Jun 1 PMID: 29856934.
132. Abraham EP, Chain E. An enzyme from bacteria able to destroy penicillin. Nature. 1940;146:837. https://doi.org/10.1038/146837a0.
133. Khameneh B, Iranshahy M, Soheili V, Fazly Bazzaz BS. Review on plant antimicrobials: a mechanistic viewpoint. Antimicrob Resist Infect Control. 2019;16(8):118. https://doi.org/10.1186/s13756-019-0559-6. PMID: 31346459; PMCID: PMC6636059.

134. Zipfel PF, Skerka C. From magic bullets to modern therapeutics: Paul Ehrlich, the German immunobiologist and physician coined the term 'complement.' Mol Immunol. 2022;150:90–8. https://doi.org/10.1016/j.molimm.2022.08.002. Epub 2022 Aug 23 PMID: 36027818.
135. Dassanayake MK, Khoo TJ, An J. Antibiotic resistance modifying ability of phytoextracts in anthrax biological agent Bacillus anthracis and emerging superbugs: a review of synergistic mechanisms. Ann Clin Microbiol Antimicrob. 2021;20(1):79. https://doi.org/10.1186/s12941-021-00485-0. PMID: 34856999; PMCID: PMC8641154.
136. Radulović NS, Blagojević PD, Stojanović-Radić ZZ, Stojanović NM. Antimicrobial plant metabolites: structural diversity and mechanism of action. Curr Med Chem. 2013;20(7):932–52 PMID: 23210781.
137. Reiter J, Levina N, van der Linden M, Gruhlke M, Martin C, Slusarenko AJ. Diallylthiosulfinate (Allicin), a volatile antimicrobial from garlic (allium sativum), kills human lung pathogenic bacteria, including MDR strains, as a Vapor. Molecules. 2017;22(10):1711. https://doi.org/10.3390/molecules22101711. PMID:29023413;PMCID:PMC6151386.
138. Matile PH. The mustard oil bomb: compartmentation of the myrosinase system. Biochem Physiol Pflanz. 1980;175(8–9):722–31.
139. Yoshimoto N, Saito K. S-Alk(en)ylcysteine sulfoxides in the genus Allium: proposed biosynthesis, chemical conversion, and bioactivities. J Exp Bot. 2019;70(16):4123–37. https://doi.org/10.1093/jxb/erz243.
140. Commission Implementing Regulation (EU) 2024/2878 of 8 November 2024 approving the basic substance Allium fistulosum, processed, in accordance with Regulation (EC) No 1107/2009 of the European Parliament and of the Council, and amending Commission Implementing Regulation (EU) No 540/2011.
141. Magryś A, Olender A, Tchórzewska D. Antibacterial properties of Allium sativum L. against the most emerging multidrug-resistant bacteria and its synergy with antibiotics. Arch Microbiol. 2021;203(5):2257–68. https://doi.org/10.1007/s00203021-02248-z. Epub 2021 Feb 27. PMID: 33638666; PMCID: PMC8205873.
142. Cho H, Park JY, Kim DH, Han J, Natesan K, Choi MS, Lee SK, Kim J, Cho K, Ahn BO. Understanding the defense mechanism of Allium plants through the onion isoallicinomics study. Front Plant Sci. 2024;11(15):1488553. https://doi.org/10.3389/fpls.2024.1488553. PMID: 39722878; PMCID: PMC11668612.
143. Imai S, Tsuge N, Tomotake M, Nagatome Y, Sawada H, Nagata T, Kumagai H. Plant biochemistry: an onion enzyme that makes the eyes water. Nature. 2002;419(6908):685. https://doi.org/10.1038/419685a. PMID: 12384686.
144. Xu Z, Zhang H, Yu H, Dai Q, Xiong J, Sheng H, Qiu J, Jiang L, Peng J, He X, Xin R, Li D, Zhang K. Allicin inhibits Pseudomonas aeruginosa virulence by suppressing the rhl and pqs quorum-sensing systems. Can J Microbiol. 2019;65(8):563–74. https://doi.org/10.1139/cjm-2019-0055. Epub 2019 Apr 22 PMID: 31009577.
145. Jakobsen TH, Warming AN, Vejborg RM, et al. A broad range quorum sensing inhibitor working through sRNA inhibition. Sci Rep. 2017;7:9857. https://doi.org/10.1038/s41598-017-09886-8.
146. Yazdani M, Beihaghi M, Ataee N, Zabetian M, Khaksar S, Nasrizadeh H, Chaboksavar M. Anti-quorum sensing effects of SidA protein on Escherichia coli receptors: in silico analysis. J Biomol Struct Dyn. 2024;1–12. https://doi.org/10.1080/07391102.2024.2322632.
147. Jin Z, Zhao S, Li H, Ouyang Q, Tao N. Identification and validation of garlic (allium sativum) metabolites as quorum sensing inhibitors of bacillus cereus targeting the PlcR receptor: an in silico and in vitro study. Foodborne Pathog Dis. 2024. https://doi.org/10.1089/fpd.2024.0098. Epub ahead of print. PMID: 39435711.
148. Li N, Zhang J, Yu F, Ye F, Tan W, Hao L, Hu X. Garlic-Derived quorum sensing inhibitors: a novel strategy against fungal resistance. " drug design. Development and Therapy. 2024;18:6413–26. https://doi.org/10.2147/DDDT.S503302.
149. Khandagale K, Roylawar P, Randive P, et al. Isolation and expression profiling of insecticidal lectins from wild alliums against onion thrips (thrips tabaci lindeman). Proc Natl Acad Sci India, Sect B Biol Sci. 2022;92:451–459. https://doi.org/10.1007/s40011-022-01360-4.

150. Wan Q, Li N, Du L, Zhao R, Yi M, Xu Q, Zhou Y. Allium vegetable consumption and health: an umbrella review of meta-analyses of multiple health outcomes. Food Sci Nutr. 2019;7(8):2451–70. https://doi.org/10.1002/fsn3.1117. PMID: 31428334; PMCID: PMC6694434.
151. Kothari D, Lee WD, Kim SK. Allium flavonols: health benefits, molecular targets, and bioavailability. Antioxidants (Basel). 2020;9(9):888. https://doi.org/10.3390/antiox9090888. PMID: 32961762; PMCID: PMC7555649.
152. Shang A, Cao SY, Xu XY, Gan RY, Tang GY, Corke H, Mavumengwana V, Li HB. Bioactive compounds and biological functions of garlic (Allium sativum L.). Foods. 2019;8(7):246. https://doi.org/10.3390/foods8070246. PMID: 31284512; PMCID: PMC6678835.
153. Rendu F, Daveloose D, Debouzy JC, Bourdeau N, Levy-Toledano S, Jain MK, Apitz-Castro R. Ajoene, the antiplatelet compound derived from garlic, specifically inhibits platelet release reaction by affecting the plasma membrane internal microviscosity. Biochem Pharmacol. 1989;38(8):1321–8. https://doi.org/10.1016/0006-2952(89)90339-0. PMID: 2539823.
154. Ferri N, Yokoyama K, Sadilek M, Paoletti R, Apitz-Castro R, Gelb MH, Corsini A. Ajoene, a garlic compound, inhibits protein prenylation and arterial smooth muscle cell proliferation. Br J Pharmacol. 2003;138(5):811–8. https://doi.org/10.1038/sj.bjp.0705126. PMID: 12642382; PMCID: PMC1573737.

Chapter 10
Processing, Varieties, Natural Products and Food Systems

Contents

10.1 Recipe. Stovies . . . 290
10.2 Production of Secondary Natural Products Resulting from Biotic and Abiotic Stress . . . 292
10.3 Post-Harvest Production of Secondary Natural Products . . . 293
10.4 Processing Parameters—pH, Heat, Ionic Strength, Pressure—Effect on Secondary Natural Products . . . 297
10.5 Microbial Processing (Fermentation) and Secondary Natural Products . . . 302
10.6 Processing and Storage—Effects on Secondary Natural Products . . . 304
10.7 Ultra Processed Food (UFP)—Effect on Secondary Natural Products. A Novel Processing Index . . . 305
References . . . 307

Abstract The Hanseatic League was an important trading network during the twelfth-seventeenth centuries. In addition to trading goods and maintaining standards, the League also helped to spread ideas and concepts—in particular in northern Europe during the Enlightenment. The League was in part protectionist and countered the great empires of the time (Spain, France and England) and was, therefore a barrier against economic stress. In the case of organisms, they also have to deal with biotic and abiotic stress and have developed various responses. One of these is the production of heat shock proteins (HSPs) while another is the upregulation of pathways leading to the production of secondary natural products. Stress responses are interlinked and have significant consequences. They are also linked to post-harvest events including processing and storage. In addition, some secondary natural products are inhibitors of autolytic enzymes and can, therefore, play a role in the maintenance of food structural integrity. The role of antioxidants both in maintaining foods and in subsequent health benefits for consumers has been extensively studied however results have been inconsistent and methodologies may not be sufficiently sensitive to detect minor perturbations (positive and negative). The role of ultraprocessed food (UPF) in health and disease has been widely discussed however the most popular system (NOVA) is contentious. The possibility of using a system based on the metabolism of some secondary natural products as a multivariate indicator of the level of processing could be pursued.

B. Hanley, *Secondary Natural Products in Foods and Food Systems*,
https://doi.org/10.1007/978-3-032-00292-1_10

Keywords Hanseatic league · Biotic and abiotic stress · Processing · HSPs · UPF

10.1 Recipe. Stovies

The use of vegetables to reduce the amount of meat required to make a dish has a substantial history and has been adopted by many cultures, globally. Examples only from Europe include variations on the theme of hotpot, stoemp, Irish stew, cassoulet, garbure, goulash, hasenpfeffer and bigos to name but a few. They generally comprise a small amount of meat, fat of some sort, vegetables (often potatoes) and stock. Long cooking periods is also a feature of this type of cooking. It is, in essence, a slow, long way of processing meat and vegetables. It is often seasoned with herbs that tend to contain flavour compounds that are sufficiently heat stable and involatile to survive lengthy heating and do not evaporate. It tends to be a winter dish and uses older vegetables. Starch is the main carbohydrate present in potatoes (*Solanum tuberosum*) and many other root vegetables. One of the advantages of this type of vegetable is that the starch functional qualities are mostly maintained on storage, although the molecular structure of starch may change in time (see below) [1] however there are a number of factors that may affect the initial starch quality including agronomic conditions and processing. In addition, the amylose content of starch decreases on storage [2] and sprouting and cold-induced sweetening can affect potato quality if they are stored at low temperatures for prolonged periods and the use of amylase inhibitors is recommended to prevent the hydrolysis of starch [3]. One of the key determinants of potato quality, indeed is the maintenance of starch content. One of the effects of sprouting in potatoes is the conversion of starch to sugar—reduction in the starch and an increase in the sucrose, fructose, and glucose contents of the tubers—and this results in a loss of weight and firmness of the stored tubers [4]. For that reason, a number of sprout supressing treatments including ethylene [5] and other treatments [6] are used. In some cases, sprouting is a necessary pre-requisite to planting and therefore is encouraged by treatments that increase phytohormone production [7] and encourage growth. In addition to macronutrients and vitamins, potatoes also contain a number of secondary natural products including toxic glycoalkaloids [8], anthocyanins [9] and phenolic compounds [10]. Indeed, varietal choice of potatoes and correlation of this with beneficial phytochemicals (anthocyanins and phenolics) may be a reason why genetic modification of tomatoes to produce fruit containing anthocyanins may not ultimately be successful as a health and nutrition strategy [11].

The Hanseatic League was an important trading alliance during the 12th to the seventeenth centuries. The League, which started in what is now Germany, was a confederation of independent merchant guilds (Hansa), each with its own set of regulations and rules. They recognised the benefits in working together through common principles. They also formed alliances for defence against larger countries and this quasi-autonomous structure continued to be influential over several hundreds of years [12]. The Hanseatic League was more of a protectionist concept than an EU-style free trade area however towns such as Bruges and Leiden with specialist trade opportunities benefited from the quality standards that were applied by the League. After the break up of the Hanseatic League due to changing trading conditions—particularly to the Americas and the Far East—it continued to have influence on the intellectual and academic links between the members of the League for a period after the League as a trading entity ceased to exist. For example Hermann Boerhaave (1668–1738) was professor of chemistry at Leiden and was a considerable influence on the development of chemistry in Europe in general and in Edinburgh (a member of the Hanseatic League) in particular. Many chemists and physicians who set up the Edinburgh Medical School (the oldest Medical School in the UK) studied with Boerhaave at Leiden. Another legacy of the Hanseatic League was the spread of particular cooking practices including the development of meat and vegetable stews throughout northern Europe. The influence of this—and practices associated with the storage of food for long sea voyages (salting, pickling and smoking—all of which will be considered later) continue alongside the use of herbs and spices. The link between the Hanseatic League and Scotland suggests the Scottish version of potato and vegetable stew—stovies—is a good place to begin an exploration of food processing and the effect of secondary natural products. In addition, the cooking of the Hanseatic towns of Northern Germany (particularly Bremen, Hamburg and Lubeck) includes dishes like Bremer Labskaus, a mixture of meat, potatoes and onions that is, in a culinary sense, close to stovies. While some aspects of food processing were already mentioned in Chap. 8, the current chapter will extend and complement the earlier discussion and will have a different emphasis.

Stovies—The Hanseatic League of Cooking

Stovies are a simple way of cooking leftover meat or, in this case, sausages. Start with a knob of meat fat (or lard) or (if you prefer) some oil (two tablespoons or so)in a skillet or heavy pan. Add a chopped onion and a chopped leek. Although both onion and leek are members of the Allium family, they have different alk(en)yl cysteine sulphoxides and produce different flavour compounds after hydrolysis by the alliinase enzyme that is released after chopping [13]. **Fry the onion and leek for 5–10 min until they are translucent. This also inactivates the alliinase enzyme and, while you may get some higher homologues of S compounds formed from initially released hydrolysis products, fresh hydrolysis does not take place. Some of the volatiles will evaporate so fry gently so as to retain some of the low molecular, volatile compounds but the higher molecular weight compounds will remain anyway, dissolved in the fat. The distribution of flavour compounds will be affected by the frying temperature. Split open the sausages** [3] **and form the meat into balls and fry with the onion/leek mixture. Add sliced or diced potatoes and other root vegetables (carrot, parsnip etc.) cut to the same size as the potatoes. Add stock to just cover the vegetables. Simmer for 45 min and check that the potatoes are cooked. Add a little more stock as required and cook for another 15 min if necessary. Add freshly ground pepper. Serve in a shallow bowl and add salt once served. The salt is purely there for flavouring so by coating it over the surface, you use less salt than if you add it**

earlier in the cooking process. If you want to crisp the potatoes, you can put the skillet under a grill and brown carefully.

10.2 Production of Secondary Natural Products Resulting from Biotic and Abiotic Stress

The level of production of secondary natural products can be a consequence of stress. Biotic and abiotic stress can result in the production of a range of stress-related responsive elements whose function is to enable the system to better respond to the stress. Abiotic stress is that which is introduced by environmental factors (temperature, pH, radiation etc.) while biotic stress is introduced by a biological agent (disease, biologically-derived toxic challenge, damage by competing organisms or by being the target of predation or consumption). Clearly processing can result in abiotic stress however some techniques (for example, fermentation) may cause biotic stress. One of the most significant (and common) mechanisms for dealing with stress is the production of heat shock proteins (HSPs). These proteins are produced in response to a number of stress signals. The first examples were those in which the stressing agent was heat and hence the responsive elements were referred to as HSPs. One of the consequences of heat (or other stress events) is the denaturation and/or misfolding of proteins and a primary function of HSPs was the refolding or even the de novo folding of proteins as they are produced [14]. In addition to abiotic stress such as heat, it has also been found that some naturally occurring compounds can potentiate a heat shock response. For example, glutamine is known to increase expression of HSP 70 (the '70' refers to the molecular weight of the HSP) [15]. Glutamine is also implicated in the expression of other HSPs and the induction of a stress response to disease and injury [16]. A number of other amino acids have been tested in vitro and some of these have been found to induce an HSP response. One of the key initiators of HSP response in eukaryotes is the production of Heat Shock Factor (HSF1). The role of HSFs is to bind to transcription sequences for HSPs and to promote their expression [17]. Amino acid deprivation has been found to reduce the expression of HSF1 and may, therefore, reduce the protective stress response. A number of other biotic factors have been found to increase HSP production including some bioactive peptides, dietary fibre and certain probiotics (or something that they produce). Some secondary natural products have also been found to influence HSP expression [18]. These include phenolics such as resveratrol [19]. In addition, some S-containing natural products (compounds from the genus *Allium* and isothiocyanates from *Brassica* species have been suggested to operate by activating HSF1 [20] and upregulating HSP 70 production.

In addition to the induction of an HSP protective response, secondary natural products can be produced at elevated levels in response to biotic or abiotic stress and this may induce the initiation of a protective cascade. Many of these appear to have, in addition to an effect on protective mechanisms such as the production of HSPs, an antiinflammatory action and may be a source of novel pharmacological agents [21]. In the case of plants, unlike mammals they cannot move to alleviate stress caused by inducers so their response has to be triggered by a range of factors and be effective against a wide number of potential hazards [22]. Many of these responses are mediated by transcription factors. In plants, a range of inducible transcription factors have been found whose expression can mediate a number of responses. Modulation of plant protective factors such as secondary natural product accumulation, may be a mechanism to prevent damage (feeding deterrence, anti-microbial activity etc.), they may have a regulatory function or may be a shunt pathway for the storage of complex molecular structures that can be catabolised as and when required for primary metabolite production. These tri-functional properties have been referred to as a 'functional trichotomy' [23]. The concept that secondary natural products just have a single function is outdated and overly simplistic however attempting to verify polyfunctionality in the context of different properties of organisms may lead to false assumptions and parallels being drawn which are reductionist rather than accurate. It remains unclear if the observed apparent functional diversity is a planned evolutionary outcome with specific and distinct advantages or is a random consequence of changes to primary protective (and evolutionarily advantageous) function as appears to be the case for many of the unintentional pharmaceutical uses of some medicinal natural products. The existence of a particular set of compounds with biological activities may not be anything more than serendipitous. The functional classification of secondary natural products can tempt natural product chemists (and others including plant physiologists) to try to identify patterns of use leading to polyfunctional evolutionary advantage. This may be a consequence of the combination of reductionist chemistry being combined with the more holistic approach of systems biology. This synthesis of different traditions may produce insights but it may also provide patterns that have a false coherence. Secondary natural products are produced at elevated levels due to biotic and abiotic stress and, in some cases, these products are physiologically beneficial for the producing organism however the understanding of how secondary natural products, stress response, primary metabolism and other effects are linked remains fragmentary and partial.

10.3 Post-Harvest Production of Secondary Natural Products

The study of the post-harvest production of secondary natural products should take into account the pre-processing steps (including harvesting and storage) but it should exclude the actual processing parameters that will be examined in more detail in Sect. 10.4. The post-harvest process includes the cessation of growing, the separation of different parts of the organism and the pre-processing storage of the harvested material. Many post-harvest processes are enzymatic and lead to deterioration in the quality of the harvested material [24] and in some cases these can lead to the transformation of natural products following autolysis and the production of a range of secondary natural products including isothiocyanates, thiols, cyanogenic aglycones and others. Some of these will be discussed as illustrative examples.

Secondary natural products can be defined either as intrinsic or extrinsic. The intrinsic substances are those which derive from the harvested material (as noted above) while the extrinsic compounds are those which arise from sources other than the harvested material. Many of the extrinsic substances arise from microbial (bacterial, fungal) contamination and the production of mycotoxins and other bioactive entities. Some of the intrinsic compounds have anti-microbial properties and prevent microbial growth or spoilage [25]. Post harvest senescence is a complex process involving interlinked cascades of reactions [26] some of which can be seen to provide a balance between anti-microbial activity and oxidation [27] with intrinsic reactive oxygen species being quenched by endogenous antioxidants. One of the most significant classes of intrinsic natural products are the range of phenolic (and related) antioxidants [28]. In some cases these are added while in others they arise from the harvested product itself [29]. In a wider sense, the preservation of antioxidant levels improves the perceived healthpromoting properties of some products [30]. Harvested products remain, at least for a period of time, metabolically active. The degradation of living material can cause the break down of cell walls and of cellular structures and lead to loss of product quality and browning of the product [31] through the action of polyphenol oxidases and peroxidases. Attempts to minimise the consequences of oxidation are an important aspect of post-harvest management [32]. Some secondary natural products are undesirable from a health, processing or quality perspective. The reduction of the levels of cyanogenic glycosides in cropped plant material has been the subject of considerable interest [33] and the manipulation of metabolic cycling pathways may be of value [34]. Breeding and strain selection has been used to develop an understanding of biosynthetic pathways and genomics that may, for example, be able to contribute to the production of Solanaceae species with low glycoalkaloid content [35] and 'double low' (low erucic acid and low glucosinolate) rapeseed [36]. On a molecular level, production of basic helix-loop-helix (bHLH) proteins, which are known transcription factors, has been observed and found to correlate with the reduction of post-harvest deterioration and production of cyanogenic glucosides [37]. Molecular level understanding will increase the throughput of identification of varieties with desirable traits and this will facilitate the manipulation of biosynthetic pathways as a promising approach to preserving post-harvest quality. Gene editing technology shows promise as a 'minimally invasive' technique [38]. An alternative approach is to reduce the levels of specific, less desirable secondary natural products (as was instanced with the examples of glycoalkaloids, erucic acid and glucosinolates previously) by strain development or by genetic manipulation [39]. In this case a natural mutant with a particular characteristic was identified. A more deterministic approach is to increase the rate of removal of a specific compound by accelerating its catabolism *in* vivo. In the case of caffeine, the rate limiting step in the catabolic degradation is the first demethylation of the methyl xanthine molecule (Fig. 10.1).

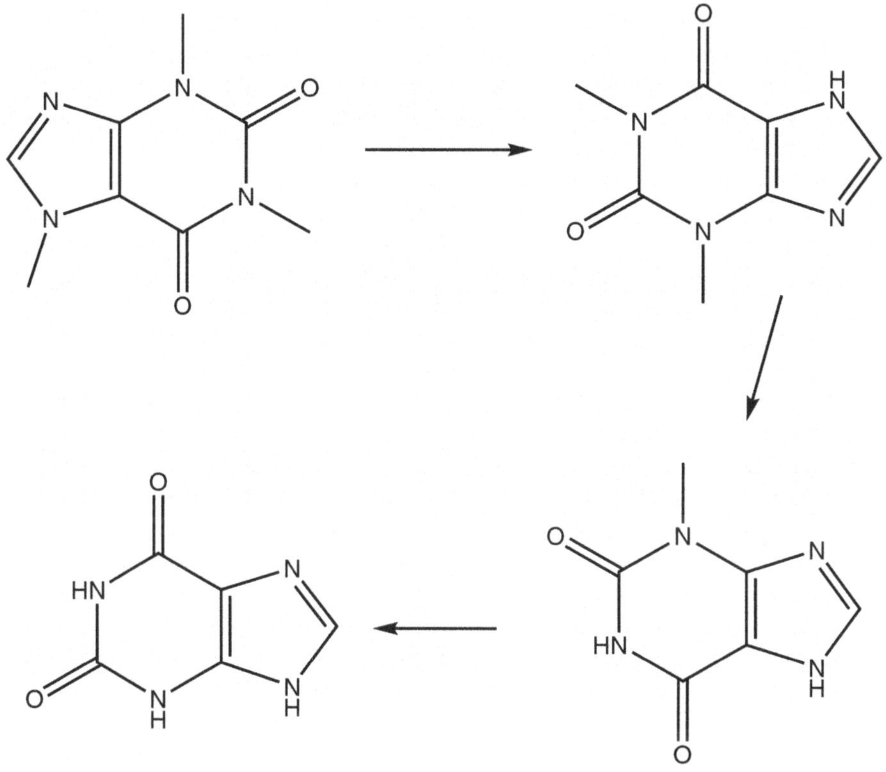

Fig. 10.1 Catabolism of caffeine

One strategy for the post-harvest manipulation of caffeine levels is the incorporation of bacterial or fungal genes to convert caffeine via theophylline and theobromine to xanthine [40] (which then enters the purine degradation pathway) [41]. In the cases of sulphides from *Allium* species and isothiocyanates (and other post hydrolysis bioactive constituents) from *Brassica* species, the formation of the bioactive secondary natural products is dependent upon the degree of tissue disruption that occurs during harvesting and post-harvest storage. In both cases the (relatively) inactive precursors are stored separately from the enzyme that hydrolyses them to give the biologically active products (although, note that a cytosolic alliinase enzyme has recently been found and therefore tissue disruption is not obligatory for production of some sulphides). The effect of pre and post-harvest conditions on the production of biologically active compounds in garlic has been reviewed [42]. Much of the interest in the bioactive components of vegetables have centred upon medicinal and pharmacological uses however a number of components have also been found to protect the post-harvest stored product from later deterioration [43] and if these can be combined with a beneficial health effect then there may be polyfunctional opportunities for crop manipulation.

Extrinsic secondary natural products are also present (or in some cases synthesised de novo) in post-harvest crops. In many cases, these are bioactive, and often toxic, entities produced by microbial species. The most significant of these are the group of mycotoxins. Their origin, production and elimination have been extensively studied [44]. In addition, mycotoxins have been the subject of extensive biosynthetic studies by natural product chemists and others. In general conditions that favour the growth of microbial species (high humidity, warmth) will lead to the production of mycotoxins [45]. One area of particular interest is that of the actual or potential interactions between different mycotoxins [46] or between mycotoxins and intrinsic secondary natural products. An example of this latter effect could be the cumulative effect of phytoestrogens and mycoestrogens [47]. Since mycotoxins are extrinsic natural products, they are produced in actively growing microbial species. This means that an understanding of those factors that impact upon their biosynthesis and the flux of metabolites can help in the design of post-harvest conditions and the development of treatments (chemical inhibitors) that may lessen the toxic load that is consumed [48]. It is not simply the production of different mycotoxins that is of concern. Although estrogens were mentioned previously, the combination of mycotoxins and intrinsic natural products has been relatively little studied in terms of synergistic effects [49] although bioassay-based investigations (including metabolic effects) would appear to be potentially a fruitful area for investigation [50]. Among the effects noted for mycotoxins are impacts on the immune system [51] and this may also affect susceptibility to other toxic entities including viruses and bacteria [52]. In addition, some natural products have been reported to have immunomodulatory effects [53] and an interaction with mycotoxins may be feasible. Host species (humans and animals) have a significant microbial population in the intestine (microbiome) and elsewhere. The possibility of cross-talk between endogenous microbial species and exogenous, microbially-produced toxins has been suggested [54] and this is likely to be of significance. One area of growing interest, which has a basis in the biological activity of compounds, is the development of bioanalytical methods for the analysis of mycotoxins—particularly in clinical samples [55]. Functional assays that provide direct information on the consequences of exposure to multiple secondary natural products will be of particular value. Because of the target organ-oriented nature of mycotoxicosis and the activation by cytochrome enzymes of many members of the class [56] it is likely that organoid cultures will prove to be particularly useful and will help to differentiate between effects on different types of cell [57].

Secondary natural products may be produced or modified by post-harvest events. The post-harvest storage of fresh material is often associated with deterioration (see above) and, while this is not an inevitable consequence, the loss of homeostasis, the degradation of biological structures and the release of autolytic principles will tend to lead to a loss of integrity and structure. Some foods are processed by addition of other components (salt, acid, heating to inactivate spoilage enzymes etc.), however these will be discussed in Sect. 10.4. Underground anaerobic fermentation is common in some cultures but this is still a comparatively rare form of storage. The effect of cooking and burying on protein has been studied [58] from an archaeological standpoint and it seems that the presence of other food components may have a protective effect since proteins bind strongly to the minerals in ceramic vessels and are not readily released. The process is, however, non-linear and this may lead to bias in the material isolated.

10.4 Processing Parameters—pH, Heat, Ionic Strength, Pressure—Effect on Secondary Natural Products

The effect of various types of processing on secondary natural products has been studied. Some of the processes used can be considered in terms of the crossing of an interface between one state and another. In a living matrix, secondary natural products can be considered to be in a state of changing equilibria. Within this state they have the potential to be changed if conditions change. The nature of the change will depend on the type of change experienced. For example, mechanical disruption, as has been noted previously, can enable constituents that are physically separated to mix and to react together. If one of the components is an enzyme and the other is a substrate for the enzyme then a catalysed reaction can take place. In this case, the activation energy for the reaction is lowered, this constitutes a change in conditions and the reaction can proceed. The interface between enzyme and substrate which keeps them apart is, literally, removed. Rather than mechanical disruption, if energy in the form of heat is introduced then an interface may be disrupted however, above a certain temperature, the enzyme structural integrity is lost (it is denatured) and the reaction cannot take place. These two examples—mechanical disruption and heat—may both lead to disruption of an interface, but with different consequences. In the case of mechanical disruption, the movement of chemicals and enzymes across the interface between different parts of material to permit mixing leads to a reaction. In the case of heating, if the energy input is sufficient to disrupt the enzyme structure and the interface between an ordered (active) and disordered (inactive) enzyme is lost then catalysis by the enzyme will not take place even if the substrate and enzyme share a location and his can affect the cocktail of compounds present.

The development of processing was initially guided by the need to prevent microbial (and other) spoilage and to preserve food. As noted in Sect. 10.3, microbial contamination can lead to the production and accumulation of microbial toxic natural products. In addition, foods may also contain anti-nutrients and removal of these can be an important consequence of processing. Finally, living systems (including foods) contain enzymes (and other biological macromolecules) and deactivation/denaturation of these can help to preserve some desirable characteristics and prevent degradation. While listing the effects of different types of processing on specific secondary natural products may provide some information, this will always be an incomplete exercise since there are a large number of processes and enormous numbers of natural products that can be affected both in isolation and in combination with other members of the class. It is, therefore, more valuable to take some specific examples and to examine these as examples which may help to guide studies of other members. The choice of which examples to use is guided by importance and impact and comprise the following:

- S-containing natural products from *Allium* and *Brassica* species and asparagusic acid.
- Enzymes associated with secondary natural products.
- Anti-nutrients—phytates and saponins.
- Flavones and related compounds.

The main S-containing natural products from *Brassica* species and the genus *Allium* occur as relatively biologically inert compounds (glucosinolates and alk(en)yl-Lcysteine sulphoxides respectively. When subjected to typical processing conditions (heat, addition of salt, changes in pH), they will form a range of breakdown products [59] including nitriles, disulphides, isothiocyanates and thiocyanate ion. In both cases, under conditions in which the co-occurring enzymes (myrosinase and alliinase) are present, a range of compounds can be formed after tissue disruption to enable access of the enzyme to the vacuole where the precursor is stored. Heat and some other processing conditions can inactivate myrosinase [60] and alliinase [61]. Perhaps the most significant distinguishing feature of the enzymic breakdown of glucosinolates is that the aglucone undergoes a Lossen-type rearrangement to give an isothiocyanate (Fig. 10.2).

While other compounds can be formed [62], the formation of an isothiocyanate is indicative of hydrolysis by myrosinase. Since many of the biological properties (both beneficial and deleterious) of *Brassica* species can be attributed to the isothiocyanates or products arising from them, processing (if it inactivates myrosinase) may have the effect of diminishing the biological potential of *Brassica* species. In addition, while some intestinal bacteria are able to hydrolyse glucosinolates (i.e. have an enzyme with thioglucosidase capability), the extent of processing will also have an effect on the bioavailability of glucosinolates and their breakdown products [63]. The metabolism of glucosinolates in vivo may also be affected by the presence of mammalian sulfatases [64] and intestinal microbial sulphatases [65]. These enzymes may be important since desulpho glucosinolates are not hydrolysed by myrosinase and may, indeed, be thioglucosidase inhibitors [66]. Another interesting natural S-containing compound is asparagusic acid (1,2-dithiolane-4-carboxylic acid). It occurs exclusively in asparagus and is derived, biosynthetically from isobutyric acid. The use of a range of processing techniques to improve the bioavailability of asparagusic acid and other phytochemicals from asparagus has been noted [67].

A range of enzymes are associated with secondary natural products. As noted above, both glucosinolates and S-alk(en)yl-L-cysteine sulphoxides are hydrolysed by cooccurring enzymes and, in the case of glucosinolates, the biological relevance of sulphatases has been described. In general, processing disrupts enzyme activity either directly as a result of denaturation or as a consequence of the release of inhibitors. In addition, some cofactors are labile and this may also affect enzyme activity [68]. In addition to the effect of processing on naturally occurring enzymes in food systems, the use of enzymes themselves in food processing is an area of

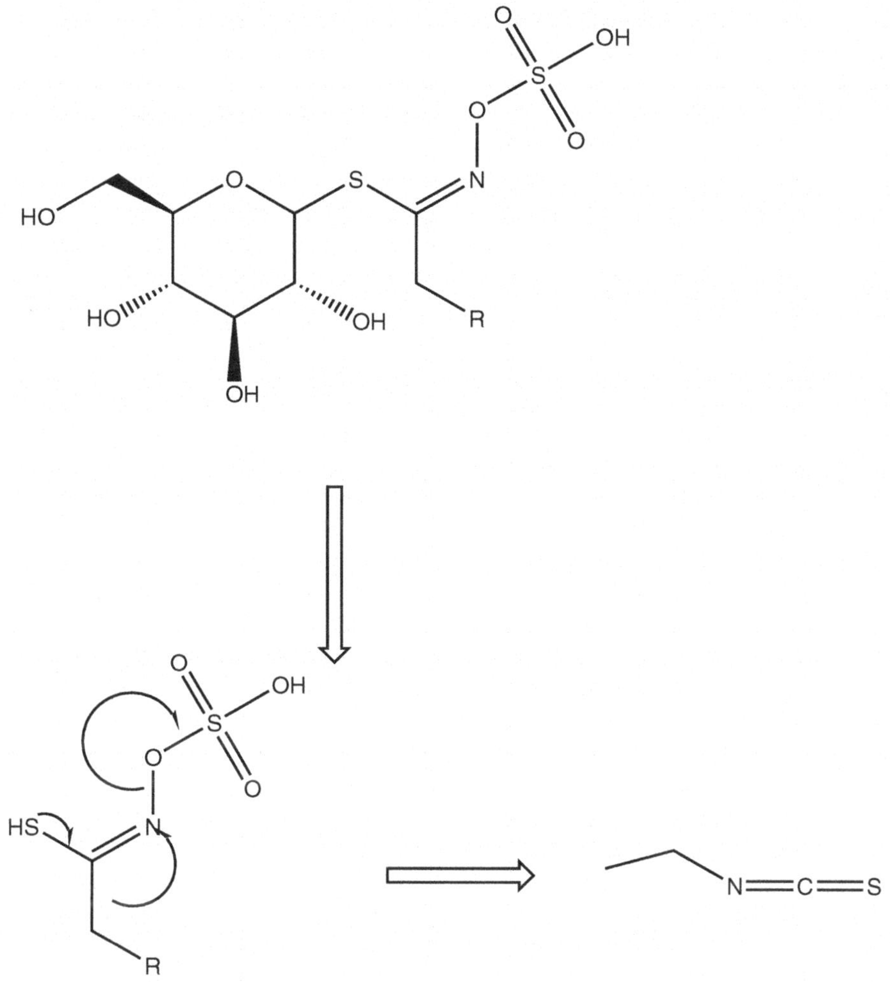

Fig. 10.2 The lossen rearrangement

substantial interest [69]. In addition to the use of enzymes directly in food processing [70], some enzyme inhibitors that occur naturally (for example, phenols) or are added (polyphenol oxidase and protease inhibitors) are also important modulators of enzyme activity and are, therefore significant components of food processing systems [71].

The presence of antinutrients such as phytates and saponins can affect the nutritional profile of foods. While the term 'antinutrients' is convenient, the whole concept of antinutrients has been questioned [72]. This raises the question of whether antinutrients are components that inhibit the uptake of nutrients (see phytates below), break down the nutrients themselves (for example, thiaminase) or interfere with nutrient-dependent physiological pathways (for example, phytoestrogens).Most foods (even individual ingredients) contain a mixture of constituents and, while the reductionist, chemistry approach is probably the most convenient definition (for example, phytates are defined as being made up of six phosphate groups, attached to an inositol ring), functional definitions (for example, phytoestrogens) are also used. The phosphate groups in phytates are able to bind cations—some of which are nutritionally important. This specific capability is one of the main reasons why phytates are defined as antinutrients. They are found in grains and legumes however if diets contain sufficient minerals to counteract the chelating properties of phytates then, nutritionally, they do not pose a risk. In addition a range of processing techniques (including soaking, fermentation, sprouting, germinating, and cooking) will reduce the levels of phytates significantly. While phytates are generally considered to be a deleterious component of the diet, they often occur in foods that are rich in fibre and in prebiotics. In addition phytates can help to regulate iron absorption and may play a role in immune regulation. The effects of phytate, as is the case with many specific dietary components, is dependent upon other factors in the diet, the physiological effect that is measured (and its importance to the target group and the degree and type of processing carried out. In the case of saponins, their role as factors affecting nutrition has been documented [73]. They consist of a sugar moiety (hydrophilic attached to a lipophilic group. They are, therefore amphipathic and are active at surfaces. They have a range of positive properties including being antimicrobial and anti-inflammatory agents. Saponins are, however, generally bitter and therefore are disfavoured for food use. They have also been defined as antinutrients [74] and attempts have been made to remove them by processing [75]. While some positive aspects of saponin consumption have been noted [73], overall they are considered to be antinutritional factors [76]. While bitterness does not map precisely onto mammalian toxicity [77] and the bitter rejection response is not linked in quantitative terms to toxicity [78], the fact that all saponins are bitter might suggest that avoidance of food components containing antinutrient saponins may at least have some evolutionary justification. The occurrence and ingestion of thaminase represents another type of antinutrient. Different forms of thiaminase occur but one of the most studied examples cleaves the pyrimidine and thiazole ring [79]. Thiaminase activity has been implicated in thiamine deficiency—particularly in marine organisms and from dietary factors [80] and it is therefore a functional antinutrient.

The processing of flavones and related compounds is a useful example of the interaction of processing with secondary natural products. Flavones are members of the class of flavonoids (Fig. 10.3).

They have been extensively studied and their occurrence and biological properties have been investigated and documented [81]. They can be hydroxylated or glycosylated and this affects their physiological properties. As members of the flavonoid class of natural products, flavones are defined in terms of their chemical structure. Many of the studies of flavones have grouped them together with other flavonoids [82] however flavones have come under increasing individual scrutiny. It has been noted that flavones (and their glycosides) are relatively stable to processing. Much of

Fig. 10.3 Structure of flavones

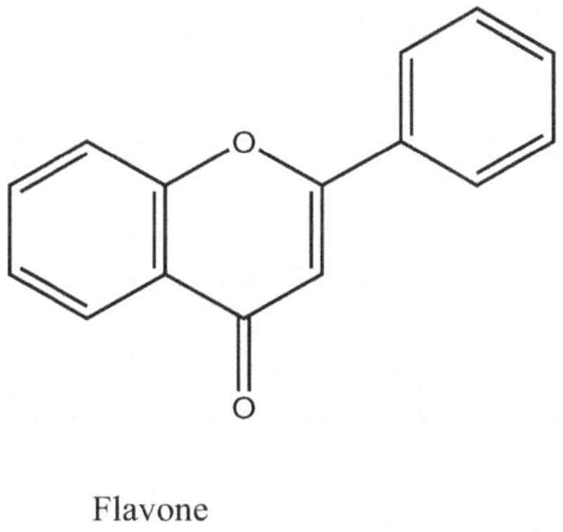

the interest (and therefore many of the studies) have been stimulated by the antioxidative properties of flavones and, while it is difficult to translate measured in vitro antioxidant activity into physiological effects, the production of theories—often validated by putative mechanistic studies—has fuelled many speculative studies [83]. Many of these studies have started from the assumption that physiological processes (respiration, catabolism etc.) will give rise to free radical formation (particularly reactive oxygen species—ROS) and the introduction of antioxidants will, therefore, help to protect against damage. The equivalence between endogenous and exogenous antioxidants remains a matter of debate [84] and even the role of ROS in the promotion of defence elements has received comparatively little attention and is poorly understood [85] but despite this, the promotion of consumption of antioxidants as protection against cellular damage leading to inflammatory diseases and clinical endpoints such as cancer persist [86] although much sceptism remains and clinical trials demonstrating a link between antioxidant intake and disease endpoints remain unconvincing [87]. The link between circulating antioxidant levels and specific clinical endpoints associated with cardiovascular disease, for example [88] suggests that the 'more is better' argument often attributed to antioxidants may be overly simplistic [89]. This has not prevented studies that begin with the supposition that preservation of antioxidant activity of flavones is desirable. Using this index, the effect of processing using extrusion, baking, high-pressure treatment and yeast fermentation was assessed and fermentation was recommended based upon its capacity to retain the antioxidant activity of extracted flavones [90]. The possibility of glycosidases from foods being responsible for the production of flavone aglycones has been noted [91] and this may improve the bioavailability of active species however it should be noted that the use of reductionist, pharmacological principles may not be the most suitable paradigm for assessing the consequences of processing upon the complex mixture of bioactive compounds present in foods.

10.5 Microbial Processing (Fermentation) and Secondary Natural Products

As noted above (Sect. 10.4) in the case of flavones, fermentation may be the least biologically disruptive method of processing. Fermentation is defined as an anaerobic process in which a carbon source (normally a carbohydrate) is converted into a range of other products including alcohols, acids or gases. While Pasteur (and most scientists who followed him) regarded fermentation as a purely anaerobic process, aerobic fermentation has become an important process, in the production of antibiotics, for example [92]. Yeasts, bacteria, fungi and plants are all capable of carrying out fermentation. The use of fermentation to carry out the production of natural products is the foundation of industrial biotechnology [93]. While many natural products only occur in small amounts and/or are part of a complex mixture of chemical structures, the use of recombinant DNA technology has enabled considerable amounts of biologically useful small molecule material to be produced. Different types of fermentation can be employed with different host organisms providing the biological apparatus to be used to support the fermentative process [94] In addition, biological macromolecules such as recombinant proteins, vaccines, enzymes and biologically important products such as insulin are able to be produced in fermentation tanks. In addition to the production of bioactive molecules, fermentation is also an important source of foods [95] and of preservation both in terms of prevention of microbial growth and spoilage and the generation of novel characteristics such as flavours and textures. Fermentation has also been used to act as a source of specific bioactive dietary components that are used a food supplements [96]. In the context of the use of fermentation and its effect on secondary natural products, much of the interest is in the use of fermentation to carry out specific biotransformations. This can include the utilisation of the following:

- Natural or recombinant enzymes to carry out reactions
- Whole organisms to either carry out individual biotransformations or a series of reactions to give a desired product.

Both groups have been extensively studied so, rather than providing a comprehensive review, specific examples will be provided to illustrate the main principles involved. While much of the emphasis has been placed on the specificity of fermentations for specific endpoints, mixed fermentations have also been used to access novel biosynthetic pathways [97] and AI-guided fermentation has increased the scope and applicability of fermentation at large scale both for pharmaceutical [98] and food use [99]. As was seen in the case of flavones, generic claims about health benefits based on attributes such as antioxidative properties are used in the promotion of fermentation [100] but these should be treated with caution.

Single enzyme biotransformations have grown in popularity in industrial fermentations—particularly for enantiospecific reactions [101]. Such 'chemoenzymatic synthesis' has a number of advantages over synthetic chemistry approaches [102] including specificity and the generally mild conditions that are able to be used.

While reactions using enzymes can be slow to proceed to completion compared with organic synthesis and enzymes are often incompatible with organic solvents [103], the specificity of the reactions catalysed can mean that side reactions are minimised (so isolation of the required products from complex reaction mixtures is not required), eventual yields are high and enzymes can be stabilised by immobilisation [104]. Such approaches can also be used with whole cells. Despite these advances, the differences in solubility of enzymes (generally aqueous compatible) and non-polar organic molecules means that synthetic organic chemists are often reluctant to take a biological route. The use of enzymes is often reaction and product specific [105] and is, therefore, philosophically different from synthetic chemistry in which the ability of reagents to carry out the same reaction on a range of starting materials is seen as an advantage although immobilisation has in the past been seen as a converging viable option for both [106]. In simple terms, the challenge of carrying out the synthesis of secondary natural products can proceed from two strands of thought—the 'disconnection' or retrosynthetic approach is one in which a molecule is broken down into constituent parts (disassembled) which are then reassembled to give the desired product. This approach can be used both with traditional synthetic reagents [107]—sometimes aided by AI approaches—and in biocatalysis [108]. The alternative approach—which is perhaps more favoured by natural product chemists due to familiarity with biologically coherent transformations, is the biomimetic approach in which actual or plausible intermediates are postulated and the molecule synthesised using a route that is biologically derived [109]. The first biomimetic synthesis is considered to be that of tropinone by Robinson in 1917 [110]. The 'added value' of a biomimetic approach is that in addition to resulting in the synthesis of a desired compound, it can also test the plausibility of a putative biosynthetic scheme. An alternative approach to the use of biomimetic chemistry to construct a putative biosynthetic pathway using chemical reagents in place of enzymes, is the construction of molecules which mechanistically mimic an enzyme active site and thereby carry out individual biomimetic reactions [111]. One of the challenges in the use of enzymes, rather than enzyme mimics in synthesis is the need for cofactors. A cofactor is an entity which provides a specific element required for an enzymically catalysed reaction to proceed. The definition of cofactors has been further clarified to distinguish between cofactors (often metal ions whose function is to bind to the active site and enable the reaction to proceed) and coenzymes (organic molecules that are able to transfer functional groups or electrons to a substrate or intermediate and are cycled between an active and an inactive state). The cycling of coenzymes is an important aspect of the effective use of enzymes to catalyse reactions [112]. The value of carrying out individual reactions in a biological pathway has been appreciated and some of the advantages mentioned, synthetic biology and engineering biology have been developed to apply engineering principle to biological pathway exploitation and design [113]. This makes use of complete pathways to synthesise biological molecules and to monitor the reactions in a bioreactor [114]. The essence of synthetic biology is the Design-Build-Test-Learn paradigm [115] and the application of this systematic concept leads to the development of ways to interrogate biological systems rather than individual reactions. The concept of modular design is, in some ways, a biomimetic mimic of the polyketide biosynthetic system [116]. The principles of bioreactors is not only applied to the production of high value pharmaceuticals—as exemplified by the production of penicillin by deep tank fermentation by Pfizer. Interestingly one of the first successes of Pfizer was in the food rather than the pharmaceutical area—the production of citric acid for use in a range of products including soft drinks. The later developed technology and company expertise in the use of deep tank fermentation enabled them to develop the necessary technology for the commercial production of penicillin [117]. There are currently a number of bioreactor approaches used for the production of foods and food ingredients. While meat production has attracted considerable interest, other applications for the use of bioreactors have grown including for fuels and bulk protein [118] as well as high value products such as long chain polyunsaturated fatty acids [119]. While bioreactor design and the capability to monitor its outputs has grown in sophistication, a number of challenges remain both in the food and pharmaceutical areas and this will likely fuel the application of engineering to biological systems. Perhaps a signifier of success will be more apparent when procedures are developed de novo for biological systems and then applied to other engineering challenges rather than purely the application of tried and tested engineering paradigms to biology.

10.6 Processing and Storage—Effects on Secondary Natural Products

The effects of processing and storage on secondary natural products remains an area of considerable interest. As noted previously, molecules can be considered from the context of those which are produced by the stored material itself (endogenous products) and those produced by other environmental factors (exogenous factors such as mycotoxins).

While secondary natural products can occur at elevated levels due to biotic and abiotic stress induced both biosynthesis, considerable attention has focused upon the control of the production of another class of naturally-derived products—food processing contaminants and these have been the subject of considerable regulatory consideration. Most processing contaminants occur as a consequence of the thermal treatment of food and subsequent reactions between its constituents [120]. The most significant of these include acrylamide, advanced glycation end products, heterocyclic aromatic amines, furan, polycyclic aromatic hydrocarbons, N-nitroso compounds, monochloropropane diols and their esters. Measures to control their presence are generally focused upon modification of the processing conditions although this has been combined, in some cases, with using raw materials that contain lower levels of the (generally primary) constituents that are precursors of the toxic entities. Much of the studies have concentrated upon the kinetics of formation of the products and therefore on manipulations that impact upon this. There are fewer concerns over the production of compounds that derive from the effects of processing on secondary natural products since the starting materials generally occur in lower concentrations. In a small number of cases (for example goitrogens and naturally occurring estrogens), a diverse range of structures give rise to a specific physiological effect.

In the case of xenoestrogens, the effects may not necessarily be deleterious but are dependent upon gender, age, health status and other factors [121]. Both estrogens and anti-estrogens [122] can occur in foods and processing may increase their bioavailability [123]. This may also be the case with other bioactive compounds. In addition the potential synergistic effects of bioactive compounds released from processed foods has been relatively little studied.

10.7 Ultra Processed Food (UFP)—Effect on Secondary Natural Products. A Novel Processing Index

The concept of ultra processed foods (UPFs) was introduced as a label to try to define foods that are subjected to high degrees of 'industrial processing' and contain ingredients that are not found or used in kitchens [124]. While the trend towards highly processed foods has been extensively reported, the system that was proposed (called the NOVA system) for classifying foods [125] is riddled with inconsistencies [126] and, in addition, it fails to take a holistic view of diet. Conceptually the idea of classifying food on the basis of degree of processing and to then create the manifestation of the system based upon the nomenclature of some ingredients is confusing both to scientists and, perhaps more importantly, to consumers. Various other bodies have also attempted to explain UPF and UPF guidelines including national regulatory bodies however statements such as "The nutrition literature and authoritative reports increasingly recognise the concept of ultra-processed foods (UPF), as a descriptor of unhealthy diets" [127] create further confusion by trying to define UPF in the context of health rather than ingredients or processing. One of the challenges, therefore with a process-based classification system is that it does not define food in terms of its constituents (and possible synergies between components) but it defines such foods in terms of energy density, effects on the gut microbiome and a range of health endpoints [128]. The problem with such an approach is that it takes little account of potential confounders and begins with a proposition that the researchers then seek to find statistical evidence to support. This is the combination of poor experimental design and poor data analysis that is the sign of bad science [129]. In addition, as terms like UPF gain traction through the media and the apparent plausibility of the concept, people are more likely to be influenced by the incomplete and partial narrative they have been told than by scientific evidence [130].

If food is to be defined in terms of the degree of processing that it is subjected to then 'biomarkers of processing' should be developed to enable the comparison of different foods containing different ingredients would be a logical place to start. Such a system—based upon secondary natural products—will be more complex that the simplistic NOVA-based system but it will contain a scientific basis rather than one based on subjective perception. It will also provide a nutritional basis for the classification of food rather than one with no substantive intellectual basis.

As noted previously, foods contain a number of components with variable levels of sensitivity to processing conditions. These can be used as a measure of the degree of processing that has been carried out and multivariate analysis would be a useful defining principle.

Level One Processing

The first level of processing is the non thermal disruption of structure. Depending on the level of disruption, enzymes will be released that can hydrolyse secondary natural products to give compounds such as isothiocyanates and nitriles and thiols and related compounds. In addition glycoside hydrolases will facilitate the production of aglycones that can be analysed directly or subjected to further hydrolysis to release hydrogen cyanide. This will be the minimum process level.

Level Two Processing

The second level of processing will be the application of heat and/or pressure. This combination is known to largely inactivate food enzymes such as myrosinase [131]. The levels of hydrolysis products formed will be a biomarker for the level of processing and the processing history of the food. In addition, the differing labilities of different enzymes (alliinases, glycoside hydrolases) means that measurement of the different products formed will be a biomarker for the different levels of processing within the category.

Level Three Processing

The third level of processing will reflect the effect on the secondary natural products themselves. Foods contain a large number of secondary natural products with varying levels of stability to the conditions (thermal, ionic strength, pH, pressure) commonly used during processing. By measuring the levels (and, importantly, the ratios) of these different compounds and their breakdown products and subjecting this data to multivariate analysis, it will be possible to determine the levels of processing that has been carried out. Much of the information concerning the stability of secondary natural products to different conditions is already available from previous studies therefore this approach will be largely a collation exercise.

An alternative to the three level approach outlined above is the measurement of the integrity of the cell wall. Microbial and plant cell walls have varying levels of robustness and the release of certain components upon acid hydrolysis or treatment with enzymes has been used to characterise the components of cell walls. A similar approach may be able to be used to define the level of degradation as a consequence of processing [132]. It may also be possible to measure the bioaccessibility of some components that are released as a consequence of cell wall degradation [133].

References

1. Brążkiewicz K, Pobereżny J, Wszelaczyńska E, Bogucka B. Potato starch quality in relation to the treatments and long-term storage of tubers. Sci Rep. 2025;15(1):4144. https://doi.org/10.1038/s41598-025-85657-0. PMID: 39900680; PMCID: PMC11791064.
2. Labuschagne M, Phalafala L, Osthoff G, van Biljon A. The influence of storage conditions on starch and amylose content of South African quality protein maize and normal maize hybrids. J Stored Prod Res. 2014;56:16–20. https://doi.org/10.1016/j.jspr.2013.11.004.
3. Zhang H, Hou J, Liu J, et al. Amylase analysis in potato starch degradation during cold storage and sprouting. Potato Res. 2014;57:47–58. https://doi.org/10.1007/s11540-014-9252-6.
4. Haider MW, Nafees M, Iqbal R, et al. Postharvest starch and sugars adjustment in potato tubers of wide-ranging dormancy genotypes subjected to various sprout forcing techniques. Sci Rep. 2023;13:14845. https://doi.org/10.1038/s41598-023-37711-y.
5. Dai H, Fu M, Yang X, Chen Q. Ethylene inhibited sprouting of potato tubers by influencing the carbohydrate metabolism pathway. J Food Sci Technol. 2016;53(8):3166–74. https://doi.org/10.1007/s13197-016-2290-0. Epub 2016 Jul 26. PMID: 27784911; PMCID: PMC5055881.
6. Daniels-Lake BJ, Prange RK. The canon of potato science: 41. Sprouting Potato Res. 2007;50:379–82. https://doi.org/10.1007/s11540-008-9065-6.
7. Yin M, Hu D, Yu X, Wang Y, Song S, Wang C, Hu Q, Wen Y. Polyacrylamide regulated phytohormone balance and starch degradation to promote seedpotato sprouting and emergence. Plants (Basel). 2024;13(19):2796. https://doi.org/10.3390/plants13192796. PMID: 39409666; PMCID: PMC11478544.
8. Mensinga TT, Sips AJ, Rompelberg CJ, van Twillert K, Meulenbelt J, van den Top HJ, van Egmond HP. Potato glycoalkaloids and adverse effects in humans: an ascending dose study. Regul Toxicol Pharmacol. 2005;41(1):66–72. https://doi.org/10.1016/j.yrtph.2004.09.004. Epub 2004 Dec 10 PMID: 15649828.
9. Mattoo AK, Dwivedi SL, Dutt S, Singh B, Garg M, Ortiz R. Anthocyanin-rich vegetables for human consumption-focus on potato, sweetpotato and tomato. Int J Mol Sci. 2022;23(5):2634. https://doi.org/10.3390/ijms23052634. PMID: 35269776; PMCID: PMC8910313.
10. Akyol H, Riciputi Y, Capanoglu E, Caboni MF, Verardo V. Phenolic compounds in the potato and its byproducts: an overview. Int J Mol Sci. 2016;17(6):835. https://doi.org/10.3390/ijms17060835.
11. Butelli E, Titta L, Giorgio M, et al. Enrichment of tomato fruit with healthpromoting anthocyanins by expression of select transcription factors. Nat Biotechnol. 2008;26:1301–8. https://doi.org/10.1038/nbt.1506.
12. Von der Deutschen Hansa. Eine Historische Skizze (Classic Reprint). Harry Denicke ISBN 10: 0282631496/ISBN 13: 9780282631499.
13. Mondy N, et al. Aroma analysis of fresh and preserved onions and leek by dual solid-phase microextraction-liquid extraction and gas chromatography-mass spectrometry. J Chromatogr A. 2002;963(1–2):89–93.
14. Santoro MG. Heat shock factors and the control of the stress response. Biochem Pharmacol. 2000;59:55–63. https://doi.org/10.1016/S0006-2952(99)00299-3.
15. Singleton KD, Wischmeyer PE. Glutamine protection against sepsis and lung injury is dependent on heat shock protein 70 expression. Am J Regul Integr Comp Physiol. 2007;292:1839–45. https://doi.org/10.1152/ajpregu.00755.2006.
16. Xue H, Sawyer MB, Field CJ, Dieleman LA, Murray D, Baracos VE. Bolus oral glutamine protects rats against CPT-11-induced diarrhea and differentially activates cytoprotective mechanisms in host intestine but not tumor. J Nutr. 2008;138:740–6. https://doi.org/10.1093/jn/138.4.740.
17. Trinklein ND, Murray JI, Hartman SJ, Botstein D, Myers RM. The role of heat shock transcription factor 1 in the genome-wide regulation of the mammalian heat shock response. Mol Biol Cell. 2004;15(3):1254–61. https://doi.org/10.1091/mbc.e0310-0738. Epub 2003 Dec 10. PMID: 14668476; PMCID: PMC363119.

18. Moura CS, Lollo PCB, Morato PN, Amaya-Farfan J. Dietary nutrients and bioactive substances modulate heat shock protein (hsp) expression: a review. Nutrients. 2018;10(6):683. https://doi.org/10.3390/nu10060683. PMID: 29843396; PMCID: PMC6024325.
19. Liu LL, He JH, Xie HB, Yang YS, Li JC, Zou Y. Resveratrol induces antioxidant and heat shock protein mRNA expression in response to heat stress in blackboned chickens. Poult Sci. 2014;93:54–62. https://doi.org/10.3382/ps.2013-03423.
20. Zhang Y, Ahn YH, Benjamin IJ, Honda T, Hicks RJ, Calabrese V, Cole PA, Dinkova-Kostova AT. HSF1-Dependent Upregulation of Hsp70 by SulfhydrylReactive Inducers of the KEAP1/NRF2/ARE Pathway. Chem Biol. 2011;18:1355–61. https://doi.org/10.1016/j.chembiol.2011.09.008.
21. Yeshi K, Crayn D, Ritmejerytė E, Wangchuk P. Plant secondary metabolites produced in response to abiotic stresses has potential application in pharmaceutical product development. Molecules. 2022;27(1):313. https://doi.org/10.3390/molecules27010313. PMID: 35011546; PMCID: PMC8746929.
22. Rabeh K, Hnini M, Oubohssaine M. A comprehensive review of transcription factor-mediated regulation of secondary metabolites in plants under environmental stress. Stress Biology. 2025;5:15. https://doi.org/10.1007/s44154-024-00201-w.
23. Erb M, Kliebenstein DJ. Plant secondary metabolites as defenses, regulators, and primary metabolites: the blurred functional trichotomy. Plant Physiol. 2020;184(1):39–52. https://doi.org/10.1104/pp.20.00433. Epub 2020 Jul 7. PMID: 32636341; PMCID: PMC7479915.
24. Shezi S, Ngcobo MEK, Khanyile N, Ncama K. Physio-Metabolic mechanisms behind postharvest quality deterioration in broccoli (Brassica oleracea var. Italica) and Swiss Chard (Beta vulgaris L. var. Cicla): a review. Plants (Basel). 2024;13(22):3174. https://doi.org/10.3390/plants13223174. PMID: 39599383; PMCID: PMC11597650.
25. Padhi S, Routray W. Effects of natural materials on food preservation and storage. In: Parameswaranpillai J, Jayakumar A, Radhakrishnan EK, Siengchin S, Radoor S, editor. Natural materials for food packaging application; 2023 https://doi.org/10.1002/9783527837304.ch14.
26. Li C, Yu W, Liao W. Role of nitric oxide in postharvest senescence of fruits. Int J Mol Sci. 2022;23(17):10046. https://doi.org/10.3390/ijms231710046. PMID: 36077446; PMCID: PMC9456340.
27. Meitha K, Pramesti Y, Suhandono S. Reactive oxygen species and antioxidants in postharvest vegetables and fruits. Int J Food Sci. 2020;10(2020):8817778. https://doi.org/10.1155/2020/8817778. PMID: 33381540; PMCID: PMC7749770.
28. Papoulias E, Siomos AS, Koukounaras A, Gerasopoulos D, Kazakis E. Effects of genetic, pre- and post-harvest factors on phenolic content and antioxidant capacity of white asparagus spears. Int J Mol Sci. 2009;10(12):5370–80. https://doi.org/10.3390/ijms10125370. PMID: 20054475; PMCID: PMC2801999.
29. Petcu CD, Tăpăloagă D, Mihai OD, Gheorghe-Irimia RA, Negoiță C, Georgescu IM, Tăpăloagă PR, Borda C, Ghimpețeanu OM. Harnessing natural antioxidants for enhancing food shelf life: exploring sources and applications in the food industry. Foods. 2023;12(17):3176. https://doi.org/10.3390/foods12173176. PMID: 37685108; PMCID: PMC10486681.
30. Villa-Rodriguez JA, Palafox-Carlos H, Yahia EM, Ayala-Zavala JF, Gonzalez-Aguilar GA. Maintaining antioxidant potential of fresh fruits and vegetables after harvest. Crit Rev Food Sci Nutr. 2015;55(6):806–22. https://doi.org/10.1080/10408398.2012.685631. PMID: 24915343.
31. Serra S, Anthony B, Boscolo Sesillo F, Masia A, Musacchi S. Determination of post-harvest biochemical composition, enzymatic activities, and oxidative browning in 14 apple cultivars. Foods. 2021;10(1):186. https://doi.org/10.3390/foods10010186. PMID: 33477632; PMCID: PMC7831482.
32. Mahajan PV, Caleb OJ, Singh Z, Watkins CB, Geyer M. Postharvest treatments of fresh produce. Philos Trans A Math Phys Eng Sci. 2014;372(2017):20130309. https://doi.org/10.1098/rsta.2013.0309. PMID: 24797137; PMCID: PMC4006172.

33. Cutler AJ, Conn EE. The synthesis, storage and degradation of plant natural products: cyanogenic glycosides as an example. In: Creasy LL, Hrazdina G, editors. Cellular and subcellular localization in plant metabolism. Recent advances in phytochemistry, vol. 16. Springer:Boston, MA;1982. https://doi.org/10.1007/978-1-4757-4727-0_8.
34. Pičmanová M, Neilson EH, Motawia MS, Olsen CE, Agerbirk N, Gray CJ, Flitsch S, Meier S, Silvestro D, Jørgensen K, Sánchez-Pérez R, Møller BL, Bjarnholt N. A recycling pathway for cyanogenic glycosides evidenced by the comparative metabolic profiling in three cyanogenic plant species. Biochem J. 2015;469(3):375–89. https://doi.org/10.1042/BJ20150390. Epub 2015 Jun 11 PMID: 26205491.
35. Akiyama R, Watanabe B, Nakayasu M, et al. The biosynthetic pathway of potato solanidanes diverged from that of spirosolanes due to evolution of a dioxygenase. Nat Commun. 2021;12:1300. https://doi.org/10.1038/s41467-02121546-0.
36. Hu J, Chen B, Zhao J, et al. Genomic selection and genetic architecture of agronomic traits during modern rapeseed breeding. Nat Genet. 2022;54:694–704. https://doi.org/10.1038/s41588-022-01055-6.
37. An F, Xiao X, Chen T, Xue J, Luo X, Ou W, Li K, Cai J, Chen S. Systematic analysis of bHLH transcription factors in cassava uncovers their roles in postharvest physiological deterioration and cyanogenic glycosides biosynthesis. Front Plant Sci. 2022;16(13): 901128. https://doi.org/10.3389/fpls.2022.901128. PMID: 35789698; PMCID: PMC9249602.
38. Kumari C, Sharma M, Kumar V, Sharma R, Kumar V, Sharma P, Kumar P, Irfan M. Genome editing technology for genetic amelioration of fruits and vegetables for alleviating post-harvest loss. Bioengineering (Basel). 2022;9(4):176. https://doi.org/10.3390/bioengineering9040176. PMID: 35447736; PMCID: PMC9028506.
39. Maluf MP, da Silva CC, de Oliveira Mde P, Tavares AG, Silvarolla MB, Guerreiro O Jr. Altered expression of the caffeine synthase gene in a naturally caffeine-free mutant of Coffea arabica. Genet Mol Biol. 2009;32(4):802–10. https://doi.org/10.1590/S1415-47572009005000090. Epub 2009 Dec 1. PMID: 21637458; PMCID: PMC3036905.
40. Dash SS, Gummadi SN. Catabolic pathways and biotechnological applications of microbial caffeine degradation. Biotechnol Lett. 2006;28:1993–2002. https://doi.org/10.1007/s10529-006-9196-2.
41. Fox IH. Degradation of purine nucleotides. In: Kelley WN, Weiner IM editors. Uric acid. Handbook of experimental pharmacology, vol. 51. Springer:Berlin, Heidelberg;1978. https://doi.org/10.1007/978-3-642-66867-8_5.
42. Martins N, Petropoulos S, Ferreira IC. Chemical composition and bioactive compounds of garlic (Allium sativum L.) as affected by pre- and post-harvest conditions: a review. Food Chem. 2016;211:41–50. https://doi.org/10.1016/j.foodchem.2016.05.029. Epub 2016 May 6. PMID: 27283605.
43. Chowdhury NN, Islam MN, Jafrin R, Rauf A, Khalil AA, Emran TB, Aljohani ASM, Alhumaydhi FA, Lorenzo JM, Shariati MA, Simal-Gandara J. Natural plant products as effective alternatives to synthetic chemicals for postharvest fruit storage management. Crit Rev Food Sci Nutr. 2023;63(30):10332–50. https://doi.org/10.1080/10408398.2022.2079112. Epub 2022 May 25 PMID: 35612470.
44. Tola M, Kebede B. Occurrence, importance and control of mycotoxins: a review. Cogent Food & Agriculture. 2016;2(1). https://doi.org/10.1080/23311932.2016.1191103.
45. Awuchi CG, Ondari EN, Ogbonna CU, Upadhyay AK, Baran K, Okpala COR, Korzeniowska M, Guiné RPF. Mycotoxins affecting animals, foods, humans, and plants: types, occurrence, toxicities, action mechanisms, prevention, and detoxification strategies—a revisit. Foods. 2021;10(6):1279. https://doi.org/10.3390/foods10061279. PMID: 34205122; PMCID: PMC8228748.
46. Alassane-Kpembi I, Schatzmayr G, Taranu I, Marin D, Puel O, Oswald IP. Mycotoxins co-contamination: methodological aspects and biological relevance of combined toxicity studies. Crit Rev Food Sci Nutr. 2017;57(16):34893507. https://doi.org/10.1080/10408398.2016.1140632. PMID: 26918653.

47. Chen X, Uzuner U, Li M, Shi W, Yuan JS, Dai SY. Phytoestrogens and mycoestrogens induce signature structure dynamics changes on estrogen receptor α. Int J Environ Res Public Health. 2016;13(9):869. https://doi.org/10.3390/ijerph13090869. PMID: 27589781; PMCID: PMC5036702.
48. Wang W, Liang X, Li Y, Wang P, Keller NP. Genetic Regulation of Mycotoxin Biosynthesis. J Fungi (Basel). 2022;9(1):21. https://doi.org/10.3390/jof9010021. PMID: 36675842; PMCID: PMC9861139.
49. Ghazi T, Arumugam T, Foolchand A, Chuturgoon AA. The impact of natural dietary compounds and food-borne mycotoxins on DNA methylation and cancer. Cells. 2020;9(9):2004. https://doi.org/10.3390/cells9092004. PMID: 32878338; PMCID: PMC7565866.
50. Heikkinen T, Küblbeck J, Rysä J. Metabolic disruption by mycotoxins: focus on metabolic endpoints steatosis, adipogenesis and glucose metabolism in vivo and in vitro. Arch Toxicol. 2025. https://doi.org/10.1007/s00204-025-03957-w.
51. Oh S Y, Mead PJ, Sharma BS, Quinton VM, Boermans HJ, Smith TK, Swamy HV, Karrow NA. Effect of Penicillium mycotoxins on the cytokine gene expression, reactive oxygen species production, and phagocytosis of bovine macrophage (BoMacs) function. Toxicol In Vitro. 2015;30(1 Pt B):446–53. https://doi.org/10.1016/j.tiv.2015.09.017. Epub 2015 Sep 21. PMID: 26394380.
52. Cleary D. Effects of mycotoxins on macrophages and their possible clinical implications. In: Gupta S, Pathak YV, editors. Macrophage targeted delivery systems. Springer:Cham; 2022. https://doi.org/10.1007/978-3-030-841645_23.
53. Beigoli S, Boskabady MH. The molecular basis of the immunomodulatory effects of natural products: a comprehensive review. Phytomedicine. 2024;135: 156028. https://doi.org/10.1016/j.phymed.2024.156028. Epub 2024 Sep 10 PMID: 39276685.
54. Xia D, Mo Q, Yang L, Wang W. Crosstalk between mycotoxins and intestinal microbiota and the alleviation approach via microorganisms. Toxins (Basel). 2022;14(12):859. https://doi.org/10.3390/toxins14120859. PMID: 36548756; PMCID: PMC9784275.
55. Karsauliya K, Yahavi C, Pandey A, Bhateria M, Sonker AK, Pandey H, Sharma M, Singh SP. Co-occurrence of mycotoxins: a review on bioanalytical methods for simultaneous analysis in human biological samples, mixture toxicity and risk assessment strategies. Toxicon. 2022;30(218):25–39. https://doi.org/10.1016/j.toxicon.2022.08.016. Epub 2022 Aug 29 PMID: 36049662.
56. Wen J, Mu P, Deng Y. Mycotoxins: cytotoxicity and biotransformation in animal cells. Toxicol Res (Camb). 2016;5(2):377–87. https://doi.org/10.1039/c5tx00293a. PMID: 30090353; PMCID: PMC6062401.
57. Gao Y, Meng L, Liu H, Wang J, Zheng N. The Compromised Intestinal Barrier Induced by Mycotoxins. Toxins (Basel). 2020;12(10):619. https://doi.org/10.3390/toxins12100619. PMID: 32998222; PMCID: PMC7600953.
58. Evans M, Hagan R, Boyd OJ, Bondetti M, Craig OE, Collins MJ, Hendy J. The impact of cooking and burial on proteins: a characterisation of experimental foodcrusts and ceramics. R Soc Open Sci. 2024;11(9): 240610. https://doi.org/10.1098/rsos.240610. PMID: 39416716; PMCID: PMC11482021.
59. Abdel-Massih RM, Debs E, Othman L, Attieh J, Cabrerizo FM. Glucosinolates, a natural chemical arsenal: More to tell than the myrosinase story. Front Microbiol. 2023;5(14):1130208. https://doi.org/10.3389/fmicb.2023.1130208. PMID: 37089539; PMCID: PMC10114928.
60. Van Eylen D, Oey I, Hendrickx M, et al. Behavior of mustard seed (Sinapis alba L.) myrosinase during temperature/pressure treatments: a case study on enzyme activity and stability. Eur Food Res Technol. 2008;226:545–53. https://doi.org/10.1007/s00217-007-0569-0.
61. Rabinkov A, Zhu XZ, Grafi G, Galili G, Mirelman D. Alliin lyase (Alliinase) from garlic (Allium sativum). Biochemical characterization and cDNA cloning. Appl Biochem Biotechnol. 1994;48(3):149–71. https://doi.org/10.1007/BF02788739. PMID: 7979352.

62. Eisenschmidt-Bönn D, Schneegans N, Backenköhler A, Wittstock U, Brandt W. Structural diversification during glucosinolate breakdown: mechanisms of thiocyanate, epithionitrile and simple nitrile formation. Plant J. 2019;99(2):329–43. https://doi.org/10.1111/tpj.14327. Epub 2019 Apr 29. PMID: 30900313; PMCID: PMC6850609.
63. Barba FJ, Nikmaram N, Roohinejad S, Khelfa A, Zhu Z, Koubaa M. Bioavailability of glucosinolates and their breakdown products: impact of processing. Front Nutr. 2016;16(3):24. https://doi.org/10.3389/fnut.2016.00024. PMID: 27579302; PMCID: PMC4985713.
64. Mashima R, Nakanishi M. Mammalian sulfatases: biochemistry, disease manifestation, and therapy. Int J Mol Sci. 2022;23(15):8153. https://doi.org/10.3390/ijms23158153. PMID: 35897729; PMCID: PMC9330403.
65. Baldelli S, Lombardo M, D'Amato A, Karav S, Tripodi G, Aiello G. Glucosinolates in human health: metabolic pathways, bioavailability, and potential in chronic disease prevention. Foods. 2025;14:912. https://doi.org/10.3390/foods14060912.
66. Bourderioux A, Lefoix M, Gueyrard D, Tatibouét A, Cottaz S, Arzt S, Burmeister WP, Rollin P. The glucosinolate-myrosinase system. New insights into enzymesubstrate interactions by use of simplified inhibitors. Org Biomol Chem. 2005;3(10):1872–9. https://doi.org/10.1039/b502990b. Epub 2005 Apr 14. PMID: 15889170.
67. Yu Q, Fan L. Improving the bioactive ingredients and functions of asparagus from efficient to emerging processing technologies: a review. Food Chem. 2021;358:129903. https://doi.org/10.1016/j.foodchem.2021.129903. Epub 2021 Apr 20. PMID: 33933971.
68. Scanlon MG, Henrich AW, Whitaker JR. Factors affecting enzyme activity in food processing. Woodhead Publishing; 2018, pp. 337–365. https://doi.org/10.1016/B978-0-08-100722-8.00014-0.
69. Fernandes P. Enzymes in food processing: a condensed overview on strategies for better biocatalysts. Enzyme Res. 2010;29(2010): 862537. https://doi.org/10.4061/2010/862537. PMID: 21048872; PMCID: PMC2963163.
70. Simpson BK, Rui X, XiuJie J. Enzyme-assisted food processing. In: Boye J, Arcand Y, editors, Green technologies in food production and processing. Food Eng Series. Springer:Boston, MA; 2012. https://doi.org/10.1007/978-14614-1587-9_13.
71. Whitaker JR. Inhibitors of enzymes in biological materials used for foods. In: Jelliffe EFP, Jelliffe DB, editors, Adverse effects of foods. Springer:Boston, MA; 1982. https://doi.org/10.1007/978-1-4613-3359-3_3.
72. Petroski W, Minich DM. Is there such a thing as "anti-nutrients"? a narrative review of perceived problematic plant compounds. Nutrients. 2020;12(10):2929. https://doi.org/10.3390/nu12102929. PMID: 32987890; PMCID: PMC7600777.
73. Timilsena YP, Phosanam A, Stockmann R. Perspectives on saponins: food functionality and applications. Int J Mol Sci. 2023;24(17):13538. https://doi.org/10.3390/ijms241713538. PMID: 37686341; PMCID: PMC10487995.
74. Thompson LU. Potential health benefits and problems associated with antinutrients in foods. Food Res Int. 1993;26:131–49. https://doi.org/10.1016/09639969(93)90069-U.
75. Ridout CL, Price KR, Dupont MS, Parker ML, Fenwick GR. Quinoa saponins—analysis and preliminary investigations into the effects of reduction by processing. J Sci Food Agric. 1991;54:165–76. https://doi.org/10.1002/jsfa.2740540202.
76. Samtiya M, Aluko RE, Dhewa T. Plant food anti-nutritional factors and their reduction strategies: an overview. Food Prod Process and Nutr. 2020;2:6. https://doi.org/10.1186/s43014-020-0020-5.
77. Nissim I, Dagan-Wiener A, Niv MY. The taste of toxicity: A quantitative analysis of bitter and toxic molecules. IUBMB Life. 2017;69(12):938–46. https://doi.org/10.1002/iub.1694. Epub 2017 Nov 11 PMID: 29130618.
78. Glendinning JI. Is the bitter rejection response always adaptive? Physiol Behav. 1994;56(6):1217–27. https://doi.org/10.1016/0031-9384(94)90369-7. PMID: 7878094.
79. Kreinbring CA, Remillard SP, Hubbard P, Brodkin HR, Leeper FJ, Hawksley D, Lai EY, Fulton C, Petsko GA, Ringe D. Structure of a eukaryotic thiaminase I. Proc Natl Acad Sci U S A. 2014;111(1):137–42. https://doi.org/10.1073/pnas.1315882110. Epub 2013 Dec 18. PMID: 24351929; PMCID: PMC3890816.

80. Edwards KA, Randall EA, Wolfe PC, Angert ER, Kraft CE. Dietary factors potentially impacting thiaminase I-mediated thiamine deficiency. Sci Rep. 2023;13(1):7008. https://doi.org/10.1038/s41598-023-34063-5. PMID: 37117240; PMCID: PMC10147936.
81. Hostetler GL, Ralston RA, Schwartz SJ. Flavones: food sources, bioavailability, metabolism, and bioactivity. Adv Nutr. 2017;8(3):423–35. https://doi.org/10.3945/an.116.012948. PMID: 28507008; PMCID: PMC5421117.
82. Wang J, Wang B, Chen C, Dong J, Zhang H. Utilization and separation of flavonoids in the food and medicine industry: current status and perspectives. Separations. 2024;11:349. https://doi.org/10.3390/separations11120349.
83. Losada-Barreiro S, Sezgin-Bayindir Z, Paiva-Martins F, Bravo-Díaz C. Biochemistry of antioxidants: mechanisms and pharmaceutical applications. Biomedicines. 2022;10(12):3051. https://doi.org/10.3390/biomedicines10123051. PMID: 36551806; PMCID: PMC9776363.
84. Kozlov AV, Javadov S, Sommer N. Cellular ROS and antioxidants: physiological and pathological role. Antioxidants (Basel). 2024;13(5):602. https://doi.org/10.3390/antiox13050602. PMID: 38790707; PMCID: PMC11117742.
85. Salehi B, Martorell M, Arbiser JL, Sureda A, Martins N, Maurya PK, Sharifi-Rad M, Kumar P, Sharifi-Rad J. Antioxidants: positive or negative actors? Biomolecules. 2018;8(4):124. https://doi.org/10.3390/biom8040124. PMID: 30366441; PMCID: PMC6316255.
86. Griffiths K, Aggarwal BB, Singh RB, Buttar HS, Wilson D, De Meester F. Food antioxidants and their anti-inflammatory properties: a potential role in cardiovascular diseases and cancer prevention. Diseases. 2016;4(3):28. https://doi.org/10.3390/diseases4030028. PMID: 28933408; PMCID: PMC5456284.
87. Dewell A, Tsao P, Rigdon J, Gardner CD. Antioxidants from diet or supplements do not alter inflammatory markers in adults with cardiovascular disease risk. A pilot randomized controlled trial. Nutr Res. 2018;50:63–72. https://doi.org/10.1016/j.nutres.2017.10.017. Epub 2017 Nov 2. PMID: 29540273; PMCID: PMC5858717.
88. Yang R, Lv M, Yang X, et al. A Mendelian randomized study of circulating antioxidants in the diet and risk of cardiovascular disease. Sci Rep. 2025;15:10341. https://doi.org/10.1038/s41598-025-94369-4.
89. Janciauskiene S. The beneficial effects of antioxidants in health and diseases. Chronic Obstr Pulm Dis. 2020;7(3):182–202. https://doi.org/10.15326/jcopdf.7.3.2019.0152. PMID: 32558487; PMCID: PMC7857719.
90. Xia G, Li X, Zhang Z, Jiang Y. Effect of food processing on the antioxidant activity of flavones from Polygonatum odoratum (Mill.) Druce. Open Life Sci. 2021;16(1):92–101. https://doi.org/10.1515/biol-2021-0010. PMID: 33817302; PMCID: PMC7874555.
91. Hostetler GL, Riedl KM, Schwartz SJ. Effects of food formulation and thermal processing on flavones in celery and chamomile. Food Chem. 2013;141(2):1406–11. https://doi.org/10.1016/j.foodchem.2013.04.051. Epub 2013 Apr 25. PMID: 23790931; PMCID: PMC3850025.
92. Zhgun AA. Pharmaceutical fermentation: antibiotic production and processing. Fermentation. 2024;10:419. https://doi.org/10.3390/fermentation10080419.
93. Pham JV, Yilma MA, Feliz A, Majid MT, Maffetone N, Walker JR, Kim E, Cho HJ, Reynolds JM, Song MC, Park SR, Yoon YJ. A review of the microbial production of bioactive natural products and biologics. Front Microbiol. 2019;20(10):1404. https://doi.org/10.3389/fmicb.2019.01404. PMID: 31281299; PMCID: PMC6596283.
94. RamyaSree B, Jose PA, Divakar K. Fermentative production of secondary metabolites from bioengineered fungal species and their applications. In: Hesham AL, Upadhyay R, Sharma G, Manoharachary C, Gupta V, Editors, Fungal biotechnology and bioengineering. Fungal biology. Springer: Cham; 2020. https://doi.org/10.1007/978-3-030-41870-0_11.
95. Voidarou C, Antoniadou M, Rozos G, Tzora A, Skoufos I, Varzakas T, Lagiou A, Bezirtzoglou E. Fermentative foods: microbiology, biochemistry, potential human health benefits and public health issues. Foods. 2020;10(1):69. https://doi.org/10.3390/foods10010069. PMID: 33396397; PMCID: PMC7823516.
96. Rusu AV, Trif M, Rocha JM. Microbial secondary metabolites via fermentation approaches for dietary supplementation formulations. Molecules. 2023;28:6020. https://doi.org/10.3390/molecules28166020.

97. Pettit RK. Mixed fermentation for natural product drug discovery. Appl Microbiol Biotechnol. 2009;83:19–25. https://doi.org/10.1007/s00253-009-1916-9.
98. Hamdy R, et al. Artificial intelligence and discovery of microbial natural products. In: Cruz JN, editors, Drug discovery and design using natural products. Springer: Cham; 2023. https://doi.org/10.1007/978-3-031-35205-8_3.
99. Yadav H, Jain S, Rastamanesh R, Bomba A, Catanzaro R, Marotta F. Fermentation technology in the development of functional foods for human health: where we should head. Ferment Technol. 2012;1(1):e102. https://doi.org/10.4172/2167-7972.1000e102. Epub 2012 Jan 30. PMID: 31565543; PMCID: PMC6764775.
100. Sarıtaş S, Portocarrero ACM, Miranda López JM, Lombardo M, Koch W, Raposo A, El-Seedi HR, de Brito Alves JL, Esatbeyoglu T, Karav S, et al. The impact of fermentation on the antioxidant activity of food products. Molecules 2024; 29:3941. https://doi.org/10.3390/molecules29163941.
101. Valentová K. Biotransformation of natural products and phytochemicals: metabolites, their preparation, and properties. Int J Mol Sci. 2023;24(9):8030. https://doi.org/10.3390/ijms24098030. PMID: 37175731; PMCID: PMC10178883.
102. Chong G. Guiding chemoenzymatic synthesis. Nat Chem Biol. 2025;21:465. https://doi.org/10.1038/s41589-025-01876-6.
103. Hinzmann A, Adebar N, Betke T, Leppin M, Gröger H. Biotransformations in pure organic medium: organic solvent-labile enzymes in the batch and flow synthesis of nitriles. Eur J Org Chem. 2019;2019:6911–6. https://doi.org/10.1002/ejoc.201901168.
104. Khan MR. Immobilized enzymes: a comprehensive review. Bull Natl Res Cent. 2021;45:207. https://doi.org/10.1186/s42269-021-00649-0.
105. Swizdor A, Panek A, Milecka-Tronina N, Kołek T. Biotransformations utilizing β-oxidation cycle reactions in the synthesis of natural compounds and medicines. Int J Mol Sci. 2012;13(12):16514–43. https://doi.org/10.3390/ijms131216514. PMID: 23443116; PMCID: PMC3546705.
106. Ley S, Baxendale I. New tools and concepts for modern organic synthesis. Nat Rev Drug Discov. 2002;1:573–86. https://doi.org/10.1038/nrd871.
107. Kreutter D, Reymond JL. Multistep retrosynthesis combining a disconnection aware triple transformer loop with a route penalty score guided tree search. Chem Sci. 2023;14(36):9959–69. https://doi.org/10.1039/d3sc01604h. PMID: 37736648; PMCID: PMC10510629.
108. de Souza ROMA, Miranda LSM, Bornscheuer UT. A retrosynthesis approach for biocatalysis in organic synthesis. Chem Eur J. 2017; 23:12040.
109. Razzak M, De Brabander J. Lessons and revelations from biomimetic syntheses. Nat Chem Biol. 2011;7:865–75. https://doi.org/10.1038/nchembio.709.
110. Robinson R. LXIII.—a synthesis of tropinone. J Chem Soc Trans. 1917;111:762–768.
111. Biomimetic chemistry: biology as an inspiration. J Biol Chem. 284(3):1337–42.
112. Robinson PK. Enzymes: principles and biotechnological applications. Essays Biochem. 2015;59:1–41. https://doi.org/10.1042/bse0590001. Erratum in: Essays Biochem. 2015;59:75. PMID: 26504249; PMCID: PMC4692135.
113. Castle SD, Stock M, Gorochowski TE. Engineering is evolution: a perspective on design processes to engineer biology. Nat Commun. 2024;15(1):3640. https://doi.org/10.1038/s41467-024-48000-1. PMID: 38684714; PMCID: PMC11059173.
114. Mao C, Mao Y, Zhu X, Chen G, Feng C. Synthetic biology-based bioreactor and its application in biochemical analysis. Crit Rev Anal Chem. 2024;54(7):2467–84. https://doi.org/10.1080/10408347.2023.2180319. Epub 2023 Feb 20 PMID: 36803337.
115. Garner KL. Principles of synthetic biology. Essays Biochem. 2021;65(5):791–811. https://doi.org/10.1042/EBC20200059. PMID: 34693448; PMCID: PMC8578974.
116. Grininger M. Enzymology of assembly line synthesis by modular polyketide synthases. Nat Chem Biol. 2023;19:401–15. https://doi.org/10.1038/s41589-023-01277-7.

117. Bud R. Innovators, deep fermentation and antibiotics: promoting applied science before and after the Second World War. Dynamis. 2011;31(2):323–41. https://doi.org/10.4321/s0211-95362011000200004. PMID: 22332462.
118. Palladino F, Marcelino PRF, Schlogl AE, José ÁHM, Rodrigues RCLB, Fabrino DL, Santos IJB, Rosa CA. Bioreactors: applications and innovations for a sustainable and healthy future—a critical review. Appl Sci 2024;14:9346. https://doi.org/10.3390/app14209346.
119. Guo D-S, Ji X-J, Ren L-J, Li G-L, Huang H. Improving docosahexaenoic acid production by Schizochytrium sp. using a newly designed high-oxygen-supply bioreactor. AIChE J. 2017;63:4278–86. https://doi.org/10.1002/aic.15783.
120. Zahir A, Khan IA, Nasim M, Azizi MN, Azi F. Food process contaminants: formation, occurrence, risk assessment and mitigation strategies - a review. Food Addit Contam Part A Chem Anal Control Expo Risk Assess. 2024;41(10):1242–74. https://doi.org/10.1080/19440049.2024.2381210. Epub 2024 Jul 22 PMID: 39038046.
121. Patisaul HB, Jefferson W. The pros and cons of phytoestrogens. Front Neuroendocrinol. 2010;31(4):400–19. https://doi.org/10.1016/j.yfrne.2010.03.003. Epub 2010 Mar 27. PMID: 20347861; PMCID: PMC3074428.
122. Paramasivam A, Murugan R, Jeraud M, Dakkumadugula A, Periyasamy R, Arjunan S. Additives in processed foods as a potential source of endocrinedisrupting chemicals: a review. J Xenobiot. 2024;14(4):1697–1710. https://doi.org/10.3390/jox14040090. PMID: 39584955; PMCID: PMC11587131.
123. Tham DM, Gardner CD, Haskell WL. Clinical review 97: potential health benefits of dietary phytoestrogens: a review of the clinical, epidemiological, and mechanistic evidence. J Clin Endocrinol Metab. 1998;83(7):2223–35. https://doi.org/10.1210/jcem.83.7.4752. PMID: 9661587.
124. Monteiro CA, Cannon G, Levy RB, Moubarac JC, Louzada ML, Rauber F, Khandpur N, Cediel G, Neri D, Martinez-Steele E, Baraldi LG, Jaime PC. Ultraprocessed foods: what they are and how to identify them. Public Health Nutr. 2019;22(5):936–41. https://doi.org/10.1017/S1368980018003762. Epub 2019 Feb 12. PMID: 30744710; PMCID: PMC10260459.
125. Monteiro CA, Cannon G, Moubarac J-C, et al. The UN decade of nutrition, the NOVA food classification and the trouble with ultra-processing. Public Health Nutr. 2018;21:5–17.
126. Braesco V, Souchon I, Sauvant P, Haurogné T, Maillot M, Féart C, Darmon N. Ultra-processed foods: how functional is the NOVA system? Eur J Clin Nutr. 2022;76(9):1245–53. https://doi.org/10.1038/s41430-022-01099-1. Epub 2022 Mar 21. PMID: 35314769; PMCID: PMC9436773.
127. Elizabeth L, Machado P, Zinöcker M, Baker P, Lawrence M. UltraProcessed foods and health outcomes: a narrative review. Nutrients. 2020;12(7):1955. https://doi.org/10.3390/nu12071955. PMID: 32630022; PMCID: PMC7399967.
128. Zinöcker MK, Lindseth IA. The Western diet–microbiome–host interaction and its role in metabolic disease. Nutrients. 2018;10:E365.
129. Smaldino PE, McElreath R. The natural selection of bad science. R Soc Open Sci. 2016;3(9): 160384. https://doi.org/10.1098/rsos.160384. Erratum in: R Soc Open Sci. 2023;10(9):rsos231026. 10.1098/rsos.231026. PMID: 27703703; PMCID: PMC5043322.
130. Michal AL, Zhong Y, Shah P. When and why do people act on flawed science? Effects of anecdotes and prior beliefs on evidence-based decisionmaking. Cogn Res. 2021;6:28. https://doi.org/10.1186/s41235-02100293-2.
131. Okunade OA, Ghawi SK, Methven L, Niranjan K. Thermal and pressure stability of myrosinase enzymes from black mustard (Brassica nigra L. W.D.J Koch. var. nigra), brown mustard (Brassica juncea L. Czern. var. juncea) and yellow mustard (Sinapsis alba L. Subsp Maire) seeds. Food Chem. 187:485–90. ISSN 0308–8146 https://doi.org/10.1016/j.foodchem.2015.04.054.
132. Costa G, Plazanet I. Plant cell wall, a challenge for its characterisation. Adv Biol Chem. 2016;6:70–105. https://doi.org/10.4236/abc.2016.63008.

133. Holland C, Ryden P, Edwards CH, Grundy MM. Plant cell walls: impact on nutrient bioaccessibility and digestibility. Foods. 2020;9(2):201. https://doi.org/10.3390/foods9020201. PMID: 32079083; PMCID: PMC7074226.

Chapter 11
Conclusions and the Future of Natural Product Research

Contents

11.1 Natural Products in Foods—Context and Importance 318
11.2 The Complexity of Secondary Natural Products in Foods 321
11.3 Epicurus, Empiricism and Democritus 323
11.4 Secondary Natural Products in Health and Disease 325
11.5 Added Value and Added Benefits of Secondary Natural Products 328
11.6 Optimisation of Food Constituents 330
11.7 The Future of Secondary Natural Products in Foods and Food Systems 331
References 332

Abstract While the primary purpose of food is to provide nutrition and a source of macro and micronutrients, secondary natural products are important in the differentiation between foods including taste, flavour, protection of the host species, additional health benefits for consumers and response to biotic and abiotic stress. Much of the non-pharmacological emphasis on secondary natural products in foods has focused on the removal of anti-nutritive components such as phytates, glucosinolates, saponins, cyanogenic glycosides and others. Optimisation of beneficial secondary natural products has been comparatively little studied since the starting point of optimisation of intake levels has been difficult to define or achieve. Secondary natural products are important in both health and disease—even though methods to determine how individual components (and mixtures of compounds) act on whole organisms. Part of the challenge is that secondary natural products have a different value from the perspective of the producing and the consuming species and this conflict lies at the core of how they are investigated. The classification of quorum sensing molecules as being members of the class of semiochemicals is important both from an organisational standpoint and as a way of thinking about them. Optimisation of levels of secondary natural products will require a redefining of what 'optimal' means and a better understanding of complex living systems and mixtures of compounds. Current paradigms designed to investigate the relevance and importance of food and food system derived secondary natural products have identified some of the challenges but appear to be failing to provide coherent solutions. This does not mean that the identified challenges are insoluble but perhaps the traditional reductionist, isolative and synthetic approaches are inappropriate.

B. Hanley, *Secondary Natural Products in Foods and Food Systems*,
https://doi.org/10.1007/978-3-032-00292-1_11

Keywords Health · Disease · Optimisation · Semiochemicals · Complexity

11.1 Natural Products in Foods—Context and Importance

While the primary purpose of food is to provide the macro and micronutrients necessary for health [1], secondary natural products provide significant levels of differentiation between foods. These include taste, flavour, protection of the host species, additional health benefits for consumers and response to biotic and abiotic stress [2]. Traditionally, the chemical examination of natural products has followed the 'analysis/synthesis' paradigm. Products have been isolated from natural sources, subjected to analysis (either of the whole entity of after controlled breakdown into constituent parts), compared with known structures from related sources then, as a final step in the confirmatory scheme, synthesised in order to absolutely confirm the proposed structure. The use of biomimicry and the understanding of the degree of relatedness of natural products led chemists to hypothesise about how molecules were built in nature and this branched out into the study of biosynthesis. Knowing how small molecules were assembled would only satisfy one segment of the curiosity of chemists about how the natural world functioned. While the use of the enzymes by living systems was known for millennia (for example in the development of fermentation), the concept of catalysis was not developed until the eighteenth century. As with much empirical chemistry, the phenomenon of catalysis was observed before the process itself was named by Berzelius in 1836 [3]. The first enzymes to be discovered and characterised catalysed hydrolytic rather than synthetic reactions. This is perhaps consistent with the 'analysis first then synthesis' paradigm that prevailed in natural product chemistry, but it may also simply be a function of the interests of key figures of the time such as Pasteur and the ease with which hydrolytic systems could be studied. The history of the development of enzymes as a distinct entity (separable but derived from living systems) has been described [4]. It was reasonable that the reductionist paradigms that applied to secondary natural products should also apply to enzymes and other biological macromolecules such as carbohydrates, lipids and, in particular, to proteins. By the early part of the twentieth century, the reductionism that characterises chemistry led to the isolation and characterisation of individual enzymes including urease, pepsin, trypsin, and chymotrypsin. These were activity-guided exercises rather than investigations of physicochemical properties The two principal drivers for the use of enzyme-catalysed processes were (as mentioned previously) the production of bulk chemicals such as citric acid and the utility of biological systems in the production of biologically active entities such as ephedrine [5]. As purification of enzymes led to their structural elucidation, it became clear that, unlike DNA, the structures of proteins did not fit a single model (there was no protein equivalent to the ubiquitous double helix of DNA) and the structure of individual protein and enzymes would have to be 'solved' uniquely. Following the structural insights provided by reductionist isolation and analysis of individual enzymes, the interest in proteins (and enzymes) shifted towards understanding the biological complexity of the DNA/RNA/protein process. Despite this, the development of radioisotope tracer techniques facilitated the study of biosynthetic pathways of proteins [6] and other molecules and reductionism remained as an important driver [7] as knowledge of the existence of primary, secondary and tertiary structures of proteins (and the relationship between them) began to be grasped as the way they were constructed in vivo was deconvoluted. Research focussed on readily available and biologically important materials which were accessible in abundance (unlike many secondary natural products. Both haem [8] and chlorophyll [9] have been studied from the context of their core structure, their biosynthesis and their role in the evolution of key energetic pathways of living systems (their biological function)—a functionality/structure duality [10]. The study of the biosynthesis of individual molecules, the pathways that operated and the way the synthesis was controlled by living systems (including the relationship between genes and protein primary structure) later provided important and useful information about how biosynthetic systems might be manipulated.

In the context of foods, manipulation of secondary natural products has tended to focus on the removal of deleterious (anti-nutritional) factors and the maximisation of yield, ease of harvest and resistance to disease and pests. Initially this was carried out by crossing compatible varieties in order to maximise favourable agronomic traits. Examples of these are the introduction of leafless (and semi-leafless) pea varieties [11] and the study of their agronomic characteristics [12] and the variety choice and breeding characteristics of wheat [13]. Beyond agronomy and developing strains with favourable secondary natural product profiles became prominent with the development of the 'double zero' rapeseed variety [14]. The adaptations were designed to improve the yield, ease of harvesting or nutritional profile of the crop. Other breeding programmes targeted disease resistance and aspects such as growth rate and agronomic characteristics (see above). While breeding programmes have led to the introduction of varieties with greatly improved characteristics, these are limited and genetic manipulation [15], gene editing [16] and molecular marker assisted programmes [17] have been developed (and adopted with varying degrees of success). The primary driver for many new and manipulated varieties is the improvement in agronomic properties (in the widest sense). In terms of secondary natural products in foods, many of the studies (as noted previously in the case of rapeseed [14]) have been oriented towards the elimination of secondary natural product profiles and compounds with undesirable properties. There are some exceptions in which optimisation of production of certain secondary natural products is desirable—for example the elucidation of the taxol biosynthetic pathway may enable the development of synthetic biology approaches to this important pharmaceutical [18, 19] and the development of similar biosynthetic strategies may be employed in the production of other valuable secondary natural products. Such strategies focus upon molecules of known pharmacological efficacy. In the case of most non-pharmacological secondary natural products, the products have a comparatively low intrinsic value since they do not explicitly cure a disease state. This is often because the health benefits are relatively ill defined in terms of clinical targets and dietary (or dietary supplement) levels are sufficiently high to remove a major driving force for the development of enhanced means of production. An example of this is the focus on anthocyanins in tomatoes

[20]. While genetic modification and breeding programmes have led to the production of tomato plants whose fruits contain enhanced levels of certain anthocyanins, it is unclear if nutritional supplementation from a genetically modified source will be of particular benefit [21]. In addition less than 2% of ingested anthocyanins are absorbed and they are completely eliminated from the body in less than 6 h [22]. The European Food Safety Authority (EFSA) has rejected a number of claims relating to anthocyanins and beneficial physiological effects [23]. In the absence of a specific disease or nutritional deficiency target, the rationale for optimisation of production of secondary natural products is difficult to sustain. The application of pharmaceutical level evidence provision to secondary natural products in foods that have long term or chronic effects, and are likely to be subjected to confounders, is flawed. In addition, many of the claimed effects relate to subjects who are not suffering from a curable disease state (unlike the target population for pharmaceuticals who are in a non-equilibrium, disease state) and therefore endpoints that rely on a poorly constructed concept of 'health' are misleading and likely to be unquantifiable. In addition, it has been suggested that a single, all-encompassing definition of 'health' is untenable and a multidimensional approach is required [24]. The ability to adapt and self-manage has also been suggested [25] and this introduces the concept of healthy homeostasis and resilience.

Natural products in foods provide both diversity of characteristics and of use with diverse chemical structures giving the potential for polyfunctionality.

In many cases, the evolutionary pressure to develop structures with particular properties is not reflected in the primary purpose for which the molecules are subsequently employed by organisms that consume the food and this can create challenges in developing a structure/properties investigation strategy. In addition, the principle of emergence in chemistry [26] in which diverse chemical motifs are combined in natural products to give either compounds with sets of defined properties or, on occasion, novel properties will, together with artificial intelligence approaches, enable new chemical spaces (linked to biological properties) to be defined in a controlled and predictable fashion [27]. Developing the link between certain specific structural motifs and cellular responses (including phenotype) has the potential to combine reductionism (identifying common structural motifs common to a number of natural products) with a better understanding of the multiplex nature of natural products and their synergies [28]. The importance of emergent properties of natural products will become increasingly apparent as a route towards the study of the interactions of individual and groups of natural products with complex biological systems. There also exists the possibility of combining heterologous gene pathways into different hosts to enhance the production of mixtures of natural products [29] with novel emergent properties. In addition to the rise of emergent properties, natural products may also change the environment such that different properties may develop. If these are linked to genetic transcriptional regulators then a cascade of responsive emergent properties may be created that will vary according to conditions and temporal cues. Control of these properties could enable different groups of natural products to be produced in response to the external environment. Pre-harvest production of antifeedants, for example, may be followed by the synthesis of putatively beneficial components such as anthocyanins or other natural products with an individual or combined emergent property that will be of benefit to the consumer. Indeed 'emergent engineering' could become an important strategy for adding value to food products through the manipulation of secondary natural products.

11.2 The Complexity of Secondary Natural Products in Foods

Natural products in foods and food systems exist in a bilateral exchange between the host organism and the consuming species. These are often in conflict since, from a host standpoint, the purpose of many natural products is to lessen predation or provide a growth or survival advantage while the consuming species either develops methods to minimise the deleterious impacts of natural products or makes use of their ancillary properties. Some specific examples will illustrate the underlying paradigm. In addition to the different biological effects, there is also a temporal dimension. If natural products are synthesised to discourage consumption of the host by predators then the effect must be virtually immediate. This suggests that a sensory or acutely toxic effect (including enzyme inhibition) is likely to be most effective. Longer term effects (for example carcinogenicity unless it is a side effect of a more generic DNA-damaging effect resulting in cell cycle arrest, is not an acutely toxic response to ingestion of a toxin) are less likely to be an effective way for a host species to gain a survival advantage. Many of the effects of natural products, however, are not apparent in the consuming species in the short term and only become realised in the longer term. This is particularly the case for beneficial effects (for example the health promoting properties of many phenolic compounds). In addition, many species have effective methods for accelerating the excretion of xenobiotics including the inducible expression of specific xenobiotic transporter proteins [30]. If natural products are classified on the basis of their short term, toxic potential then some of the most potent (and well regulated) bioactive metabolites are those produced by microbes. In some cases, their effect is generic while for others it is specifically targeted against organisms which are in direct competition for space in the ecosystem or for resources [31]. An example of the former is 3-nitropropanoic acid which is an inhibitor of succinate dedhydrogenase and is, therefore, active at a primary metabolic level [32]. It functions in humans and animals primarily as a neurotoxin and is a physiological mimic of Huntington's disease [33]. Metabolism of the toxin is an important route to its biological deactivation and this may occur as a result of the presence of a decarboxylase enzyme as well as conjugation to amino acids. More specific natural toxins produced by microbes include penicillins and cephalosporins which prevent bacterial cell wall formation and tricothecenes which act on eukaryotic ribosomes and inhibit translation.

While there has been considerable interest in the anti-feedant properties of some plant compounds—particularly in the context of predation—and these will be discussed in more detail later, plants also contain a range of specific antimicrobial compounds that act to deter or prevent microbial colonisation [34]. In addition to compounds that act directly on specific metabolic features of pathogenic microorganisms and whose biosynthesis can be stimulated by damage to the plant or the detection of specific, microbially-derived elicitors, there is also a growing interest in the study of inhibitors of the microbial quorum sensing system [35] with concomitant disruption of microbial cell–cell communication. Features such as virulence and biofilm formation are controlled by microbial quorum sensing so its disruption is a distinct and direct challenge to microbial survival and proliferation. Antimicrobial plant compounds have been classified as phytoalexins [36] and phytoanticipins [37]. The difference between the two classes is governed by their biosynthesis and how they are produced. The production of phytoalexins is a de novo event induced by the detection of a pathogen while phytoanticipins are formed from pre-existing entities and are not produced by transcription or translation in response to a pathogenic challenge. The definition is based upon mode of production rather than biological activity is, therefore, different to the classical secondary natural toxin definition that derives either from biological activity (as with the specific enzyme inhibitor, 3-nitropropanoic acid) or producing species (as with aflatoxins from *Aspergillus flavus*) or chemical attributes (the alkali-like alkaloids). The definition of a compound as a phytoalexin or a phytoanticipin can be somewhat fluid and this is best illustrated by examples. For example, the production of certain glucosinolates is constitutive and damage to a plant leads to hydrolysis of the thioglucoside to give a range of isothiocyanates which have both antimicrobial and insecticidal activity [38] and glucosinolates and their products are, therefore, phytoanticipins. In some cases specific pathway activation and the formation of glucosinolates in response to fungal infection has also been observed [39] and this is clearly a phytoalexin response. While the definition of plant protective factors may appear to be a somewhat esoteric area of investigation, it has a practical outcome and is important when considering either the production of disease-resistant plant varieties or strategies for gene editing to optimise the resistance of plants to microbial attack. In both cases, successful plant growth and yields can be dependent on the optimisation of endogenous plant protection factors.

As noted previously, the production of many secondary natural products has a different function in the host compared to the consuming species. For example, he beneficial attributes of many phytochemicals has been noted [40] however this would appear to be somewhat serendipitous since the disease mechanism that the compounds elicit protection against either do not occur in the producing species or confer no growth or survival advantage [41]. The enigma is, therefore, why do plants (and other organisms) produce such a range of chemically distinct species with a complex biological profile? The principal advantage for the producing species in the production of secondary metabolites is that they confer some growth or survival advantage and are, therefore, worth the biosynthetic investment [42]. In addition to protection against environmental stresses (biotic and abiotic), plant secondary metabolites may also affect important factors such as colour and odour and thereby have an impact on pollination, reproduction and growth [43]. As noted previously, there are a number of ways of classifying secondary natural products including structural characteristics (for example, glucosinolates, alkaloids), biological activity (estrogens, antimicrobials) or as either storage or components of biological cycling of materials. One other possible definition is to consider compounds as belonging to one of two classes based upon biological activity (function) rather than chemical identity (structure). The first category comprises those whose biological activity confers the same advantage to the producing and the non-target consuming species. This category could include many antibiotics which can be used by the producing organism as a microbial defence and by other organisms (for example, humans) to protect against the consequences of microbial infection [44] Another example would be anti-parasitic compounds [45] which have the same function in different circumstances and offer the same benefit to different organisms. The second group would be defined as those compounds whose biological profile for the consuming species is different to that of the producing organism. An example of this would be vinyl oxazolidinethione which contains a distinct moiety (NHC = S) which enables it to bind to a specific polymorphic form of a bitter receptor [46] and to therefore act as a feeding deterrent to any species that contains this form of the bitter receptor.

This aspect of the molecule is of greatest value to the producing species since it directly deters predation. The compound also has the trivial name 'goitrin' and the S form is, indeed, goitrogenic. Such activity, while being significant for consuming species that produce thyroid hormones, confers no advantage for the producing species. The effect is simply too long-term to confer any particular advantage to the producer. By considering the polyvalent activities of secondary natural products, it is possible to build up a more complete view of their derivation and the association of compounds with the environment in which they exist. Similar considerations affect other secondary natural products such as isothiocyanates. Their immediate effect is, because of their pungency, to deter predation while longer term effects (stimulation of detoxication enzyme systems) may be of longer term benefit to predating species.

11.3 Epicurus, Empiricism and Democritus

Many of the properties of secondary natural products are associated with their nonmedicinal properties (such as, as demonstrated previously, bitterness). While the disease and health properties of substances will be considered elsewhere (Sect. 11.4), the non-pharmacological attributes of secondary natural products will be considered first in the context of their philosophical basis and the role of empiricism.

One way of considering secondary natural products is from the standpoint of different disciplines and their interaction (or crossover point). These can often be incompatible both in their attributes and in their approach. For example, chemistry is reductionist in attribute and empirical in approach and seeks therefore to develop iterations that bring only a part of the system into focus rather than the system as a whole (which is generally too complex to deal with holistically). Biology, on the other hand, can be focused on whole living systems (attribute) even if precise understanding is not possible and is, therefore, much more holistic (approach). The question to be answered is what the value is in developing a profound knowledge of each discipline individually in order to gain knowledge of the molecular/living system interaction. Each one, individually, can only give a particular perspective so the added value lies in the boundary region between them where both perspectives are visible. In outlook and attributes, chemistry and biology resemble two immiscible liquids with appropriate thermodynamic properties. The interface between the two liquids represents the boundary where exchange can take place. In the case of chemistry and biology, the boundary region is where the exchange of knowledge can take place. If chemistry is concerned primarily with molecules and their interactions and biology with systems then the interaction between a molecule and the specific part of the system it interacts with (with consequent impacts on the whole system) is the interface. To carry the analogy further, a eutectic system is a homogenous mixture in which the melting point of the mixture is lower than that of the constituents. The phase change can, therefore, take place with a lower energy input than that which would be necessary in a non-eutectic mixture. In the case of biology and chemistry, under a specific set of conditions (with the optimal mixture of the two approaches), information and knowledge will flow more freely between them and will then more readily define the attributes of both the molecules and the biological system. The activation energy for the acquisition of knowledge will be less if both chemistry and biology are used than if each is used individually. A crucial aspect of this approach is the incorporation of an iterative exchange between the disciplines and, therefore, some degree of shared principle. Each will contain elements that are detailed and unique but the intellectual eutectic point represents a Janus-like window between the two specialisms.

The philosophical meeting point between Epicurus and Democritus provides an alternative way of considering secondary natural products that is both molecular and systemic. The common root for the philosophies of both Epicurus and Democritus is the concept of atomism. For Democritus, the principles of atomism are concerned both with indivisible elements (atoms) and the void between them. This is an important concept since, if atoms exist and move then the void (that which is not) must also exist so the one implies the other. Epicurus also accepted the existence of atoms, which could not be seen, and a void. However Epicurus did not concentrate on the development of theoretical models to explain the macroscopic consequences of the existence and properties of atoms and voids but, rather like a chemist dealing with whole, observable systems, he takes the ultra-empirical approach of considering what he can perceive as being the most important parameter rather than the whole system. This approach assumes that nature—both on a molecular and on a systemic level, is consistent. Observations at one level are able to be translated to the other. While there are various complexities to the world views expressed by Democritus and Epicurus, the essential truth that molecular and macroscopic properties are related (even if we do not fully understand how) is an important concept in the better understanding of the relationship between secondary natural products and the living systems with which they interact. In short, the observed macromolecular behaviour of systems is consistent with the molecular interactions that are postulated but cannot be directly observed. Secondary natural products in foods are consumed and act at a molecular level which is then observed in terms of clinical endpoints. The challenge is to determine the chain of causality between molecular interactions and outcomes which can be very small and prone to the influence of confounders. From both an Epicurean (molecular) and a Democritean (systemic) standpoint, there are significant unknown factors that make arriving at definitive conclusions about the link between secondary natural products and health & disease particularly challenging. This relationship is examined in more detail in the next part of this chapter (Sect. 11.4).

11.4 Secondary Natural Products in Health and Disease

In contrast to essential macro and micronutrients, the effect of secondary natural products on health and disease, as has been noted previously, can be difficult to determine [1]. There are a number of confounding factors and these can be categorised as follows:

- The nature of the chemical species and their physiological role.
- The presence of multiple bioactive entities—both agonists and antagonists.
- The bioavailability and metabolism of bioactive constituents.
- The nature (chronic), levels (often sub-millimolar) and specificity of bioactive agents and the effects of long-term exposure. The latter is particularly significant for secondary natural products in foods since, in many cases, there will be a lifetime of exposure.
- The correlation between molecular and clinical effects (already discussed in Sect. 11.3).

The nature of secondary natural products is that they are often present in small amounts. Their production requires the application of biosynthetic capability and they are often produced by the diversion of primary products, which have a defined biological function, into secondary metabolite production. This may be a consequence of a direct interaction between primary and secondary metabolites. For example, glucosinolates are derived from amino acids [47] and the biosynthesis and metabolic engineering of glucosinolate production has consequences for amino acid stores and regulation. Similarly the cross-talk between primary and secondary metabolism has been noted in the biosynthesis of polyketides [48]. While such interactions are holistically important and may provide some information about the temporal expression of certain pathways, determination of which structures are actually present in a system is essential in devising schemes for secondary natural product isolation and this can be challenging. One of the most important approaches to determining which structures are present in a biological system is the use of quantitative structure–activity relationships (QSAR) methodology. This can help to define the types of molecules that are present based on a biological profile. In essence QSAR is a mathematical tool that links aspects of the physicochemical features of a molecule with its biological properties and vice versa. In practical terms, QSAR models can help to guide the development of a fractionation scheme to isolate a purer compound from a complex mixture that includes the products of both primary and secondary metabolism. One of the simplest examples is the isolation of alkaloids. These are defined as molecules with an ionisable nitrogen atom therefore they can be extracted into an acidic aqueous solution then, after neutralisation, back extracted into an apolar solvent. The problem is that, while some structural features can enable broad classes of compounds with similar physicochemical (and structural) characteristics to be isolated, separation of the individual components generally requires fractionation based on chromatographic characteristics [49] often combined with an assay that is specific for the biological properties of the compound of interest. There are two problems with the fractionation/assay approach. The first is that the biological properties that are of interest to a natural product chemist are rarely those that bear a relationship to the reason why the compound is being synthesised by the host species (this was referred to previously in Sect. 11.1). For example, histamine receptors are important mediators in eukaryotic systems [50] however a number of plants have been found to contain molecules with histamine antagonist properties [51]. It is unclear what the physiological role of histamine antagonists in a plant would be and therefore developing strategies to find novel histamine agonists or antagonists can only be an exercise in action at a distance rather than examining some aspect of the producing plant itself. There is no aspect of a given plant (or other biological system) that will flag up the presence of histamine receptor ligands. Many bioactive compounds synthesised by living organisms would not a priori seem to be implicated in a useful biological trait for the producing organism. The second problem is in part a consequence of the first. The amounts of the bioactive compound with interesting and useful properties are frequently not optimised with the result that only small mounts of the compound of interest are formed and its biological properties will be unoptimized. This means that a putative lead compound may have only relatively weak activity and this activity may even be lost during the fractionation procedure if it is the consequence of the presence of more than one distinct chemical entity. While an approach that relies upon an assay of the biological activity of a fraction from the host, producing species may be superficially attractive since it is more likely to result in the isolation of an optimised compound, there may be no correlation between the activity that is of value to the host and that which is the focus of the activities of the natural product chemist. For example, caffeine is bitter (like many alkaloids) and can act as an anti-feedant to reduce predation by insects [52], however, for humans it is a stimulant and is purported to impact upon cognitive function [53]. Caffeine also has some insecticidal activity (phosphodiesterase inhibition) however this is dependent upon the physicochemical environment [54] and insects that predate upon the coffee plant have *Pseudomonas* bacteria in their gut which contain an enzyme that demethylates caffeine thereby suppressing its residual insecticidal activity [55]. While genetic determinants of liking (or, at least, non-aversion) to caffeine have been suggested [56] and a correlation between certain aspects of social cognition and the genetics of caffeine bitterness perception has been suggested, any causative link remains unproven and controversial [57] and the physiological benefit of caffeine consumption as indicated by bitterness perceived by a non-target species, remains unproven.

The nature of the biological activity of secondary natural products—and in particular the differentiation between the desired effect for the producer and the consumer, is further exacerbated by the presence of a number of chemical entities with apparently conflicting biological activities (for example, agonists and antagonists). These activities will be irrelevant to the producing organism but will be significant for the consuming species. One of the clearest examples of agonist and multiple antagonist activity relating to secondary natural products from foods are bitter compounds [58] which may demonstrate different types of activity and, furthermore, this may manifest in different physiological effects depending on the location of the bitter receptor [59]. In addition, mixtures of compounds that bind to different taste receptors can act synergistically to provide complex taste effects. The relationships between different bitter compounds that bind to bitter receptors is difficult to elucidate purely in terms of structure–activity relationships. As noted above, the extra-oral presence of taste receptors at a range of sites and their link to the immune response of the consumer [60] suggests that the biological consequences of a particular receptor may be governed both by the ligands that bind to it and where the receptor is located [61]. A range of agonists and antagonists from natural sources have been isolated and studied [62] however it is likely that, in many cases, the receptor-ligand binding activity is an evolutionarily unanticipated consequence of the biosynthesis of the secondary natural product.

One of the most significant challenges to the realisation of the biological potential of secondary natural products is the uptake and bioavailability of specific chemical entities. This was mentioned previously in the context of QSAR and physicochemical properties (see above). Measurement of indices such as hydrophobicity/hydrophilicity have been used for the in silico assessment of drug molecules [63] but other measures that can affect absorption efficiency include metabolism and excretion. In addition, many natural products occur as hydrolysable entities (for example, glycosides) and have to be converted into the active form after ingestion. In addition to limiting the efficacy of natural products in an in vivo environment, limited absorption can create a disconnect between clinical effects and in vitro bioassay-guided optimisation of QSAR. The use of pharmacological approaches to optimise bioavailability has been suggested [64] however this type of approach may not be suitable for the chronic effects found for many secondary natural products that occur in foods. Natural products in foods are not directly analogous to the bioactive constituents of foods [65] and a purely pharmacological strategy may not be appropriate.

The chronic exposure levels of secondary natural products in foods is an important differentiating factor between foods and medicinal products. In addition, the effects of secondary natural products may be a consequence of exposure to multiple elements. For example, goitrogenic agents derived from plants of the genus *Brassica* includes both vinyl oxazolidinethione (goitrin) and thiocyanate ion. Both are produced as a consequence of the hydrolysis of glucosinolates but they are produced from different precursors, are structurally different and have very different mechanisms of action. Synergies between them are likely to exist and these are more likely to occur since both active agents derive from the same plant species. Many secondary natural products in foods have been associated with chronic and long term conditions such as cancer [66], osteoporosis [67] and reduction in blood pressure and hence reduction in the risk of cardiovascular disease [68]. The most significant challenges to the development of a chain of causation between dietary secondary natural products and clinical endpoints include the effect of confounders, the lack of predictive, mechanistic biomarkers that enable studies to be carried out in shorter periods of time than a whole lifetime and maintenance of a particular dietary regime over an extended period of time. This means that claims for the effective treatment of diseases with secondary natural products from foods should be treated with caution.

The difficulty in establishing the correlation between molecular and clinical effects has already been discussed (Sect. 11.3) and, while improvements in model systems have been developed, these remain an experimental and clinical challenge.

11.5 Added Value and Added Benefits of Secondary Natural Products

As already noted (Sect. 11.4) the added value of natural products is dependent upon the organism (producer or consumer) and its perspective in relation to the product. In the case of the producing organism, the secondary natural product confers some specific advantages. This can either be a deterrent to consumption or it can act to reduce the growth of competing organisms or pathogenic microbiota. The second advantage (particularly in the case of plants) is the attraction of organisms involved in, for example, pollination or seed dispersal [69]. The range of attractants and deterrents produced by one organism that affect another organism (either of the same type or different) are called semiochemicals. There are a range of subclasses of these including pheromones and allelochemicals. One particular class of semiochemicals, which have been referred to earlier, are quorum sensing molecules. At first thought of as being microbially derived and active, it has been suggested that these molecules (and inhibition of their action) may make a significant contribution to inter-kingdom signalling [70] and, while quorum sensing remains key specifically to prokaryotic microbial responses to the environment [71], the discovery of a peptide that regulates virulence in a fungal species suggests that eukaryotic organisms can make use of a similar mechanism [72]. It may be tempting to consider the collective action mediated by quorum sensing mechanisms to be part of a pathway towards differentiation however microbial communities in environments such as biofilms have been studied and, while some degree of interdependence is demonstrated, this does not imply any level of specific purpose leading to differentiation. Modulation of bacterial quorum sensing by eukaryotes [73] and the use of quorum sensing by unicellular eukaryotes such as yeast [74] suggests that quorum sensing and modulation of signalling pathways is a key aspect of all living systems. One interesting aspect of inter-kingdom communication using quorum sensing is the interaction of an important bacterial quorum sensing molecule (competence stimulating peptide) with a bitter taste receptor (T2R14) in gingival epithelial cells leading to stimulation of the immune response [75]. Another example is modulation induced by one of the most common autoinducer molecules produced by Gram negative bacteria (N-(3-oxododecanoyl)-L-homoserine lactone) (AHL-12). This molecule will bind to the eukaryotic bitter receptor T2R38 and has been implicated in the induction of enhancement of phagocytosis, increased expression of adhesion receptors and induction of chemotaxis [76] Indeed T2R38 appears to be an essential component of the early response to bacterial infiltration through its interaction with bacterial quorum sensing molecules [77]. The extra-oral role of bitter taste receptors has been mentioned previously and is the subject of frequent reviews [78]. In the context of food-related natural products, bitter taste receptors have also been found to use advanced glycation end-products (produced by the glycation of amino acids in proteins and found as processing contaminants) [79] as ligands [80]. The increasing realisation of the role of G-protein coupled receptors (GPCRs) such as bitter receptors in different organs and their use of secondary natural products as ligands, may stimulate further investigations into the biological significance of a range of secondary natural products. While GPCRs do not occur in prokaryotes (although homologues have been found in bacteria [81]) and it has been suggested that they may have evolved from prokaryotic genes [82]. If this is the case then prokaryotic organisms may have a hitherto unappreciated functional use for secondary natural products from a range of sources (endogenous, prokaryotic and eukaryotic).

While the more complete understanding of the functional properties of secondary natural products is important, simply having a fuller understanding of the structures that are present is an important starting point [83]. In the past, this has involved the natural product chemist in the optimisation of extraction and separation of individual components (some of which are only present in small amounts). There are a number of other issues (loss of bioactivity during fractionation, lost synergistic effects between different components). The determination of all the molecular components of a system is the basis for metabolomics. The principle of metabolomics is the analysis of all the metabolites that are present in a living system [84]. All the metabolites are analysed (generally using multiple techniques) and a vast array of data generated in which an analytical pattern is correlated with types of biological activity. Such data-intensive methods are scientifically and philosophically different to the classical, reductionist analysis/separation and then synthesis cycle that was formerly used in assessing the importance of natural products. While much of the interest has been on the development of methods to understand and exploit useful pharmacological properties, metabolomic approaches may also be adopted as part of the development of new, alternative methods (NAMs) to assess the toxicity of secondary natural products and their metabolites.

11.6 Optimisation of Food Constituents.

The optimisation of food constituents in terms of secondary natural products is complex. It crucially depends on what characteristics are defined as most useful (and important) and how that assessment is made. The idea that natural products have fixed properties is, as has been demonstrated, simplistic. Even between the creator and the consumer of a product, there can be significant differences in the effect of natural products. In addition, the longer term effects of some of these may not be apparent. Compounds that bind to taste receptors in the oral cavity and give rise to aversion may, upon binding to the same receptors situated elsewhere, promote the immune system and help in preventing infection. They may also interact with completely different biological systems to stimulate excretion and detoxication. To use an analogy of the philosophy of Nietzsche, true benefits may accrue from something that is, at first encounter, unpleasant. In other words "A little poison now and then: that maketh pleasant dreams. And much poison at last for a pleasant death." In this particular case, not all that is bitter is to be avoided at all costs. Similarly some sweet compounds may have deleterious effects so the avoidance of bitterness and the attraction to sweetness can, in the longer term, be disadvantageous. Not all bitter substances are bad and not all sweet substances are good. This means that the optimisation of intake of secondary natural products—even on a single compound basis, is not a trivial undertaking.

As noted previously, much of the interest in developing food systems has focused upon yield, processing (including harvesting) characteristics and resistance to pests. Of these, only the last of them has a direct link with secondary natural products and, even then, it is important to balance toxicity and taste with the capacity to resist biotic and abiotic stress. This is particularly important since the levels of production of protective secondary natural products can be increased under stress conditions along with other responses [85]. These other responses can, as mentioned previously, include heat shock proteins [86] and cytokinin production [87] among others.

Some of the challenges in deciding on the optimal levels of secondary natural products are the quantification of the levels of compounds, the beneficial causal effect that is required and the length of exposure. In addition, it is rare for a producing species to contain just a single secondary natural product so potential synergies and antagonistic effects must also be considered. This is particularly the case when pharmacologically active substances are the object of attention [88] however in the case of most food substances where chronic effects are observed, determining the effect due to a single component in the presence of significant confounders is also challenging.

In the case of overtly toxic entities (for example specific dietary toxins, mutagens and compounds with organ-specific toxic effects such as neurotoxins), the optimal level of consumption is the consequence of, where possible, complete elimination of the entity from the diet. There is significant regulatory apparatus to define steps that should be taken and this is supported by considerable analytical expertise [89] however this can lead to situations where regulatory limits are set based upon analytical limits of detection rather than a toxicological assessment. The optimal level for the intake of beneficial secondary natural products is much more difficult to ascertain. It relies on quantification of the principle that health is more than just the absence of disease and on the establishment of a link between a specific component and a positive clinical outcome. In practice, this is difficult to achieve and is, for the most part, the subject of very generic recommendations for specific populations [90] or relates to specific disease states [91]. There has been considerable interest in the concept of personalised (or stratified) nutrition [92] and for recommendations to be made based upon predispositions (often genetic), lifestyles or other specific personal attributes (age, gender etc.) however such approaches are expensive and, once again, are subject to significant confounders. In terms of the health of the general population, they are unlikely to be significantly impactful.

11.7 The Future of Secondary Natural Products in Foods and Food Systems

The tradition of natural product research being fuelled by the search for bioactive principles of pharmacological significance has underwritten much of the progress that has been made as a result of the application of reductionist chemical principles [93]. It has been realised, more recently, that many organisms have the biosynthetic capacity to produce many more bioactive compounds than were previously thought to occur [94]. Genome mining has been a fruitful area for exploration and this has largely been driven by the advances in understanding of the mechanisms of disease progression and the potential for the discovery of new pharmacologically active agents. In the case of food-related biological systems, there has not been the same impetus and many traditional methods of, for example, plant trait analysis have been employed to investigate factors that may influence particular attributes [95]. Many studies have concentrated on those aspects that may produce immediate results—agronomy, disease resistance and yield. In those instances where secondary metabolites have been 'optimised', this has often been with the aim on reducing toxin or anti-nutritional component levels [96]. In addition to traditional breeding, advanced molecular techniques have also been developed to reduce anti-nutritional factors while maintaining desirable qualities such as disease and pest resistance and yield [97].

The prevalence of chronic diseases has been a feature of recent history and this has been linked to dietary factors [98] and to lifestyle [99]. It has been estimated that the global cost of chronic disease is in the region of trillions of dollars per annum [100] so, in addition to the human cost, there is a significant economic impact to be considered. The potential of secondary natural products to reduce the burden of chronic disease outcomes has been the subject of considerable speculation however these are often of a generic nature and deal with patterns of behaviour rather than specific components [101] and while attempts have been made to link secondary natural products with specific aspects of some diseases (particularly cancer and inflammatory conditions [102]), these have often been piecemeal in nature and take insufficient account of confounders. In addition the nature of chronic diseases is that they are long term conditions and the application of techniques developed for short term, pharmacological testing are unlikely to provide a foundation for success [103]. The in vitro testing procedures that have been developed are oriented towards short term outcomes although longer term culture systems have been investigated [104] however the emphasis has remained on toxicity testing rather than benefit identification [105].

The development of optimised food and food systems requires that certain milestones must be identified and achieved. These include the following:

1. The development of a clear understanding of the mechanistic relationship between secondary natural products in foods and food systems and health/disease. This includes both in vitro and in vivo techniques using biomarkers that are predictive of eventual clinical outcomes in order to obviate the need for 'full lifetime' experiments.
2. A better understanding of the synergistic and competing effects of secondary natural products at realistic exposure levels and timeframes.
3. A strategy for the development of optimisation of desirable secondary natural product biosynthesis at different growth/developmental stages. For example, the production of anti-feedant compounds in the early stage of growth of plants followed by the production of beneficial secondary natural products at the time of (or post) harvest.

Current paradigms designed to investigate the relevance and importance of food and food system derived secondary natural products have identified some of the challenges but appear to be failing to provide coherent solutions. This does not mean that the identified challenges are insoluble but perhaps the traditional reductionist, isolative and synthetic approaches are inappropriate.

References

1. Inui T, Hanley AB, Tee E-S, Nishihira J, Tontisirin K, Van Dael P, Eggersdorfer M. The role of nutrition in ageing Asia: what can be implemented with the existing insights. Nutrients 2021;13:2222.
2. Drasar PB, Khripach VA. Growing importance of natural products research. Molecules. 2019;25(1):6. https://doi.org/10.3390/molecules25010006. PMID: 31861374; PMCID: PMC6983153.
3. Smith JK. History of catalysis. In: Horváth I, editors. Encyclopedia of catalysis. 2010. https://doi.org/10.1002/0471227617.eoc098.pub2.
4. Heckmann CM, Paradisi F. Looking back: a short history of the discovery of enzymes and how they became powerful chemical tools. ChemCatChem. 2020;12(24):6082–102. https://doi.org/10.1002/cctc.202001107. Epub 2020 Oct 1. PMID: 33381242; PMCID: PMC7756376.
5. Hildebrandt G, Klavehn W, Manufacture of Laevo-1-Phenyl-2Methylaminopropanol-1, 1934, US1956950 A.
6. Rittenberg D. The state of the proteins in animals as revealed by the use of isotopes. Symp Quant Biol. 1941;9:283–9.

7. Wooley JC, Ye Y. A historical perspective and overview of protein structure prediction. In: Xu Y, Xu D, Liang J, editors. Computational methods for protein structure prediction and modeling. Biological and medical physics biomedical engineering. Springer:New York, NY; 2007. https://doi.org/10.1007/978-0-387-68372-0_1.
8. Kořený L, Oborník M, Horáková E, Waller RF, Lukeš J. The convoluted history of haem biosynthesis. Biol Rev Camb Philos Soc. 2022;97(1):141–62. https://doi.org/10.1111/brv.12794. Epub 2021 Sep 2 PMID: 34472688.
9. Blankenship RE. Early evolution of photosynthesis. Plant Physiol. 2010;154(2):434–8. https://doi.org/10.1104/pp.110.161687. PMID: 20921158; PMCID: PMC2949000.
10. Hendry GA, Jones OT. Haems and chlorophylls: comparison of function and formation. J Med Genet. 1980;17(1):1–14. https://doi.org/10.1136/jmg.17.1.1. PMID: 6988593; PMCID: PMC1048480.
11. Janusauskaite D. Productivity of three pea (Pisum sativum L.) varieties as influenced by nutrient supply and meteorological conditions in boreal environmental zone. Plants. 2023;12:1938. https://doi.org/10.3390/plants12101938.
12. Shen Y, Johnson EN, Syrovy LD, Warkentin TD, De Devini S, Shirtliffe SJ. Evaluation of yield and agronomic performance of leafed and semi-leafless pea blends. Agron J. 2022;114:2762–73. https://doi.org/10.1002/agj2.21134.
13. Mourad AMI, Alomari DZ, Alqudah AM, Sallam A, Salem KFM. Recent advances in wheat (Triticum spp.) breeding. In: Al-Khayri J, Jain S, Johnson D, editors, Advances in plant breeding strategies: cereals. Springer: Cham; 2019. https://doi.org/10.1007/978-3-030-23108-8_15.
14. Bell JM. Nutrients and toxicants in rapeseed meal: a review. J Anim Sci. 1984;58:996–1010. https://doi.org/10.2527/jas1984.584996x.
15. Karalis DT, Karalis T, Karalis S, Kleisiari AS. Genetically Modified Products, Perspectives and Challenges. Cureus. 2020;12(3): e7306. https://doi.org/10.7759/cureus.7306. PMID: 32313747; PMCID: PMC7164548.
16. Gaj T, Sirk SJ, Shui SL, Liu J. Genome-Editing Technologies: Principles and Applications. Cold Spring Harb Perspect Biol. 2016;8(12): a023754. https://doi.org/10.1101/cshperspect.a023754. PMID: 27908936; PMCID: PMC5131771.
17. Hasan N, Choudhary S, Naaz N, Sharma N, Laskar RA. Recent advancements in molecular marker-assisted selection and applications in plant breeding programmes. J Genet Eng Biotechnol. 2021;19(1):128. https://doi.org/10.1186/s43141-021-00231-1. PMID: 34448979; PMCID: PMC8397809.
18. Jiang B, et al. Characterization and heterologous reconstitution of Taxus biosynthetic enzymes leading to baccatin III. Science. 2024;383:622–9. https://doi.org/10.1126/science.adj3484.
19. Li Y, Zhang G, Pfeifer BA. Current and emerging options for taxol production. Adv Biochem Eng Biotechnol. 2015;148:405–25. https://doi.org/10.1007/10_2014_292. PMID: 25528175.
20. Gonzali S, Perata P. Anthocyanins from purple tomatoes as novel antioxidants to promote human health. Antioxidants (Basel). 2020;9(10):1017. https://doi.org/10.3390/antiox9101017. PMID: 33092051; PMCID: PMC7590037.
21. Gonçalves AC, Nunes AR, Falcão A, Alves G, Silva LR. Dietary effects of anthocyanins in human health: a comprehensive review. Pharmaceuticals (Basel). 2021;14(7):690. https://doi.org/10.3390/ph14070690. PMID: 34358116; PMCID: PMC8308553.
22. Mazza G, Kay CD, Cottrell T, Holub BJ. Absorption of anthocyanins from blueberries and serum antioxidant status in human subjects. J Agric Food Chem. 2002;50:45–8. https://doi.org/10.1021/jf020690l.
23. EFSA Panel on Dietetic Products, Nutrition and Allergies (NDA); Scientific Opinion on the substantiation of health claims related to anthocyanins from Ribes nigrum L. and improvement of visual adaptation to the dark (ID 2750) pursuant to Article 13(1) of Regulation (EC) No 1924/2006. EFSA J. 2011;9(6):2204. [14 pp.]. https://doi.org/10.2903/j.efsa.2011.2204.
24. Leonardi F. The definition of health: towards new perspectives. Int J Health Serv. 2018;48(4):735–48. https://doi.org/10.1177/0020731418782653. Epub 2018 Jun 14 PMID: 29902944.

25. Huber M, Knottnerus JA, Green L, Horst HVD, Jadad AR, Kromhout D, et al. How should we define health? BMJ 2011;343:d4163. https://doi.org/10.1136/bmj.d4163.
26. Luisi PL. Emergence in chemistry: chemistry as the embodiment of emergence. Found Chem. 2002;4:183–200. https://doi.org/10.1023/A:1020672005348.
27. Saldívar-González FI, Aldas-Bulos VD, Medina-Franco JL, Plisson F. Natural product drug discovery in the artificial intelligence era. Chem Sci. 2021;13(6):1526–46. https://doi.org/10.1039/d1sc04471k. PMID: 35282622; PMCID: PMC8827052.
28. Herzon S. Emergent properties of natural products. Synlett. 1835;29:1823. https://doi.org/10.1055/S-0037-1610242.
29. Park D, Swayambhu G, Lyga T, Pfeifer BA. Complex natural product production methods and options. Synth Syst Biotechnol. 2021;6(1):1–11. https://doi.org/10.1016/j.synbio.2020.12.001. PMID: 33474503; PMCID: PMC7803631.
30. Miyauchi H, Moriyama S, Kusakizako T, et al. Structural basis for xenobiotic extrusion by eukaryotic MATE transporter. Nat Commun. 2017;8:1633. https://doi.org/10.1038/s41467-017-01541-0.
31. Awuchi CG, Ondari EN, Ogbonna CU, Upadhyay AK, Baran K, Okpala COR, Korzeniowska M, Guiné RPF. Mycotoxins affecting animals, foods, humans, and plants: types, occurrence, toxicities, action mechanisms, prevention, and detoxification strategies—a revisit. Foods. 2021;10(6):1279. https://doi.org/10.3390/foods10061279. PMID: 34205122; PMCID: PMC8228748.
32. Bendiksen Skogvold H, Yazdani M, Sandås EM, Østeby Vassli A, Kristensen E, Haarr D, Rootwelt H, Elgstøen KBP. A pioneer study on human 3-nitropropionic acid intoxication: Contributions from metabolomics. J Appl Toxicol. 2022;42(5):818–29. https://doi.org/10.1002/jat.4259. Epub 2021 Nov 1 PMID: 34725838.
33. Brouillet E, Jacquard C, Bizat N, Blum D. 3-Nitropropionic acid: a mitochondrial toxin to uncover physiopathological mechanisms underlying striatal degeneration in Huntington's disease. J Neurochem. 2005;95:1521–40. https://doi.org/10.1111/j.14714159.2005.03515.x.
34. González-Lamothe R, Mitchell G, Gattuso M, Diarra MS, Malouin F, Bouarab K. Plant antimicrobial agents and their effects on plant and human pathogens. Int J Mol Sci. 2009;10(8):3400–19. https://doi.org/10.3390/ijms10083400. PMID: 20111686; PMCID: PMC2812829.
35. Koh CL, Sam CK, Yin WF, Tan LY, Krishnan T, Chong YM, Chan KG. Plantderived natural products as sources of anti-quorum sensing compounds. Sensors (Basel). 2013;13(5):6217–28. https://doi.org/10.3390/s130506217. PMID: 23669710; PMCID: PMC3690052.
36. Bednarek P, Osbourn A. Plant-microbe interactions: Chemical diversity in plant defense. Science. 2009;324:746–8. https://doi.org/10.1126/science.1171661.
37. VanEtten HD, Mansfield JW, Bailey JA, Farmer EE. Two classes of plant antibiotics: Phytoalexins versus phytoanticipins. Plant Cell. 1994;6:1191–2. https://doi.org/10.1105/tpc.6.9.1191.
38. Ratzka A, Vogel H, Kliebenstein DJ, Mitchell-Olds T, Kroymann J. Disarming the mustard oil bomb. Proc Natl Acad Sci USA. 2002;99(17):11223–8. https://doi.org/10.1073/pnas.172112899.
39. Bendarek P, Pislewska-Bednarek M, Svatos A, Schneider B, Doubsky J, Mansurova M, Humphry M, Consonni C, Panstruga R, Sanchez-Vallet A, Molina A, Schulze-Lefert P. A glucosinolate metabolism pathway in living plant cells mediates broad-spectrum antifungal defenses. Science. 2009;323:101–6. https://doi.org/10.1126/science.1163732.
40. Samtiya M, Aluko RE, Dhewa T, Moreno-Rojas JM. Potential health benefits of plant food-derived bioactive components: an overview. Foods. 2021;10(4):839. https://doi.org/10.3390/foods10040839. PMID: 33921351; PMCID: PMC8068854.
41. Liu RH. Dietary bioactive compounds and their health implications. J Food Sci. 2013;78:A18–25. https://doi.org/10.1111/1750-3841.12101.
42. Chen D, Mubeen B, Hasnain A, Rizwan M, Adrees M, Naqvi SAH, Shahzad SM, Hussain M, Alharby HF, Irshad MK, et al. Role of promising secondary metabolites to confer resistance against environmental stresses in crop plants: current scenario and future perspectives. Front Plant Sci. 2022;13: 881032. https://doi.org/10.3389/fpls.2022.881032.

43. Kossel A. Ueber Schleim und Schleimbildende Stoffe. Dtsch Med Wochenschr. 1891;17:1297–9. https://doi.org/10.1055/s-0029-1206875.
44. Patel P, Wermuth HR, Calhoun C, et al. Antibiotics. [Updated 2023 May 26]. In: StatPearls [Internet]. Treasure Island (FL): StatPearls Publishing; 2025 Jan-. Available from: https://www.ncbi.nlm.nih.gov/books/NBK535443/.
45. Wink M. Medicinal plants: a source of anti-parasitic secondary metabolites. Molecules. 2012;17(11):12771–91. https://doi.org/10.3390/molecules171112771. PMID: 23114614; PMCID: PMC6268567.
46. Wooding SP, Ramirez VA, Behrens M. Bitter taste receptors: Genes, evolution and health. Evol Med Public Health. 2021;9(1):431–47. https://doi.org/10.1093/emph/eoab031. PMID: 35154779; PMCID: PMC8830313.
47. Mikkelsen MD, Petersen BL, Olsen CE, Halkier BA. Biosynthesis and metabolic engineering of glucosinolates. Amino Acids. 2002;22(3):27995. https://doi.org/10.1007/s007260200014. PMID: 12083070.
48. West AR, Bailey CB. Crosstalk between primary and secondary metabolism: Interconnected fatty acid and polyketide biosynthesis in prokaryotes. Bioorg Med Chem Lett. 2023;91:129377. https://doi.org/10.1016/j.bmcl.2023.129377. Epub 2023 Jun 14. PMID: 37328038; PMCID: PMC11239236.
49. Zhang QW, Lin LG, Ye WC. Techniques for extraction and isolation of natural products: a comprehensive review. Chin Med. 2018;17(13):20. https://doi.org/10.1186/s13020-018-0177-x. PMID: 29692864; PMCID: PMC5905184.
50. Panula P. Histamine receptors, agonists, and antagonists in health and disease. Handb Clin Neurol. 2021;180:377–87. https://doi.org/10.1016/B978-0-12820107-7.00023-9. PMID: 34225942.
51. Lim S, Jeong I, Cho J, Shin C, Kim KI, Shim BS, Ko SG, Kim B. The natural products targeting on allergic rhinitis: from traditional medicine to modern drug discovery. Antioxidants (Basel). 2021;10(10):1524. https://doi.org/10.3390/antiox10101524. PMID: 34679659; PMCID: PMC8532887.
52. Muñoz IJ, Schilman PE, Barrozo RB. Impact of alkaloids in food consumption, metabolism and survival in a blood-sucking insect. Sci Rep. 2020;10:9443. https://doi.org/10.1038/s41598-020-65932-y.
53. Fiani B, Zhu L, Musch BL, Briceno S, Andel R, Sadeq N, Ansari AZ. The neurophysiology of caffeine as a central nervous system stimulant and the resultant effects on cognitive function. Cureus. 2021;13(5): e15032. https://doi.org/10.7759/cureus.15032. PMID: 34150383; PMCID: PMC8202818.
54. Araque P, Casanova H, Ortiz C, Henao B, Pelaez C. Insecticidal activity of caffeine aqueous solutions and caffeine oleate emulsions against Drosophila melanogaster and Hypothenemus hampei. J Agric Food Chem. 2007;55(17):6918–22. https://doi.org/10.1021/jf071052b. Epub 2007 Jul 21 PMID: 17658827.
55. Ceja-Navarro J, Vega F, Karaoz U, et al. Gut microbiota mediate caffeine detoxification in the primary insect pest of coffee. Nat Commun. 2015;6:7618. https://doi.org/10.1038/ncomms8618.
56. Cornelis MC, van Dam RM. Genetic determinants of liking and intake of coffee and other bitter foods and beverages. Sci Rep. 2021;11(1):23845. https://doi.org/10.1038/s41598-021-03153-7. PMID: 34903748; PMCID: PMC8669025.
57. Kapellou A, Pilic L, Mavrommatis Y. Habitual caffeine intake, genetics and cognitive performance. J Psychopharmacol. 2024;39(3): 233243. https://doi.org/10.1177/02698811241303601.
58. Brockhoff A, Behrens M, Roudnitzky N, Appendino G, Avonto C, Meyerhof W. Receptor agonism and antagonism of dietary bitter compounds. J Neurosci. 2011;31(41):14775–82. https://doi.org/10.1523/JNEUROSCI.292311.2011. PMID: 21994393; PMCID: PMC6703385.
59. Tuzim K, Korolczuk A. An update on extra-oral bitter taste receptors. J Transl Med. 2021;19(1):440. https://doi.org/10.1186/s12967-021-03067-y. Erratum in: J Transl Med. 2021;19(1):478. 10.1186/s12967021-03137-1. PMID: 34674725; PMCID: PMC8529754.

60. Xi R, Zheng X, Tizzano M. Role of taste receptors in innate immunity and oral health. J Dent Res. 2022;101(7):759–68. https://doi.org/10.1177/00220345221077989. Epub 2022 Feb 22. PMID: 35191336; PMCID: PMC9218499.
61. Morini G. The taste for health: the role of taste receptors and their ligands in the complex food/health relationship. Front Nutr. 2024;30(11):1396393. https://doi.org/10.3389/fnut.2024.1396393. PMID: 38873558; PMCID: PMC11169839.
62. Daly JW. Nicotinic agonists, antagonists, and modulators from natural sources. Cell Mol Neurobiol. 2005;25:513–52. https://doi.org/10.1007/s10571-005-3968-4.
63. Stielow M, Witczyńska A, Kubryń N, Fijałkowski Ł, Nowaczyk J, Nowaczyk A. The bioavailability of drugs-the current state of knowledge. Molecules. 2023;28(24):8038. https://doi.org/10.3390/molecules28248038. PMID: 38138529; PMCID: PMC10745386.
64. Paroha S, Dewangan RP, Sahoo PK. Pharmaceutical technology for improving the bioavailability of natural products. In: Saneja A, Panda A, Lichtfouse E, editors, Sustainable agriculture reviews 43. Sustainable agriculture reviews, vol. 43. Springer: Cham; 2020. https://doi.org/10.1007/978-3-030-41838-0_1.
65. Atanasov AG, Zotchev SB, Dirsch VM, et al. Natural products in drug discovery: advances and opportunities. Nat Rev Drug Discov. 2021;20:200–16. https://doi.org/10.1038/s41573-020-00114-z.
66. Li Y, Li S, Meng X, Gan RY, Zhang JJ, Li HB. Dietary Natural Products for Prevention and Treatment of Breast Cancer. Nutrients. 2017;9(7):728. https://doi.org/10.3390/nu9070728. PMID: 28698459; PMCID: PMC5537842.
67. Bartolozzi E. The natural approach to osteoporosis. Clin Cases Miner Bone Metab. 2015;12(2):111–5. https://doi.org/10.11138/ccmbm/2015.12.2.111. Epub 2015 Oct 26. PMID: 26604935; PMCID: PMC4625766.
68. Wilburn AJ, King DS, Glisson J, Rockhold RW, Wofford MR. The natural treatment of hypertension. J Clin Hypertens (Greenwich). 2004;6(5):242–8. https://doi.org/10.1111/j.1524-6175.2004.03250.x. PMID: 15133406; PMCID: PMC8109646.
69. Wink M. Evolution of secondary metabolites from an ecological and molecular phylogenetic perspective. Phytochemistry. 2003;64(1):319. https://doi.org/10.1016/s0031-9422(03)00300-5. PMID: 12946402.
70. Coolahan M, Whalen KE. A review of quorum-sensing and its role in mediating interkingdom interactions in the ocean. Commun Biol. 2025;8:179. https://doi.org/10.1038/s42003-025-07608-9.
71. Mukherjee S, Bassler BL. Bacterial quorum sensing in complex and dynamically changing environments. Nat Rev Microbiol. 2019;17:371–82. https://doi.org/10.1038/s41579-019-0186-5.
72. Homer CM, Summers DK, Goranov AI, Clarke SC, Wiesner DL, Diedrich JK, Moresco JJ, Toffaletti D, Upadhya R, Caradonna I, Petnic S, Pessino V, Cuomo CA, Lodge JK, Perfect J, Yates JR 3rd, Nielsen K, Craik CS, Madhani HD. Intracellular action of a secreted peptide required for fungal virulence. Cell Host Microbe. 2016;19(6):849–64. https://doi.org/10.1016/j.chom.2016.05.001. Epub 2016 May 19. PMID: 27212659; PMCID: PMC5186401.
73. Rajamani S, Lee L, Smith E, Majireck M, Mohan R. Modulation of bacterial quorum sensing by eukaryotes. In: Bramhachari P, editors, Implication of quorum sensing and biofilm formation in medicine, agriculture and food industry. Springer: Singapore; 2019. https://doi.org/10.1007/978-981-32-9409-7_4.
74. Avbelj M, Zupan J, Raspor P. Quorum-sensing in yeast and its potential in wine making. Appl Microbiol Biotechnol. 2016;100(18):7841–52. https://doi.org/10.1007/s00253-016-7758-3. Epub 2016 Aug 9 PMID: 27507587.
75. Medapati MR, Singh N, Bhagirath AY, Duan K, Triggs-Raine B, Batista EL, Chelikani P. Bitter taste receptor T2R14 detects quorum sensing molecules from cariogenic Streptococcus mutans and mediates innate immune responses in gingival epithelial cells. FASEB J. 2021;35(3): e21375. https://doi.org/10.1096/fj.202000208R. PMID: 33559200.
76. Gaida MM, Dapunt U, Hänsch GM. Sensing developing biofilms: the bitter receptor T2R38 on myeloid cells. Pathog Dis. 2016;74(3):ftw004. https://doi.org/10.1093/femspd/ftw004. Epub 2016 Jan 17. PMID: 26782143; PMCID: PMC4748947.

77. Cohen NA. The genetics of the bitter taste receptor T2R38 in upper airway innate immunity and implications for chronic rhinosinusitis. Laryngoscope. 2017;127(1):44–51. https://doi.org/10.1002/lary.26198. Epub 2016 Sep 21. PMID: 27650657; PMCID: PMC5177547.
78. Shaik FA, Singh N, Arakawa M, Duan K, Bhullar RP, Chelikani P. Bitter taste receptors: Extraoral roles in pathophysiology. Int J Biochem Cell Biol. 2016;77(Pt B):197–204. https://doi.org/10.1016/j.biocel.2016.03.011. Epub 2016 Mar 23 PMID: 27032752.
79. Twarda-Clapa A, Olczak A, Białkowska AM, Koziołkiewicz M. Advanced glycation end-products (AGEs): formation, chemistry, classification, receptors, and diseases related to AGEs. Cells. 2022;11(8):1312. https://doi.org/10.3390/cells11081312. PMID: 35455991; PMCID: PMC9029922.
80. Jaggupilli A, Howard R, Aluko RE, Chelikani P. Advanced glycation endproducts can activate or block bitter taste receptors. Nutrients. 2019;11(6):1317. https://doi.org/10.3390/nu11061317. PMID: 31212814; PMCID: PMC6628017.
81. Jones AJY, Gabriel F, Tandale A, Nietlispach D. Structure and dynamics of GPCRs in lipid membranes: physical principles and experimental approaches. Molecules. 2020;25(20):4729. https://doi.org/10.3390/molecules25204729. PMID: 33076366; PMCID: PMC7587580.
82. Zhang Z, Jin Z, Zhao Y, Zhang Z, Li R, Xiao J, Wu J. Systematic study on Gprotein couple receptor prototypes: did they really evolve from prokaryotic genes? IET Syst Biol. 2014;8(4):154–61. https://doi.org/10.1049/ietsyb.2013.0037. PMID: 25075528; PMCID: PMC8687355.
83. Salem MA, Perez de Souza L, Serag A, Fernie AR, Farag MA, Ezzat SM, Alseekh S. Metabolomics in the context of plant natural products research: from sample preparation to metabolite analysis. Metabolites. 2020;10(1):37. https://doi.org/10.3390/metabo10010037. PMID: 31952212; PMCID: PMC7023240.
84. Goodacre R, Vaidyanathan S, Dunn WB, Harrigan GG, Kell DB. Metabolomics by numbers: acquiring and understanding global metabolite data. Trends Biotechnol. 2004;22:245–52. https://doi.org/10.1016/j.tibtech.2004.03.007.
85. Du B, Haensch R, Alfarraj S, Rennenberg H. Strategies of plants to overcome abiotic and biotic stresses. Biol Rev. 2024;99:1524–36. https://doi.org/10.1111/brv.13079.
86. Hu C, Yang J, Qi Z, Wu H, Wang B, Zou F, Mei H, Liu J, Wang W, Liu Q. Heat shock proteins: biological functions, pathological roles, and therapeutic opportunities. MedComm (2020). 2022;3(3):e161. https://doi.org/10.1002/mco2.161. PMID: 35928554; PMCID: PMC9345296.
87. O'Brien JA, Benková E. Cytokinin cross-talking during biotic and abiotic stress responses. Front Plant Sci. 2013;19(4):451. https://doi.org/10.3389/fpls.2013.00451. PMID: 24312105; PMCID: PMC3833016.
88. Rajčević N, Bukvički D, Dodoš T, Marin PD. Interactions between natural products—a review. Metabolites. 2022;12(12):1256. https://doi.org/10.3390/metabo12121256. PMID: 36557296; PMCID: PMC9786035.
89. Ahuja V, Singh A, Paul D, Dasgupta D, Urajová P, Ghosh S, Singh R, Sahoo G, Ewe D, Saurav K. Recent advances in the detection of food toxins using mass spectrometry. Chem Res Toxicol. 2023;36(12):18341863. https://doi.org/10.1021/acs.chemrestox.3c00241. Epub 2023 Dec 7. PMID: 38059476; PMCID: PMC10731662.
90. Jankovic N, Geelen A, Streppel MT, de Groot LC, Orfanos P, van den Hooven EH, Pikhart H, Boffetta P, Trichopoulou A, Bobak M, Bueno-deMesquita HB, Kee F, Franco OH, Park Y, Hallmans G, Tjønneland A, May AM, Pajak A, Malyutina S, Kubinova R, Amiano P, Kampman E, Feskens EJ. Adherence to a healthy diet according to the World Health Organization guidelines and all-cause mortality in elderly adults from Europe and the United States. Am J Epidemiol. 2014;180(10):978–88. https://doi.org/10.1093/aje/kwu229. Epub 2014 Oct 15. PMID: 25318818; PMCID: PMC4224363.
91. Lichtenstein AH, Appel LJ, Vadiveloo M, Hu FB, Kris-Etherton PM, Rebholz CM, Sacks FM, Thorndike AN, Van Horn L, Wylie-Rosett J. 2021 dietary guidance to improve cardiovascular health: a scientific statement from the american heart association. Circulation. 2021;144(23):e472–87. https://doi.org/10.1161/CIR.0000000000001031. Epub 2021 Nov 2 PMID: 34724806.

92. Ordovas JM, Ferguson LR, Tai ES, Mathers JC. Personalised nutrition and health BMJ 2018;361:bmj.k2173 https://doi.org/10.1136/bmj.k2173.
93. Katz L, Baltz RH. Natural product discovery: past, present, and future. J Ind Microbiol Biotechnol. 2016;43(2–3):155–76. https://doi.org/10.1007/s10295015-1723-5. Epub 2016 Jan 6 PMID: 26739136.
94. Baltz RH. Natural product drug discovery in the genomic era: realities, conjectures, misconceptions, and opportunities. J Ind Microbiol Biotechnol. 2019;46(3–4):281–99. https://doi.org/10.1007/s10295-018-2115-4. Epub 2018 Nov 27 PMID: 30484124.
95. Li J, Prentice IC. Global patterns of plant functional traits and their relationships to climate. Commun Biol. 2024;7:1136. https://doi.org/10.1038/s42003-024-06777-3.
96. Samtiya M, Aluko RE, Dhewa T. Plant food anti-nutritional factors and their reduction strategies: an overview. Food Prod Process and Nutr. 2020;2:6. https://doi.org/10.1186/s43014-020-0020-5.
97. Duraiswamy A, Sneha ANM, Jebakani KS, Selvaraj S, Pramitha JL, Selvaraj R, Petchiammal KI, Kather Sheriff S, Thinakaran J, Rathinamoorthy S, Kumar PR. Genetic manipulation of anti-nutritional factors in major crops for a sustainable diet in future. Front Plant Sci. 2023;15(13):1070398. https://doi.org/10.3389/fpls.2022.1070398. PMID: 36874916; PMCID: PMC9976781.
98. Gropper SS. The role of nutrition in chronic disease. Nutrients. 2023;15(3):664. https://doi.org/10.3390/nu15030664. PMID: 36771368; PMCID: PMC9921002.
99. Oster H, Chaves I. Effects of healthy lifestyles on chronic diseases: diet, sleep and exercise. Nutrients. 2023;15(21):4627. https://doi.org/10.3390/nu15214627. PMID: 37960280; PMCID: PMC10650398.
100. Hacker, Karen. The burden of chronic disease. Mayo Clinic Proceedings: Innovations, Quality & Outcomes, 8(1):112–119.
101. Neuhouser ML. The importance of healthy dietary patterns in chronic disease prevention. Nutr Res. 2019;70:3–6. https://doi.org/10.1016/j.nutres.2018.06.002. Epub 2018 Jul 10. PMID: 30077352; PMCID: PMC6328339.
102. Gouda NA, Alshammari SO, Abourehab MAS, Alshammari QA, Elkamhawy A. Therapeutic potential of natural products in inflammation: underlying molecular mechanisms, clinical outcomes, technological advances, and future perspectives. Inflammopharmacology. 2023;31(6):2857–83. https://doi.org/10.1007/s10787-023-01366-y. Epub 2023 Nov 11 PMID: 37950803.
103. Islam MS. Natural products and disease prevention, relief and treatment. Nutrients. 2022;14(12):2396. https://doi.org/10.3390/nu14122396. PMID: 35745128; PMCID: PMC9228901.
104. Pazos P, Fortaner S, Prieto P. Long-term in vitro toxicity models: comparisons between a flow-cell bioreactor, a static-cell bioreactor and static cell cultures. Altern Lab Anim. 2002;30(5):515523. https://doi.org/10.1177/026119290203000505.
105. Huang HJ, Lee YH, Hsu YH, Liao CT, Lin YF, Chiu HW. Current strategies in assessment of nanotoxicity: alternatives to in vivo animal testing. Int J Mol Sci. 2021;22(8):4216. https://doi.org/10.3390/ijms22084216. PMID: 33921715; PMCID: PMC8073679.

The manufacturer's authorised representative in the EU is Springer Nature Customer Service Centre GmbH, Europaplatz 3, 69115 Heidelberg, Germany. If you have any concerns regarding our products, please contact ProductSafety@springernature.com

Printed and bound by CPI Group (UK) Ltd, Croydon, CR0 4YY
01/10/2026
02225973-0009